Emerging and Advanced Technologies in Diverse Forensic Sciences

An important contribution to the professional work performed in the areas on emerging technologies, this book provides an extensive expansion of the literature base on contemporary theories and investigative techniques used in the forensic sciences.

Forensic science, as a relatively new field of research still actively identifying itself in the larger landscape of the sciences, has been sharply criticized for utilizing techniques deemed largely unscientific by subject area experts. This book presents a collective analysis and review of the existing challenges as well as directions for state-of-the-art practices found in diverse forensic settings, enabling the reader to make an informed decision about the scientific validity of forensic techniques, and emphasizes the need for a greater understanding of the use of the most appropriate methodology and procedures. The contributors address cutting-edge, developing, and even hypothetical techniques and technologies in forensics research and practice, especially as it relates to the sphere of criminal justice and law enforcement in contemporary society.

A useful work for forensics professionals, and students and scholars working in the fields of politics and technology, criminal justice, forensic psychology, police psychology, law enforcement, and forensic science.

Ronn Johnson is an Associate Professor at Creighton University School of Medicine Department of Psychiatry and a clinical psychologist at the VA Nebraska-Western Iowa Health Care System.

Emerging Technologies, Ethics and International Affairs

Series editors: Steven Barela, Jai C. Galliott, Avery Plaw, Katina Michael

This series examines the crucial ethical, legal and public policy questions arising from or exacerbated by the design, development and eventual adoption of new technologies across all related fields, from education and engineering to medicine and military affairs. The books revolve around two key themes:

- Moral issues in research, engineering and design
- Ethical, legal and political/policy issues in the use and regulation of Technology

This series encourages submission of cutting-edge research monographs and edited collections with a particular focus on forward-looking ideas concerning innovative or as yet undeveloped technologies. Whilst there is an expectation that authors will be well grounded in philosophy, law or political science, consideration will be given to future-orientated works that cross these disciplinary boundaries. The interdisciplinary nature of the series editorial team offers the best possible examination of works that address the 'ethical, legal and social' implications of emerging technologies.

New Perspectives on Technology in Society
Experimentation beyond the Laboratory
Edited by Ibo van de Poel,
Lotte Asveld and Donna Mehos

Technology, Ethics and the Protocols of Modern War
Edited by Artur Gruszczak and Paweł Frankowski

Emerging and Advanced Technologies in Diverse Forensic Sciences
Edited by Ronn Johnson

For more information about this series, please visit: https://www.routledge.com/Emerging-Technologies-Ethics-and-International-Affairs/book-series/ASHSER-1408

Emerging and Advanced Technologies in Diverse Forensic Sciences

Edited by Ronn Johnson

Routledge
Taylor & Francis Group
LONDON AND NEW YORK

First published 2019
by Routledge
2 Park Square, Milton Park, Abingdon, Oxon OX14 4RN

and by Routledge
711 Third Avenue, New York, NY 10017

Routledge is an imprint of the Taylor & Francis Group, an informa business

British Library Cataloguing-in-Publication Data
A catalogue record for this book is available from the British Library

Library of Congress Cataloging-in-Publication Data
Names: Johnson, Ronn, editor.
Title: Emerging and advanced technologies in
diverse forensic sciences / edited by Ronn Johnson.
Description: Abingdon, Oxon; New York, NY: Routledge, 2019. |
Series: Emerging technologies, ethics and international
affairs | Includes bibliographical references and index.
Identifiers: LCCN 2018020911 | ISBN 9780415789462 (hardback) |
ISBN 9781315222752 (e-book)
Subjects: LCSH: Forensic sciences—Technological innovations. |
Criminal investigation—Technological innovations.
Classification: LCC HV8073 .E5155 2019 | DDC 363.25028/4—dc23
LC record available at https://lccn.loc.gov/2018020911

ISBN: 978-0-41578946-2 (hbk)
ISBN: 978-1-31522275-2 (ebk)

Typeset in Times New Roman
by codeMantra

MIX
Paper from responsible sources
FSC® C013056

Printed and bound in Great Britain by
TJ International Ltd, Padstow, Cornwall

Contents

Figures and tables

Figures

Tables

Contributors

The editor

Dr. Ronn Johnson is an Associate Professor at Creighton University School of Medicine Department of Psychiatry and is a lead clinical psychologist at the VA Nebraska-Western Iowa Health Care System. He is a licensed and board-certified psychologist with extensive experience in academic and clinical forensic settings. Formerly, he was the Director of the Western Region of the American Board of Clinical Psychology. Dr. Johnson has served as a staff psychologist in community mental health clinics, hospitals, schools, and university counseling centers. The University of Iowa, University of Nebraska-Lincoln, University of Central Oklahoma, and San Diego State University are among the sites of his previous academic appointments. In addition, he was an adjunct professor in the Homeland Security Department at San Diego State and the California School of Forensic Studies at Alliant International University. He founded several counseling centers that currently serve racially diverse populations. These sites include the Urban Corp of San Diego County, Southern California American Indian Resource Center, Community Allies for Psychological Empowerment, and the Elim Korean Counseling Center. His practice and research areas include antiterrorism, trauma, police psychology, and transdiagnostic issues with veterans.

The contributors

Dr. John D. DeHaan has over 45 years experience in fire- and explosion-related forensic science and has been deeply involved with improving fire investigation. After earning his BS (Physics 1969), he spent 29 years with public forensic labs. He has authored six editions of Kirk's Fire Investigation since 1982 and was coauthor with Dr. David Icove in Forensic Fire Scene Reconstruction since 2004. Since 1999, he has been president of Fire-Ex Forensics, Inc., a private consulting firm that provides services to police and fire agencies, public defenders, private attorneys, and insurers in fire and explosion cases across the US and internationally.

Mr. Scott Duncan is a criminal justice instructor within the Department of Sociology, Social Work, and Criminal Justice at Bloomsburg University of Pennsylvania. Previously, Duncan spent several years as a sworn supervisor with the Metropolitan Nashville Police Department, and his current research interests include criminal investigation, missing persons, criminal justice history, and policing. He holds a B.A. from Oklahoma City University, an M.B.A. from Belmont University, an M.S. from the University of Cincinnati, and is currently a doctoral student at Nova Southeastern University.

Ms. Sarah Furnier studied Biochemistry as an undergraduate at Saginaw Valley State University before pursuing a Masters of Science at Mercyhurst University in Anthropology with a concentration in Biological and Forensic Anthropology. At present, Sarah is working towards her Ph.D. in Epidemiology at the University of Wisconsin-Madison.

Dr. Heather Garvin is a board-certified forensic anthropologist (D-ABFA). She received a B.A. in Anthropology and B.S. in Zoology from the University of Florida, where she first began to gain forensic anthropological experience at the C.A. Pound Human Identification Laboratory. She earned an M.S. in Forensic and Biological Anthropology from Mercyhurst College and a Ph.D. in Functional Anatomy and Evolution from Johns Hopkins University. Since 2012, she has been working in the Department of Applied Forensic Sciences at Mercyhurst University where she mentors undergraduate and graduate students in forensic anthropology and conducts forensic anthropological casework and research.

Dr. Randall Grometstein is a professor of criminal justice in the Behavioral Science Department at Fitchburg State University, Fitchburg, Massachusetts. She holds a B.A. from Swarthmore College, a J.D. from Boston University School of Law, and an M.S. and Ph.D. from Northeastern University, and is a member of the Massachusetts bar. Dr. Grometstein has published articles and book chapters on ethics, wrongful conviction, prosecutorial misconduct, and social construction and moral panic theory.

Dr. Wendy L. Hicks is currently the Chair of the graduate program in Criminal Justice at Ashford University. Dr. Hicks holds expertise in law-enforcement administration and white supremacy. Dr. Hicks' scholarly work has enabled her to present her research at the meetings of the European Society of Criminology, International Police Executive Symposium, American Society of Criminology, and Academy of Criminal Justice Sciences. Dr. Hicks' current research agenda is directed at extremist groups and hate crimes. Apart from her scholastic endeavors, Dr. Hicks works as a corker aboard the tall ship schooner The Bill of Rights and is a volunteer Whaler for the San Diego Natural History Museum.

Dr. Herbert H. Hill is a Regents Professor of Chemistry at Washington State University. He received a B.S. in Chemistry from Rhodes College in Memphis, TN in 1970, an M.S. in Biochemistry in 1973 from the University of Missouri in Columbia, MO, then moved to Dalhousie University with his advisor and received his Ph.D. in Chemistry in 1975 from Dalhousie University in Nova Scotia, Canada. He spent 1975–76 as a Post-doctoral Fellow in Chemistry at the University of Waterloo, Ontario, Canada. His field of study is best characterized as Analytical and Biochemistry. He has worked principally in the field of ion mobility spectrometry, more specifically on the use of ion mobility mass spectrometry (IMMS) for drug detection. Throughout his 35-year career, he has demonstrated the analytical value of IMMS, and in 1976 was the first researcher to publish research indicating that IMMS is a valuable tool for bio-analytical chemistry. In 2011, he was made a Fellow of the American Chemical Society, and in 2012, he was awarded the honor of being a Fellow of the American Association for the Advancement of Science.

Ms. Tanna M. Jacob is expected to receive her M.A. in Marriage and Family Therapy from Chapman University in Orange, CA in January 2019. She received her undergraduate degree also from Chapman University where she graduated Cum Laude with a major in psychology and a minor in sociology. She is currently an MFT trainee in the Frances Smith Center for Individual and Family Therapy in Orange, CA.

Dr. Alexandra R. Klales is the Director of the Forensic Anthropology Program and an Assistant Professor in the Sociology & Anthropology Department at Washburn University. She currently conducts forensic anthropological casework in Kansas and is an Associate Member of the American Academy of Forensic Sciences (Anthropology). Her research includes improving biological profile methods, mass fatality and fatal fire recoveries, and 3D imaging. She has been teaching forensic and physical anthropology courses at the university level since 2009, including courses on forensic science, forensic anthropology, and human skeletal biology. Lastly, she's taught short courses in forensic anthropology in the United States and internationally.

Dr. Andreas Larsson is a licensed psychologist in private practice in his home town of Stockholm, Sweden. He also lectures in psychology at Stockholm University and at Mid Sweden University. His scholarly interests are in the broad application of Acceptance and Commitment Therapy and training (ACT), which he's coauthored the only book-length review of in The Research Journey of Acceptance and Commitment Therapy (ACT). Andreas also has an interest in the experimental, behavior analytic approach to language and cognition that is related to ACT, Relational Frame Theory.

Dr. Nicholas P. Lovrich holds the rank of Regents Professor Emeritus in the School of Politics, Philosophy and Public Affairs and the honor of being a Claudius O. and Mary W. Johnson Distinguished Professorship in Political Science at Washington State University. He is on appointment as a Research Scientist in the WSU Office of Research and serves as an affiliate faculty member of the School of Public Policy at Oregon State University and is a Visiting Researcher at the University of Utah. He holds a B.A. from Stanford University (1966) and Ph.D. from the University of California, Los Angeles (1971). Lovrich is the author or coauthor/coeditor of 13 books and more than 175 peer-reviewed articles and edited book chapters.

Ms. Nichole Lovrich is currently an Assistant Public Defender in Great Falls, Montana working for the Office of State Public Defender on felony criminal cases. She received her undergraduate degree in 2006 from the University of Oregon where she graduated Summa Cum Laude with a major in political science and a minor in women and gender studies. She received her law degree from Gonzaga University School of Law in 2009. After obtaining her law license in Washington, she moved to Whitefish, Montana and worked for a brief time for the law firm of McKeon Doud, PC in Kalispell on cases involving personal injury, workman's compensation, medical malpractice, criminal law, and some family law. After obtaining her law license in Montana, she was a solo practitioner and opened her own law practice, Lovrich Law Firm, PLLC practicing in the area of family law. She received the 2014 Peer Recognition Excellence Award for her work in achieving favorable outcomes for indigent clients, including a Change of Venue in State v. Adam Sanchez, a deliberate homicide case involving the death of a Cascade County Sheriff's Deputy.

Dr. Hollianne Marshall is an Assistant Professor in the Department of Criminology at California State University, Fresno where she teaches courses within the law-enforcement curriculum. She graduated from the University of Central Florida with a Ph.D. in Sociology and a Graduate Certificate in Crime Analysis. Her research is in the areas of organized crime, urban violent crime, criminal investigation, and community policing. Her recent publications can be found in the Journal of Contemporary Criminal Justice and the Journal of Criminal Justice.

Dr. Heather McMahon is a psychologist and Certified Forensic Examiner with the Missouri Department of Mental Health. She has worked in the California Prison System and in a maximum security forensic hospital with the New York Office of Mental Health. In 2012, she was awarded the Outstanding Doctor in Forensic Psychology by Alliant University. Dr. McMahon has presented research at both the national and international level, including conferences in Scotland, Spain, England, Canada, and the Netherlands. Her research has included topics related to police hiring policies, psychological assessment, juvenile delinquency, and competency-related topics.

Dr. Jessica Mueller received her doctorate degree in clinical forensic psychology from Alliant International University, San Diego where she received the Outstanding Doctor of Psychology Award. She has presented at several conferences and lectured on methods of terrorist recruitment and online radicalization as well as emotional response following acts of terrorism. As a part of her dissertation, she created the Terrorism Emotional Arousal Measure (TEAM) with Dr. Glenn Lipson. In 2016, the Association of Threat Assessment awarded Dr. Mueller with the Dr. Chris Hatcher Memorial Scholarship. She currently works for the California Department of Corrections and Rehabilitation.

Dr. Phillip Neely is the recipient of the Doctor of Philosophy in Public Policy and Administration from Walden University and the Masters of Science in Public Administration from Central Michigan University. He received a Certificate of Management and Leadership from Georgia Law Enforcement Command College @ Columbus State University. He is an Associate Professor at Saint Leo University in Duluth, Georgia. Dr. Neely's expertise comes in the field of criminal justice and public policy.

Ms. Kori Ryan is an Assistant Professor of Behavioral Sciences at Fitchburg State University. Prior to her academic career, Dr. Ryan conducted clinical and forensic assessments and provided evidence-based interventions and with diverse populations presenting with major mental illnesses and criminal behaviors. Dr. Ryan has provided clinical and forensic services in a variety of settings, including educational, psychiatric, and correctional/community supervision settings. Dr. Ryan has presented nationally and internationally on improving multi- and interdisciplinary teaching and training for criminal justice and mental health practitioners, the intersection between the criminal justice system and mental health, clinical and forensic assessment, and clinical and forensic ethics.

Mr. Gregory Saville is an urban planner and criminologist specializing in crime prevention through environmental design. After 9 years working as a police officer he returned to graduate school and began research into the spatial patterns of crime. From 2000 to 2005, he ran a criminological research center as research professor at the University of New Haven. He currently runs AlterNation LLC Consulting in Arvada, Colorado, where he developed the SafeGrowth® crime prevention method and also spatial forensics, a technique he developed and employs as expert witness in criminal and civil cases to analyze crime scenes.

Dr. Jessica A. Tufariello is originally from Point Pleasant, NJ. Dr. Tufariello was awarded a BA from Alfred University in 2010. She then traveled to Washington State University to work on her Ph.D. under the guidance of Dr. Herbert H. Hill Jr. and later with the assistance of Dr. Nicholas P. Lovrich. The main focus of Dr. Tufariello's work was on the development of a breathalyzer for the detection of cannabis from human breath using ion mobility spectrometry. In 2016, Dr. Tufariello was awarded a Ph.D. in chemistry from Washington State University. She currently works as a scientist for Alturas Analytics, Inc. in Moscow, ID.

Acknowledgments

The editor wishes to thank his friends, families, loved ones, and students for being supportive and patient during the crafting of this work. Many people have now patiently and repeatedly copy edited and reviewed the various versions of this book. But again, I want to especially thank Tanna Jacob who devoted many days in her life to read, reread, and offer instructive remarks on this work. I am eternally indebted to my major professor and mentor Dr. Forrest Ladd from the Psychology Department of Southern Nazarene University in Bethany, Oklahoma. Dr. Ladd's Christ-centered encouragement, guidance, and cogent insights have tremendously shaped the type of psychologist that I became. Words have poor power when it comes to conveying the amount of appreciation and gratitude that I have for his unflinching support over the years.

– Ronn

Preface

The normal course of producing scientific works like this book began like every other similar endeavor. That is, the analysis of emerging technologies in forensic science started with a keen awareness of the need to articulate clear objectives. The objectives were only achievable through the recruitment of dedicated seasoned forensic professionals with special talents and knowledge. Their expertise and skills could then be reflexively applied to facilitate a firm understanding of their diverse disciplines within the context of a wide range of legal matters. While the courts recognize the relevance of various professional disciplines as well as the clarity brought by expert witnesses, that understanding is comparatively speaking rather limited when it comes to the courts. Unlike the courts, the forensic scientific base of experts is informed by professional organizations, articles disseminated through peer reviewed journals and best practice methods that are widely accepted within various disciplines. While there is some appreciation for professional backgrounds by the courts, it is often far from ideal. More often than not, the trier of facts is dependent on the experts or counter arguments made by the opposing attorneys. Despite these restrictions, the courts can become better informed through experts whose information is presented (i.e., witness qualified as an expert by knowledge, skill, experience, training, or education), in a convincing manner and often through what seems like withering cross-examination by opposing attorneys. Still, as the sheer magnitude of these types of forensic issues emerge; there is an ever increasing need to receive scientific information (i.e., evidence sufficient to support findings) that the courts can readily comprehend and then confidently rely upon while making decisions during the diverse legal matters before them.

There are two practice issues. First, the expanding the practice knowledge base requires an informed application of standard forensic scientific methods. Second, the judgments made about these practices are also important. To accomplish both tasks (i.e., expanding knowledge and making judgments about practices) a defensible sequence of events must be guided by research design. Moreover, the results from this rigorous inquiry process must be published in peer-reviewed journals. Certainly, there are times when no substantial information exists in an area of practice thus necessitating the creation of new knowledge be generated through existing forensic science methodologies. This process can be more challenging due in large part to the paucity of theories or previous work that provide a background or foundation upon which to inform that new endeavor. Nonetheless, such a research effort could potentially fuel a trend or reinforce the continued use of pre-existing practices simply because a recognized forensic expert has articulated critical issues (e.g., US Federal Rules of Evidence [FRE] and the rulings in the Daubert express the most commonly applied standards FRE Rules 701–706) that are supportive of such uses. This book exemplifies a strategic attempt to bring together a wide range of forensic professionals and have them engage in thoughtful discussions about the current and emerging technologies in their respective professional disciplines. In this case, there are explanations of the current practice status and emerging technologies found in diverse settings that reflect the overarching scientific views of the actual forensic science work.

Forensic science works exists as an expansive field with several oftentimes overlapping disciplines that must operate within a legal context. There are foundational empirically-based theories and methods are exercised by various professionals. The extent to which their opinions or work products are accepted within most courts are expected to fluctuate to some extent. However, the core of elements of what actually makes it forensic science remains pretty much the same with predictable adjustments. In this case, a distinct science covers substantial procedural elements as reflected in the wide range of forensic topics that are included in this book. That is, the subjects examined define and facilitate a firmer understanding of complexities of various forensic practice areas. For example, there are applicable laws, constructs, principles, and techniques that reflect unchanging forensic science best practices. The contributors who have collaborated on this book allow two primary goals to be reached. First, it provides readers with a review of the scope of forensic science practice encompassed in the work of a diverse group of professionals. In this case, the book provides extensive conceptual formulations as well as specific forensic methods. Second, the book informs those individuals exploring new career possibilities as to the potential professional roles that are available within diverse forensic science work settings.

Ronn Johnson, Ph.D., ABPP
VA Nebraska-Western Iowa Health Care System
Creighton University School of Medicine, Department of Psychiatry

Part I

Advanced theoretical applications relevant in diverse forensic settings

1 An overview of current and emerging technologies in diverse forensic science settings

Ronn Johnson

Forensics is the application of science to assist by providing professional discipline insights that are expected to better inform the decision-making or enforcement function for a wide range of relevant legal matters. Most people are probably more familiar with high-profile criminal acts whereby accused individuals are brought into the public's eye through various court proceedings. For example, the televised murder case of O.J. Simpson was broadcast into homes during the entire trial. Forensic science was used and misused in various ways to help the jury make sense out of all the diverse professional discipline details associated with the case that would probably go unnoticed to an uninformed juror or legal authority. Some of the data collected by crime scene analysts and police in the O.J. Simpson case was challenged for a variety of reasons. For example, parts of the crime scene evidence were kept in the pocket of one of the lead investigating detectives working on the case. This odd act alone caused a concern about the integrity of that evidence in terms of its potential of being compromised, thus significantly diminishing its legal credibility. Furthermore, this mishandling of evidence also violated Los Angeles Police Department (LAPD) policy, thereby raising serious questions about the proper chain of custody for the handling of said evidence. That is, if a police officer kept the evidence in his pocket, then how can a juror be totally confident that the data presented during trial and said to come from the evidence in question was, in fact, obtained from the actual crime scene and not contaminated with evidence from an unrelated crime scene.

A faulty or questionable chain of custody for the evidence creates a major problem for prosecuting attorneys who must work hard to convince jurors that the data had not been tampered with or degraded in any way since the time that it was first collected at the scene of the crime. Another forensic science example in this murder case centered around the LAPD criminalists while being cross-examined on the witness stand by the Simpson defense attorney Barry Scheck. Mr. Scheck raised forensically relevant questions about how DNA evidence was handled, as required by standards, in order for it to be assessed, properly collected, and then stored. The forensic science in this part of the case created major doubts about the credibility of the

LAPD professionals who were responsible for collecting and handling key crime scene evidence which impacted the outcome. Practically speaking, if a juror does not have full confidence in the collection methods used to obtain the evidence, then it often means that a case could be made that there may be insufficient evidence to find the accused guilty beyond a reasonable doubt. Doubt entered into this legal picture because the evidence in the Simpson case was viewed as biased or tainted due to the way that it was mishandled by the professionals who were legally responsible for it during the time it was collected.

The main point being expressed is that forensics science was used as a tool for determining the reliability of the procedures or methods used as well as best practices by qualified experts from various disciplines. In a broader context, many civil and criminal adjudications could not occur without the contributions afforded through forensic science. Even the decision to move forward with arrests, investigations, and all the related activities in advance of a trial are often heavily dependent on the contributions provided by forensic science. If nothing else, forensic science helps to resolve matters by better informing all the stakeholders who make up the legal community. The professionals sharing their expertise in these cases often work with police and other disciplines in order to facilitate justice for all those involved in a case. To little surprise, the approaches used in forensic science can also be dominated by confusion and/or intense debate. There are, unfortunately, occasions when the results from these forensic approaches are improperly used for a variety of motivations and reasons.

Foundational legal concepts and methods of forensic science are designed to assist in civil and criminal investigations. For example, the British Petroleum (BP) oil spill in the Gulf of Mexico resulted in several lawsuits. The US Department of Justice was one of many who filed cases against BP. During the penalty phase of the case, forensic experts engaged in professional debates regarding how much of a fine should be levied against BP based on several factors. The core of their forensic testimony in determining the fine was about whether or not BP should receive the highest-possible punishment available under guidance from the Clean Water Act. Following the spill, opposing experts on the sides of the Department of Justice (DOJ) and BP offered forensic opinions about levels of toxicity in the Gulf. The DOJ's forensic experts questioned the places (i.e., surface waters or a deep-sea plume of oil) where the BP experts collected their data, which in their opinion, failed to properly estimate the devastating consequences of the oil spill disaster. This BP case represents a valuable reference point for illustrating just how dynamic and challenging these practice areas can be when they are used to tackle the real-world issues that confront forensic science. With the rapid evolution of technologies (e.g., offshore oil drilling, smart bombs used by the military, body cams worn by police officers) in diverse settings, those forensic science professionals must assess relevant factors and be prepared to justify the science behind their decisions. For example, in case of unanticipated consequences stemming from these technologies, legal issues may be raised resulting in a re-evaluation of usage or implementation of these technologies.

Globally, the dependence of working parents on day care services allows them to fulfill their job responsibilities while at the same time feeling relatively safe in leaving their children in the hands of qualified child care providers. Yet, something as seemingly simple as a child day care can also become a caldron where forensic science is called upon to answer pressing legal questions whenever issues arise. For example, in the 1980s, there was a well-publicized child abuse case centered around McMartin Preschool in California. Here, the two owners of the day care were accused of a wide range of illegal acts (e.g., sexual abuse) during the time that the children were under their authority. Forensically, experts were used to collect data from preschool children (aged 3–5) regarding what had transpired at the preschool. For example, some of the children alleged sexual abuse. Psycho-developmentally, children at this stage are less able to verbally provide the type of details required in a potential child abuse investigation.

At some point, a decision was made to use ostensibly qualified examiners who were charged with collecting information from the children about their experiences during the time they were in attendance at the preschool. Subsequent critiques of the examiner's interviews with these children revealed that they used leading questions that significantly influenced the direction of the responses provided by the children. In addition, as part of their assessment, the examiners relied on anatomically correct dolls to secure data that would later be used in an attempt to make a case for sexual molestation against the preschool owners. However, after 6 years of criminal trials, there were no convictions and all the charges were eventually dismissed. The personal and professional lives of the accused were destroyed as a result of this highly charged public trial. At the time of this legal case, it was the longest and most expensive criminal trial in US history. Forensically, the case was problematic in several ways. First, the probative approach used by examiners was highly criticized for the manner in which the questions were framed as they tended to lead children into negative responses with respect to the accused. Second, as a psychological evaluation tool, anatomically correct dolls had not been standardized (i.e., acceptable psychometric properties) as an accepted assessment approach to use with children which meant that any of the results obtained from the use of these dolls would not reliable or valid. From a forensic psychological perspective, the case became a textbook example of exactly what not to do while conducting suspected child abuse investigation. There are other areas where science can be applied in addressing legal issues that arise in other disciplines (e.g., forensic accounting).

For example, Bernie Madoff was able to deceptively secure millions of dollars from scores of investors through what amounted to a Ponzi scheme. Over the years, he defrauded investors. Madoff was not only able to fool unsuspecting investors but the very government legal authorities who were responsible for monitoring these types of financial matters. Instead, as all Ponzi's scheme do, the one created by Madoff collapsed. The financial complexities around this criminal case involved trying to wrap minds around making an effort to determine exactly how it was done, who lost what, and who may also be responsible for perpetrating this fraudulent act. Investigating financial wrongdoing cases of this magnitude would, by default, require forensic accountants. Working in this capacity, the forensic accountants would explore issues in order to evaluate the matters at hand. Forensic accountants are trained to look beyond the numbers and deal with the multifaceted business and financial misdeeds that are attached to the case. The forensic accountants here would be charged with attending to the criminal features that are the subject of litigation. The information and interpretation of their findings would augment the work of the attorneys, judges, juries and law enforcement agencies. For in-depth assessment, fiscal records, emails, phone logs, reports, etc. become the financial skeletal remains used to assemble the forensic pieces that explain actually what transpired.

With a similar objective in mind, historically, forensic anthropologists are charged with using the science of anthropology and its various subfields in a legal context. Known to many as the Armenian Genocide is one example of a case where the expertise of forensic anthropologists has been utilized. The centennial anniversary of what has been called the Armenian Genocide took place in 2015. The details of this incident charged that somewhere between 1.5 and 3 million Armenians were killed between 1915 and 1917. The Turkish government was said to have sponsored the Armenian deaths and subsequent confiscation of their property. Furthermore, the properties belonging to those killed were seized through laws seemingly created just to facilitate the process of asset acquisition. The Turkish government has steadfastly denied this incident and has aggressively taken steps to suppress any efforts to single them out for being responsible. Yet, the charges of genocide are said to have taken place through the mass deportation and killing of hundreds of thousands of Armenians by Ottoman Turks. There are others who assess this Armenian incident as just the continuation of a preexisting war instead of an actual genocide. The core of this debate cannot ignore the historical plausibility of massive death and human suffering. In terms of science, the discipline of forensic anthropology can play a role through excavating mass graves and assessing the skeletal remains of the victims. There are scientific methods that can be applied at the sites of the mass graves where the evidence gathered can then be used to make stronger cases for proving guilt (i.e., government- or state-sponsored genocide). This type of forensic science work is vital to promoting justice as it offers documentation that meets or fails the standard definition of genocide as well as specifies the number of deaths. Any subsequent legal claims made by the families could be forensically supported through various courts that were established in order to deal with these matters.

There is another forensic science dimension to this case. Armenians or any other group of people (e.g., Native Americans), having experienced genocide or exile, are predictably at increased risk for developing the psychological by-products of cumulative or historical trauma. From a clinical forensic perspective, this particular type of trauma has unfortunately cut across generations because of the oral historical accounts that are passed down through several generations. Ethnoracial minorities experience recurrences in their intense and negative emotional reactions to what some seems like an innocuous incident because a specific unwanted circumstance is another painful reminder of egregious acts from the past. The Trail of Tears and Wounded Knee are examples of historical trauma caused through police actions. As a result, when ethnoracially contemptuous events occur, they can reflexively refresh old fears and horrors that were revealed or passed down from ancestors. Forensic science from diverse settings can assist both the courts and surviving family members with understanding what actually took place with respect to the victims during and after these racially traumatizing government or police actions.

The proliferation of video cameras and cell phones has made it possible to capture footage from police encounters as they carry out their legally authorized duties. There are several high-profile, cross-racial police incidents like Rodney King in Los Angeles, Baltimore's Freddy Gray, Ferguson's Michael Brown, and the Jamir Rice case in Cleveland. These incidents share the use of some video evidence that, at least initially, raised questions about an officer's potentially racially motivated misconduct. However, none of these cases resulted in the conviction for any of the officers involved. For example, in the Freddy Gray case, there was a prosecutor, video, and a fairly diverse jury. Yet, the courts failed to convict any of the police officers tried.

The recurring traumatized reactions from the diverse communities highlight the strong disagreement about what they assessed as actually transpiring, much less the validation of police racial bias motive. The police as part of the authorized legal process ostensibly work to achieve justice and public safety. However, the discrepancies found in the police–public perceptions of these high-profile incidents have led to some corrective actions department (i.e., universal bias trainings and DOJ consent decrees). Forensically, some of these corrective actions may have further complicated the translation of these unfortunate incidents into practices (e.g., pre-employment hiring, fitness for duty, better officer training, revisions in department policies, disciplinary actions for misconduct) that seemingly would result in the desired public safety. For example, the unarmed Eric Garner was killed by the police while he was illegally selling cigarettes on the street. Video of the incident showed one officer essentially putting the full weight of his body on Mr. Garner's neck as he laid on the ground with his hands up while restrained by several other officers at the same time. Before dying, Mr. Garner could be heard painfully screaming that he could not breathe as his neck was being compressed. None of the officers involved in the killing him were ever prosecuted. A forensic case could be made to review the personnel files of the officers involved in order to determine if previous complaints had been filed against them for excessive force. In addition, the pre-employment psychological evaluations of these same officers may reveal that there were problematic behavioral patterns that were not sufficiently examined at the time that they were hired. Moreover, it would also be prudent to determine if fitness evaluations were ordered in the aftermath of any previous excessive force complaints with these officers (e.g., especially after a sustained excessive force complaints), or a review of all of their previous psychological evaluations could uncover problems with how the psychologists approached examining these officers in areas critical to work in diverse communities. Again, the nature and quality of the pre-employment or fitness psychological evaluations are critical from a forensic science standpoint. The examiners at those previous psychological evaluations would need to have provided clear and convincing evidence that they had adequately explored any issues that would dismiss any subsequent questions about a police applicant or incumbent officer's suitability. From a forensic science perspective, there are two primary concerns at a minimum with officers involved in the types of high-profile cases. First, is there any evidence that the officer being evaluated is a negligent hire. Second, is there a forensic issue related to negligent retention after behavioral misconduct evidence emerged that should have resulted in an officer being fired, suspended, or subjected to other appropriate disciplinary actions by the department? If not, then a case could be made for negligence.

The previous overview of this field serves as a useful starting point for understanding forensic science for a variety of reasons. The aforementioned case examples illustrate the diverse settings where forensic science can be appropriately and yes, inappropriately applied. The cases also reveal that even forensic experts from opposing counsel can strongly disagree on the specific detail and technical aspects of a case in ways that are often firmly grounded in the use of scientific principles and techniques that are representative of accepted practices within their particular discipline. While this vigorous forensic expert debate can sometimes be confusing for judges and members of a jury, this process nonetheless serves a vital scientific function within the broader legal context of any case. Without such forensic insights, the outcomes in these cases would be skewed in a direction of a neonate perspective that would seriously threaten the underlying goal of achieving justice given the complexity of the issues involved in these circumstances.

The insights gleaned through the previous cases start with a sequential and extended forensic argument that secures information from diverse areas of study (e.g., anthropology, accounting, biology, chemistry, engineering, geology, mathematics, medicine, physics, and psychology). For example, in 2017, North Korea's leader, Kim Jong-Un, was accused of ordering the poisoning of his estranged half-brother using a chemical substance (e.g., a banned chemical nerve agent) that North Korea was known to have large stockpiles of and had used in the past. Forensically, toxicologists were able to successfully identify the type of poison used in the murder based on an assessment of the residual amounts found at the scene but also noted in the health effects of one of the assassins who was arrested. This case immediately became a criminal investigation because the intentional killing in this case was suitable to the courts. Moreover, the forensic side of this meant that a formal investigation was required in order to identify what evidence or pertinent facts could be used in a court of law. That is, the concept of forensics means evidence that would be applied in a court of law responsible for adjudicating the case.

Forensic science involves the use of scientific and medical techniques to the assessment (i.e., identification) and interpretation of physical or psychological evidence for legal or regulatory purposes. The techniques used in this forensic process follow generally accepted professional standards and procedures that can be appropriately validated. Most informed individuals understand that laws are crafted to ostensibly govern how we function or carry out daily life activities. When a problem emerges that cannot be readily resolved or understood by the parties involved, one of the ultimate ways to bring about resolution to the matters at hand is to pursue some type of alternative through either a civil or a criminal legal process. For example, a couple living in an old van had it catch fire as they slept near a beach where they had parked overnight. The husband was able to escape the deadly fire by kicking out a back window of the van, but the wife's charred body was later found stuck to the side entrance door. The husband reported that he awakened to smoke and flames. He also claimed that he was unable to see or rescue his wife. The husband was found to have an extensive arrest record, reputation for lying, and a chronic history of substance abuse. To no surprise, local legal authorities were quite skeptical of his accounts about what had transpired during this unfortunate fatal vehicle fire. Yet, in order to draw empirically supported conclusions in the case, the county hired a forensic fire scene expert who used a variety of assessment techniques in order to generate scientific evidence that would result in a legally defensible basis for either accepting or rejecting the accounts provided by the lone survivor. Largely, this forensic science effort was accomplished through a recreation of what actually happened at the time of the vehicle fire. Forensically, the rejection of the husband's story would mean that he would be most likely be formally charged as being responsible for his wife's homicide.

The case demonstrates how forensic science may be used to help address unresolved concerns through the application of scientific knowledge and methods within the legal system. In this specific case, the forensic science investigation was critical to helping to bring clarification to what could be a potential crime. The science work started with comprehensive interviews with the lone survivor and other individuals that were also in that area of the beach around the time of the fire incident. The fire scene details, time sequence, van construction, vehicle condition, van contents and historical functioning of vehicles of the same make, maintenance history, and year would be important information to gather from a forensic standpoint. Here, an accurate reconstruction of the fire incident is particularly relevant as this evidence would need to be credible enough to withstand the anticipated scrutiny in any subsequent legal proceedings stemming from this case.

The veracity of the lone survivor's story also hinges upon the ability of a forensic expert to essentially recreate the incident as it was reported by him and by evidence collected. These efforts may also result in the development of other plausible alternative explanations as to what transpired based on facts. Any deviation in the selection of the exact vehicle or calibration of the devices used to identify temperature from various heat sources would impact the level of confidence in the forensic data gathering sources. Moreover, since the sole victim's charred body was also evidence, was there data collected from the scene that would actually support the coroner's stated cause of death? In this case, the fire trail and composition of the interior of the van are particularly noteworthy for the information as they can potentially provide additional data about the death.

Forensic science has several applicable subspecialties that may provide relevant findings. For example, forensic chemistry is used to chemically conceptualize the information collected at the scene of this fire incident. There are several reasonable forensic questions. First, did the lungs of the victim contain credible data related to the fumes that might have been inhaled during this incident? Second, would the inhalation of the fumes contribute in some way to the victim being incapacitated or is there an alternative explanation for the cause of death? Third, was the victim incapacitated in any way before the fire? Finally, is this evidence available through other sources (e.g., blood, saliva, bones, nails, vitreous humor)? The forensic chemical report might also include information on fire residue, paints, dyes, inks, liquors, latent fingerprints, fabric, fuels, foods, and pharmaceuticals, to name a few additional potential sources of information from the actual scene. The timing of the collection of the previously identified information could be critical as some of the desirable evidence could degrade or somehow alter over time. A skilled forensic examiner would remain sensitive to the issues that are especially relevant to data collection as well as the importance of relying upon reliable analytical techniques.

Whenever there are more questions than answers, forensic science can function in a manner that facilitates a more informed understanding. For example, this process may take place using a more extensive and nuanced reconstruction of the evidence or application of various available methods of forensic analysis. This increased level of awareness can be achieved by doing something tangible, which brings greater clarity to the process by helping to learn more about what is actually being studied at the time. By default, forensic science demands the timely use of research knowledge that often improves clinical practices. Forensic science is useful because it informs the trier of fact (e.g., judge or jury) who must make legally binding decisions about the matters at hand. The decisions here must be based on the best available scientific evidence using empirically defensible or widely recognized best practice methodologies within their proper context.

The scientific foundation established in this book is critical because this information serves as an indispensable tool that ensures the continued growth in new knowledge. The timing of this forensic technology book is ideal. The process began by planning this forensic publication. The expectations were extremely high, and the challenge here was twofold. First, to provide a current and accelerated examination of select forensic technological developments by surveying the diverse professional work settings while articulating its relevance to practice. Second, to offer a concise yet wide-ranging introduction that is instructive when it comes to understanding these professions by providing dynamic insights into current and emerging forensic science trends.

Civil and criminal case issues that drive forensic science are constantly changing. This book is composed of 14 well-researched chapters that seamlessly connect and build on each other by providing illustrations of forensic developments. There are contributions from international authors with comprehensive experience in the diverse fields where forensic technology is applied. The purpose of this book is to provide a detailed overview of subspecialties that introduces and contextualizes forensic science. The book operates under a conceptual approach that is designed to make available an exhaustive and practically oriented examination of several relevant issues associated with forensic science. In addition, the blend of thematic special topics is expected to be of considerable interest to a range of cross-disciplinary professionals (e.g., attorneys, judges, criminal justice personnel, psychologists, scientific experts who testify, public safety authorities). Those individuals in training for forensic sciences are also expected to find this book particularly instructive. For example, there is a plethora of rich cases of practical applications that are used to reinforce the real-world side of forensic science as well as the challenges confronted in the work performed in various settings.

The book is divided into five parts spanning the spectrum of forensic analysis. Part I is titled "Advanced Theoretical Applications Relevant in Diverse Forensic Settings". Andreas Larsson and Ronn Johnson's chapter on Acceptance and Commitment Therapy and Training in forensic settings finds that working with psychotherapy and indeed any behavior change is always a challenge, especially if a client is coerced to attend. Acceptance and Commitment Therapy (ACT) is a third-wave behavior therapy model which emphasizes building clients' ability to find a personal purpose and use present moment focus, acceptance, and new perspectives on thoughts to cope with barriers to effective values-based action. ACT is delivered from a perspective of compassion and equality. ACT is also part of Prosocial – a model for getting groups to work more effectively together and might be useful in working with staff in forensic settings.

Ronn Johnson and Tanna Jacob's chapter assesses the global phenomenon of how Juvenile Fire Setting and Bomb Making (JFSB) continues to function as a major problem for fire service and law enforcement personnel. JFSB has garnered significant forensic psychological attention and awareness, as recent figures suggest its prevalence poses a major public safety threat. That is, the increased rates of JFSB have shown a disturbing rise in costs attached to it (e.g., deaths, property damage, and public safety responses). In terms of the clinical and forensic psychological issues involved with JFSB, one of the advanced topics is the way that risk assessments (e.g., criminogenic risks, needs rehabilitative issues, and identifies culturally responsive interventions) are used within the context of responding to these cases. In this case, the risk assessment adds critical information in terms of how public safety personnel may respond to these cases in an evidence-based way. From a clinical psychological intervention perspective, the idea of the malleability of JFSB behavior later becomes an important factor while considering the application of evidence-based interventions. This chapter introduces the advanced and emerging forensic trends in working with JFSB. Next, the focus shifts to forensic risk factors and predictors of psychopathological behavior. Then, there is a review of the myths, misdiagnoses, and missed psychiatric diagnoses with JFSBs. This section is followed by a discussion of the development of assessment tools that are specifically designed for working with this clinical forensic population. There are an examination of a sample of interventions and treatment options for JFSB. Finally, in conclusions, implications for clinical forensic psychological practice and research are highlighted.

Philip Neely's chapter examines constitutional issues of Miranda Warnings that are usually adhered to, and the background information of the interviewee as well as their relationship to the crime that has been established. Yet the instances of false confessions continue to surface despite scholarly studies, extensive research, and preventive training on this phenomenon. What induces a person to confess to a crime that is later revealed they knew nothing about has baffled law enforcement, the judicial, and laymen alike for decades. The ethical issues focus on the major developments in research addressing the link between corruption and growth, the multifaceted character of corruption, and the potential for corruption to counterbalance strides towards more significant trade openness.

In their chapter, Kori Ryan and Heather McMahon states that the fields of psychology, law, and criminal justice are inexplicably intertwined, intermingling in ways that can make service provision difficult due to issues such as changing correctional populations, legal precedent, and shifts in demographics. With these challenges comes opportunity to utilize psychologists in new and unique ways. This chapter provides an overview of how psychologists can, and do, provide a unique set of skills to address issues in correctional mental health from arrest to release. This chapter also discusses under-addressed areas where psychologists can have impactful roles, and discusses some of the barriers to implementation.

Part II is titled "Emerging Technological Advancement within Forensic Science". The chapter by Hollianne Marshall addresses the various perspectives (both theoretical and empirical) for several digital forensic technologies and how they are applied; ethical issues including boundaries in "search and seizure" of digital information, privacy rights and data storage during and after investigations, and determining intent are discussed. Much like the technologies for fingerprinting, ballistics, and DNA, digital forensic technology is also widely questioned and continues to evolve. However, the increasing ease of access to advanced technology by the general public and the lack of written law governing the use of these technologies by law enforcement led to complex issues surrounding investigation and evidentiary support.

Heather Garvin, Alexandra Klales, and Sarah Furnier describe the current uses and future applications of 3D technologies in forensic anthropological casework and research. The reader will be informed about the use of digitizers, surface scanners, CT scanners, and photogrammetry methods in the field. Limitations such as budgetary constraints, the lack of standardization, and the need for further validation are discussed, along with potential issues regarding data sharing and archiving. Although the chapter does not provide simple resolutions to these challenges, it sets the stage for discussions in future policy making and advances in the field.

John DeHaan contributed two chapters. Chapter 8 demonstrates that the major challenge in fire, explosion, accident, and crime scene investigations is to properly document all important features and dimensions. This has traditionally been carried out using a camera and a tape measure, which is time-consuming. In the late 1990s, instrument companies introduced 3D laser scanning to largely automate the tedious process of point-by-point measurements with a tape measure or a laser-based Total Station. Today's 3D laser scanners can take tens of thousands, even millions, of measurements per second creating a "point cloud" of data that can be manipulated to recreate in virtual reality all the dimensional relationships of even the largest scenes. Digital photos can be overlaid with the image data to recreate the entire scene with incredible accuracy. Dr. DeHaan's second chapter investigates the area of fire scientists and engineers who have created a world of mathematical and computer-based analytical tools to explain and predict fire growth and behaviors. These tools are widely used to plan sprinkler systems, smoke detection and control systems, and other fire safety measures. Many have provided insights into why post-fire physical indicators look the way they do and how reliably they reflect fire events that took place, leading to improvements in the accuracy of scene investigations. Fire is a very complex chemical and physical process, and fires in rooms with multiple fuel packages have not been successfully modeled to fully duplicate real test fires beyond their growth phase.

Part III is titled "Corollary Factors and Prevention Trends in Forensic Science Arenas". Gregory Saville's chapter examines how forensic architecture is of recent pedigree in the forensic sciences. Robert Lynch proposed forensic architecture as a branch within forensic science due to its difference from other forms of the field. This chapter describes the evolution of Crime Prevention Through Environmental Design (CPTED) and crime and place theories as part of forensic architecture, here titled spatial forensics. Spatial forensics and the socio-spatial patterns emerging from offender target search and selection decisions now comprise a new way to assess crime foreseeability and retroactive investigation at crime sites. A growing number of CPTED practitioners already appear at criminal and civil trials each year to assess the crime opportunities created, or prevented, by urban design. As such, spatial forensics and the CPTED principles at the foundation of its practice require a robust set of analytical procedures so that spatial forensic analysts provide reliable assessments about the crime event.

Jessica Mueller and Ronn Johnson's chapter examines lone-wolf terrorism that has been considered the fastest-growing form of terrorism, in part due to technological advances. Historically, lone wolves adopted a radical ideology on his or her own, without external influence, and then acted or attempted to act on those self-acquired beliefs. Today, lone wolves may connect with others online to seek validation of their beliefs or gather information to plan an attack. This chapter will focus on how lone wolves radicalize on the Internet and how emerging trends in technology can facilitate the radicalization process.

Part IV is identified as "Scientific Advancements in Forensic Investigations". Scott Duncan's chapter addresses Phenylketonuria (PKU) Cards as an Underutilized Resource in Forensic Investigations by noting authorities in the United States annually investigate thousands of missing persons and unidentified decedents. The lack of fingerprints, dental records, and DNA in these cases can complicate identifications, subsequently compromising criminal investigations. This chapter describes PKU or infant blood spot cards as an underutilized resource for law enforcement. Specifically, these cards contain DNA that can be used to match missing persons to recovered bodies, identify decedents from mass casualty incidents, and provide supplementary medical information pertinent to cause of death determinations. Further, card retention schedules, ethical issues, and policy implications of this investigatory approach are explored.

Finally, Part V examines "Ethical Concerns in Forensic Science". Nicholas Lovrich, Herbert Hill, Jessica Tufariello, and Nichole Lovrich's chapter "Detection of Impairing Drugs in Human Breath: Aid to Cannabis-Impaired Driving Enforcement in the Form of a Portable Breathalyzer" examines how Marijuana law liberalization has heightened concern among public safety advocates over cannabis-impaired driving. Given the rapid decline of the presence of THC in the bloodstream, there is an urgent need for the police to have point-of-contact documentation of recent acute exposure to cannabis equivalent to the portable breath test device used for alcohol impairment. Researchers at several locations are working on devices for this task, documenting the presence of THC at levels suspected of being impairing in oral fluids and breath. This will be an important new tool for the collection of forensic evidence related to driving while intoxicated/driving under the influence (DWI/DUI) offenses in due course.

Wendy Hicks' chapter on ethics addresses the questions that have arisen about the necessity for oversight for both technicians and test centers after several highly publicized misadventures at forensic science laboratories have occurred across the nation. This chapter deals specifically with ethical considerations pertaining to certification, accreditation, and state oversight. Due to the importance of forensic findings, experts have advocated for more stringent quality-control mechanisms for technicians and laboratories. Historical legal rulings coupled with contemporary case studies are used to highlight the need and lay the foundation for arguments proposing a variety of oversight mechanisms for forensic laboratories and technicians within the United States.

The text concludes with Randall Grometstein's chapter on forensic ethical implications. In this chapter, we explore the challenges faced by forensic science practitioners in the early twenty-first century, including the lack of a scientific basis for most forensic practices, the role of investigator bias, and the contribution of faulty forensic testimony to cases of wrongful conviction. This state of affairs creates ethical challenges for practitioners, who must balance the uncertainties of the physical world with the requirement of the legal system for a high degree of certainty on which to base judgments of responsibility. Acknowledging the limits of what we know and can know is an important part of the forensic practitioner's job.

This science book presents a collective analysis and review of the existing challenges and directions for state-of-the-art practices found in diverse forensic settings. The range of discourse on the topics discussed reinforces the need for a greater understanding of the use of the most appropriate methodology, which is buttressed by research supporting several forensic science procedures in regular use by highly skilled practitioners. One major challenge for forensic practitioners is the ongoing need to develop systems, processes, and techniques that promote and facilitate a culture of innovation and creativity while responding to the complex scientific questions being raised.

2 Acceptance and commitment therapy and training in forensic settings

Andreas Larsson and Ronn Johnson

The purpose of this chapter is to give and introduction to Acceptance and Commitment Therapy (ACT – said as a word "act") (Hayes, Strosahl, & Wilson, 2012) and its application in a forensic psychotherapy setting. Although part of the umbrella of Cognitive and Behavioral Therapies, ACT stands out with a precise model of psychological health based on a unique model of language and cognition and rooted in a specified philosophical account. ACT conceptualizes mental health problems as caused by a common process, this is called *psychological inflexibility,* and it is when struggling or in other ways avoiding certain experiences get in the way of living a meaningful life. Psychological inflexibility is often described using six core processes: (i) *Experiential avoidance* the propensity to avoid uncomfortable feelings even when doing so limits your choices in life, (ii) *Fusion* the pouring together of thought and the content of the thought. (iii) *Lack of values* not knowing what is personally important or living according to rules about short-term relief or the rules of others, this may lead to (iv) *Impulsivity, inaction, or persistent avoidance.* (v) *Attachment to a conceptualized self* means letting our behaviors be influenced of our ideas of who we are, either by living up to negative thoughts about not being good enough or by persevering in trying something that is unworkable to not be seen as a looser. (vi) *Loss of flexible contact with the present moment* is worrying about the past or future, unable to perceive what is going on in the present. If the readers look at these and recognize themselves it is not a mistake, in ACT, mental health problems are viewed as results of common human processes, the therapist and the client share these problematic processes. In order to help a client become more psychologically flexible the ACT therapist uses six opposing processes of *Acceptance*, *Defusion*, *Values*, *Committed action*, *Contact with the present moment*, and *Self-as-context.*

The ACT model

ACT is a trans-diagnostic approach to psychological functioning. It is based in a behavioral therapeutic paradigm theoretically and is sometimes referred to as a *third wave* of behavior therapy, Wolpean behavior therapy being the

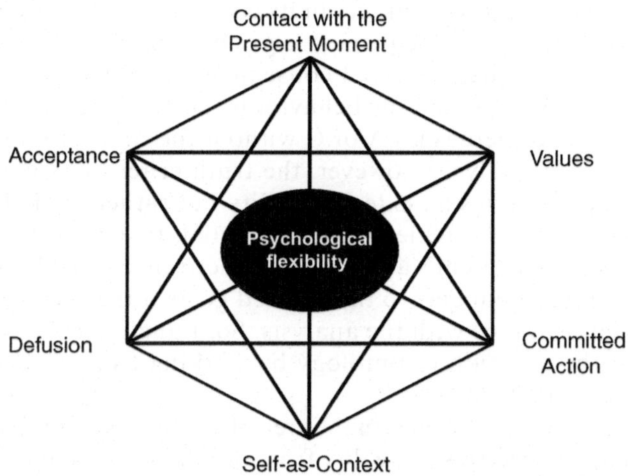

Figure 2.1 The six-sided psychological flexibility model.

first wave and traditional CBT being the *second wave* (Moran, 2008). It is a third wave because it has the behavioral analysis of the first wave but applied in a new way to the problems that the second wave showed. In a nutshell, ACT is about decreasing the negative aspects and increasing the positive aspects of the uniquely human (as far as we know) situation of thought and language (McHugh, 2011).

This chapter presents the most common conceptualization for the ACT model, the six-sided model of psychological flexibility, sometimes referred to as the hexaflex. If you feel like this is not your style, then feel free to consult a model known as the Matrix (Polk & Schoendorff, 2014), or the Open, Aware, and Active model from Steven Hayes' *Get out of your mind and into your life* (Hayes & Smith, 2005). However, the six-sided model is the most commonly used. This application sometimes seems simple, and in order for it to be used with flexibility and fidelity to the model, it is needed to go through the philosophical core of ACT (Figure 2.1).

Philosophical foundations

ACT is built from a specific philosophical foundation and is reticulately, or interdependently, related to other scientific applications, the most clearly related one being the behaviorally based paradigm for language and cognition, Relational Frame Theory – RFT (Hayes, Barnes-Holmes, Dermot, & Roche, 2001). Together they form what is called Contextual Behavior Science, which is a coherent scientific strategy.

The defined philosophical underline in ACT is called *Functional Contextualism*, and it is really a restated and clarified version of BF Skinner's *Radical Behaviorism*, one of the most misunderstood philosophies of the

20th century. Functional Contextualism is a kind of contextualism in that its unit of analysis is always an act in its context, in the traditional ACB (Antecedent-Behavior-Consequence; e.g., Ramnerö & Törneke, 2011) model the (A) antecedent and (C) consequence of a (B) behavior is the context of that behavior and the behavior is the act, without the A and C we don't get B and there is no A or C without the B. So the act and context is the unit of analysis; however, the truth criterion is not how well the model approximates an external reality but rather workability, how well the analysis aids in achieving a stated goal. In a clinical setting, it's not whether or not a client's actions are objectively caused by something but how close the client gets to their stated goals if the therapist and client work in accordance with the analysis. So, Functional Contextualism has avoided the question of ontology by working from a position of a-ontological pragmatism.

This enables some nice things in the area of working with individuals who have different perspectives on reality from the clinician, be they psychotic, political, spiritual, or religious. What matters is that the clinician and client work out what they want to work toward and how they can monitor if they are working in that direction.

Relational Frame Theory

ACT is linked with a basic science account of language and cognition known as RFT (Hayes, Barnes-Holmes, Dermot, & Roche, B, 2001). Initially, ACT was construed as a way to deal with a uniquely human phenomenon, our ability to generate and follow rules, by being reinforced when noting that we are behaving in accordance with a rule. The problem is that some rules, like "I should not feel bad," become dominant over more adaptive rules and over direct experience leading to experiential avoidance. RFT shares the philosophical assumptions of Functional Contextualism and was developed in parallel with ACT, which initially was based on a more traditional behavior analytic approach to problematic rule following that was deemed insufficient to successfully account for human adult psychopathology (Hooper & Larsson, 2015). RFT lays out our ability to build and symbolically react to relations between phenomena, including such relations themselves. Relational framing consists of three phenomenon: (i) *mutual entailment* that if we are taught one thing is like another thing, the other becomes related to the first (if I think of puppies instead of my sentence, then puppies are going to remind me of my sentence); (ii) *combinatorial entailment* that if taught that the other thing is like a third thing, the third thing now becomes related to both the one thing and the other thing (so if I used to think of my child when I thought of puppies, then my child is likely to make me think of my sentence); and third, (iii) *transformation of stimulus functions* that how I think and feel about these phenomenon will follow along in these relations but be transformed through these relations too. The examples before were of sameness, but relations can be almost anything, comparative (A is more or less than B), oppositional (A is the opposite of B), difference (A is different from B), hierarchical (A is part of B), and so on. Imagine there is this new fruit called a *jumal*, imagine biting into it – notice your mind wanting to know what a *jumal* tastes like. In addition, it's like a lemon (mutual entailment) but twice as sour (comparative framing). Notice if your mouth salivates – that is transformation of function. With these three components, RFT enables a behavioral analysis of cognition and all its associated problems that have established a new learning paradigm in behavioral psychology beyond *respondent* and *operant* learning known as *relational* learning. For book-length explorations of RFT, see Dymond and Roche (2013); Hayes, Barnes-Holmes, Dermot, and Roche (2001); and Törneke (2010), and for a book-length application of RFT to psychotherapy, see *Mastering the clinical conversation* (Villatte, Villatte, & Hayes, 2016). What changed in the development of ACT after the advent of RFT was a clearer understanding of motivation, that rule governance will happen and that it can be harnessed to work *for* the client. In ACT, the process that most clearly makes use of this is called Values.

Values

Values work in ACT means looking for what gives the client meaning or vitality in life. It is a way to organize both the clinical work and the life of the client. Values are defined as broad directions rather than concrete goals, enabling a perpetual momentum in the valued direction. This helps when patients fail to attain a goal, or are unable to, based on their current situations. Such as being incarcerated, "it is hard to cook for your children when you are in jail" – as some clients might say. But with values clarification work from an ACT perspective, we would try to get behind that goal, to look for the value behind the goal of cooking for your children. Maybe it is being present with them or being a loving parent? We would work with living that value where they are. That means asking clients how they can act while in jail in a way that moves them in their valued directions. That answer is going to be different, just like not everyone would cook, or value being a loving parent. For example, would acting in a way that increases the chance of getting out on good behavior be a way to move in the direction of being a loving parent, right now? This means valued action can always be taken and that while values work can be, values are not manualized, in order for them to work, they need to be the person's own.

Here is a highly technical definition of values from "Mindfulness for two" by Kelly Wilson and Troy Dufrene: "values are freely chosen, verbally constructed consequences of ongoing, dynamic, evolving patterns of activity, which establish predominant reinforcers for that activity that are intrinsic in engagement in the valued behavioral pattern itself" (2009).

Free choice is a tricky concept for some behavioral therapists but if we look at Skinner's writings freedom just means to be free from aversive control (Skinner, 1976), so avoidance and values are antithetical to each other. This also means that creating values when under threat is going to be difficult. This also explains why sometimes we get stuck in our values work, if the client is really working from a perspective of avoidance – which is aversive control. The most direct way to help with this is to encourage the client to imagine a place and time when they are not under aversive control, using a perfect world scenario or similar. Sometimes the aversive control is too dominant, then the processes of Acceptance and Defusion are very useful to reduce the aversive control and enable a freer choice. The rest of the definition can be said to mean that values are symbolically created and that when we act on our values the behavior and the value become one. So even when we cannot act in ways we would like to, we can choose a value that can enhance our actions.

An example of an exercise used to construct values is the tombstone exercise. In it, the client is asked to imagine visiting her or his own burial. Who would they like to see there, who has touched their life in ways that you would like them to attend? What sort of conversations would they have about the client? What sort of themes would they like to have lived? What would be their epitaph? "Here lies John Doe: He never had anxiety?" or "Here lies John Doe: He cared for his community" ask the client to choose one in a world where both are possible.

Often a life compass will be utilized; this is a list of 10 areas in life that most people can find values in: *Intimate relationships* – what sort of qualities are present in an ideal relationship with a spouse or partner. How does that enable the client to act in that area? *Family* – the kinds of relationships do you want with your parents, siblings, and others that may be part of your family. *Parenting* – the sort of parent one wants to be and the qualities one looks for in the relationships with one's children. *Friendships* – when with a close friend, what qualities are present? How is the client when they are being a good friend? *Social standing* – what societal impact does the client want to do? How does they want to act in a community? *Education/personal growth* – what part of learning something or of developing oneself is most meaningful to the client? *Physical health* – we all know physical health is important, but what makes the client want to be healthy? *Work* – what ideals and qualities would make an occupation worthwhile for a client, what to work for and how to go about it? *Recreational* – the other side of work is the time off, what recreation looks like and what makes it important. *Spiritual/religious* – qualities or meaningfulness is important to the client with regard to their religion, its practices, or spiritual life. When listing this, barriers also show up that can be approached using the other processes in the model. Oftentimes goals are reported in response to these questions about areas, such as "I want a degree" or "I want my partner back." Then we would try to look for the value in that goal, what motivates or gives meaning to that goal.

Values are something we can always move in the direction of, so for example being a caring and present partner is not something we can just check off a list and then focus on something else but requires consistent coming back to the present with parenting. On the other hand, we strive to make values work possible all the time. So if we have a client that values being a present partner but isn't able to see their partner because they are incarcerated, see if there are ways of acting with other inmates or staff that can be kind of a "practice" for when they see their partner – and thus a small step in their valued direction.

Values work is important to ACT – the entire model is aimed at living more vital lives – so stay with it, particularly with clients who have learned distrust by living in environments where they have been exploited or belittled it can take a lot of work to share closely held values. But exploring values has its own rewards as clients experience the value by talking about it, and that can increase its motivating qualities.

Committed action

Committed action is the process where the values "hit the ground" and the client acts in accordance to their values. For example, finding small things to do in everyday life that shows the importance you put in being a part of a social group, or being present in the here and now as a way to be more present as a parent when you get to see your children. Committed action is about finding and taking steps in a valued direction. When values clarification finds values that are important to the person, it is often easy to think of large steps to take, like getting an education, finding a job, getting back custody of one's children, or saving the world. However, if we want to build some self-efficacy in acting in valued directions it is important to find steps that the person can take at a fairly high rate. So breaking down larger goals into smaller, manageable steps. Studying for an hour a day for example.

Committing to a valued direction doesn't mean it is all going to be smooth sailing, anyone who's tried behavior change knows this. What committing means is sticking with it. So noticing you haven't been paying attention to your spouse even thought that was a step toward being a present and attentive partner, that is a good thing because you can now re-engage in that presence. As long as you turn back one more time than you turn away, you're on the path. If you didn't make an hour of studying, well did the minutes you spent at it feel vital?

Present moment awareness

Many clients will not easily notice when they veer from a valued life in the first part of treatment. And that is one part of the purpose of present moment awareness, or the contact with the present moment process. This will both enable noticing when impulsive or automatic behavior takes them off course and enhances the experience of travelling in a valued direction, this makes it much different from the sort of clenched teeth type accepting that many do or deciding on a behavior based on values but then suppressing internal experiences during the behaving part. Training clients to be able to be present in the present, on purpose, enables the identification of actual triggers and how our behavior impacts on our environment, as it is and not as we say it does. This enables a more effective situation-based action. It also helps in identifying urges and impulses, resulting in less reactive behavior.

In ACT, present moment awareness is sometimes trained in experiential exercises, like in mindfulness-based therapy traditions. One example can be to sit and just notice the sensations around breathing in and out for about 15 minutes a day. There are many scripts for this type of exercises, for example in the core ACT book (Hayes, Strosahl, & Wilson, 2012). For many, sitting still for 15 minutes might seem impossible, and ACT considers present moment awareness a behavior – something that can be trained. This means we can start with much smaller exercises. For example, noticing three things we can see, three things we can hear, and three things we can feel or sense, then two of each and one of each, while encouraging a non-judging stance of the focused on phenomenon. In addition, ACT does not mandate formal practice but rather encourages informal practice. Informal practices can be found in everyday activities, such as brushing teeth, doing dishes, or taking a shower. Encouraging exploring the sensations and experiences in the present. ACT also encourages clinicians to use in-session prompts for present moment awareness such as "what are you feeling as you are telling me this? Where is that located in your body?" or "can you come back to this moment here and now?"

Consistent training in present moment awareness makes it more readily accessible for the client and may introduce time into situations where they otherwise would experience an automaticity or impulsivity, introducing more choices because they are more able to notice different stimuli in their environment, both internal and external.

Self-as-context

Self-as-context is the realization that, beside any ontological stance we want to take, anything you have experienced has been from the same perspective. That there is a perspective of experiencer to the experiences, of seer to the seen, and hearer to the heard may be easy to think of, but also of thinker to the thought and feeler to the emotions inside and outside, creating a distance to them. At the same time, we are where these events take place so we are our experiences too. So if we can be both the experience and the experiencer, it is appropriate to conclude that we are more than any or even all of our experiences. Just seeing ourselves as different easily becomes "if I am different I should feel it" that that feeds the experiential avoidance or, in ACT terms "the control agenda." This has been neatly demonstrated empirically where people instructed to view themselves as more than their experience more effectively dealt with a negative experience than those just instructed to see it as different from them (Foody, Barnes-Holmes, & Barnes-Holmes, 2013). And if we are able to contact that experience of being "more than," we are much more likely to be able to take effective action in valued directions because instead of being pushed or pulled by internal events, like emotions or thoughts, the RFT analysis of self as context is that we use hierarchical (A part of B) framing of self, think of it as a tree structure on your computer hard drive if you will, where the folders higher up in a hierarchy is made up of a bunch of subfolders. Making the clients' self as more than their stories about themselves, for example, around being a convict or criminal or the place or identity as part of the offending subculture. If this enables a more flexible relating to that nefarious content, if loosing the standing in their social group is less threatening when the self is not so narrowly defined makes for a better adjustment. Even going further than that, nurturing the position of self as being the transcendent perspective from which everything has been experienced even the self-hierarchy itself can be held more lightly helping clients move in a direction of vitality, prosociality, and meaning.

Practically, self-as-context can be utilized in a number of ways. When it comes to exercises that activate this process, the most common one is the Observer exercise first mentioned in the 1999 ACT book (Hayes, Strosahl, & Wilson, 1999). This exercise invites participants to look at memories at different times during their life, noticing that they were there and that in some ways, the same I that was there then is here now – even if many things that felt like integral parts of their experience is different. They are then taken on a journey in their current life and their emotions and thoughts, continuously prompted to notice that they experience themselves as being separate from the experience. Finally, they are likened to the sky, with thoughts like clouds and the weather like emotions, able to contain all and unchanged in a fundamental way by any thunderstorm.

Another exercise is even easier: write down "I am:" five times on a piece of paper, and answer it five times in different ways, notice you are at least this many things. Now strike them out one after the other. Notice that part of you live gone. Are you still able to notice that piece missing? So then, you are not just those things! You are also the perspective that would notice them missing. And that would not change.

Acceptance

Lack of acceptance, or experiential avoidance in ACT terms means avoiding, by distraction or suppression the internal experience that one is going through at the moment. It has been defined as being "unwilling to remain in contact with particular private experiences… and takes steps to alter the form, frequency or situational sensitivity of these experiences even though doing so is not immediately necessary" (Hayes, Strosahl, & Wilson, 2012, pp. 72–73).

The experiential avoidance is sometimes very clear, like the avoidance of a socially anxious client refusing to participate in social activities or being silent in conversations; sometimes it's more opaque, like a depressed client who is staying passive as an avoidance of feeling even worse. Anger is another example of when acceptance work is effective, instead of acting on the anger, practice allowing it to arise, making space for it without acting on it opens up new avenues for action when experiencing anger. Acceptance aims to reduce the struggle with these internal experiences, making space for them and saying "yes" to having any experience that might arise while moving in valued directions.

To exemplify an acceptance exercise might be to ask the client to notice any experiences they are struggling with at the moment, anywhere in their body. Or think about something that they find difficult to cope with. Then to notice how this makes them feel, going through each part of their body and notice if there is struggle present there and what it feels like. Then the client is asked to let that struggle end. To make room for the sensation or feeling, observing the sensation as would a curious child or objective scientist. Relax any tensions around this uncomfortable sensation. Allow it to be there. Because minds typically ask us questions and work from a control agenda, questions arise about how long this needs to go on and so forth. ACT would both encourage the idea that we can only accept in the now, meaning that if we fear the future or that this event that we are trying to accept will just get worse, we can only accept what that gives rise to in the present. On the other hand, we might need to work more specifically with those thoughts using defusion strategies to allow the thoughts to simply be there in a way that they will not be a barrier to acceptance and committed action.

Defusion

The way ACT deals most directly with negative thoughts is called defusion, which is sometimes shortened into "noticing that a thought is a thought, no more, no less" because human beings live inside of our thoughts this can be quite a challenge. We know that most people deal with negative thoughts in some way or another and particularly people with mental health issues. Thanks to the wonderful work of Daniel Wegner (Wegner & Gold, 1995; Wenzlaff & Wegner, 2000), we both know that about 80% of people deal with negative thoughts by trying to suppress them, and that that is a futile exercise that often leads to an increase in the thoughts one is trying to get rid of. This has been explored from an RFT perspective (Hooper, Saunders, & McHugh, 2010; Hooper, Stewart, Duffy, Freegard, & McHugh, 2012) and is thought to be that once you try to suppress one thought you are going to have to suppress all the mutually and combinatorially related thoughts you might have – which is a futile task. Defusion means trying to relate differently to all thoughts, to become aware of the process of thinking rather than its products. As an example, a recent study (Larsson, Hooper,

Osborne, Bennett, & McHugh, 2015) utilized a defusion intervention of adding "I am having the thought that" to a negative thought – for example "I am unworthy of love" becomes "I am having the thought that I am unworthy of love" – over the course of 5 days. Compared to a standard cognitive restructuring intervention of identifying thinking errors, evaluating the evidence for and against a thought the defusion group outperformed the restructuring instruction on self-rated reductions of believability, negativity, and discomfort as well as increased willingness to have the thought. In fact, the defusion group on average no longer rated the thought as believable or uncomfortable and they were no longer unwilling to have it, indicating they might not be so controlled by struggling with the thought. If a client can stop struggling with thoughts, that often frees up a lot of time to engage in valued action. In correctional settings, it can mean defusing from rules about how society is, should or is not, like it's unfair that they belong to a discriminated against minority.

Defusion does not, as opposed to many other approaches to negative thoughts, presuppose that the thought is factually wrong. Some of the worst barriers to action are true thoughts. For example, it is true that ex-convicts have a harder time getting a job than people who have no criminal record. But is it always a helpful thought? If not, maybe just let that thought be there, you don't have to fight it or argue against it or try to distract yourself. Notice that it is a thought and get back to your committed action, writing a CV for example. It might be thoughts of what is fair or unfair, what makes hurting someone ok or not ok. But is it helping the client move in a valued direction?

Bringing the model to life

These six processes coexist and together make up ACT, although they are not the only way to parse the ACT approach to psychological health as mentioned before. Focusing on one process does not mean that the other processes are done with or not important. They are an educational tool, a prism to view clients and their problems through. In individual therapies, the therapist often uses all of the processes within a single session. If you look at Table 2.1, you see that some of the clinical examples are similar over different dimensions, meaning there are different angles of approach.

ACT differs in the goal of treatment from many other therapies in that symptom reduction is not a primary target, but valued living. Thus, the Openness processes in Table 2.1 are not primarily used to minimize the symptoms but to make the client more able to find values (based on the freely chosen part of the definition) and to relate to barriers to the committed actions. These Active engagement processes facilitates creating vitality and purpose that actually motivates the client engaging with the work. The processes of contact with the present moment and self-as-context are termed awareness here and are useful in tandem with both of the other pairs, in order to

Table 2.1 Clinical examples of the psychological inflexibility and psychological flexibility processes divided into the open, aware, and active pairs from Hayes and Smith (2005)

Inflexibility process	*Clinical example*	*ACT process*	*Clinical example*
Openness processes			
Experiential avoidance	Either closes up when talking about emotional content or acting out in order to avoid certain experiences. Narrows repertoire to avoid unpleasant experiences.	Acceptance	Letting go of the struggle with experiences.
Fusion	Thoughts standing in the way of vitality; ideas about what is possible, how the world is or should be.	Defusion	Asking client to notice the function of thoughts.
Active engagement processes			
Lack of values	Governed by avoidance, compliance, or excessive rules.	Values	Values clarification Perfect world scenario Tombstone exercise Values compass.
Impulsivity, inaction or persistent avoidance	Falling off the wagon. Acting out. Being apathetic or passive.	Committed action	Maintaining valued actions.
Awareness processes			
Attachment to a conceptualized self	Acting in line with a rigid sense of identity.	Self-as-context	Fostering a perspective of self as separate from experiences and stories about "I am."
Loss of flexible contact with the present moment	Occupied with past/future, ineffective functioning.	Contact with the present moment	Asking client to notice present-moment events and direct consequences of action.

become aware of what thoughts and emotions are barring the way and in order to both identify opportunities for committed action and to be aware of the outcomes. Perhaps what someone thinks will be a vital action will not be? Likewise, acceptance and defusion are vital to enable stepping into the present moment. Because if there is suffering in a person's life, it is going to be present in the present moment. Let's say a therapist ask someone to become more present in the room as you are talking about them valuing relating to other people in a different way. Then thoughts about what the therapist is thinking about them start to show up, and fear of rejection. The therapist can model acceptance by staying with the emotion, noticing the client emotionally closing up and asking what happens for them, encouraging contact with the present moment. As content in the form of beliefs about being unworthy of connection or similar shows up the therapist can encourage using acceptance of experiences and defusing from thoughts while gently reminding the client of the values being honored by the work being done.

Therapeutic approach

One major hurdle in working with inmates that is shared with any type of mandated counseling or behavior change work – motivation. Since the client has not chosen to be where they are, motivation is often low. Thus motivating clients to not just appear physically but psychologically and emotionally in therapy is going to be difficult, as is motivating putting in effort into their treatment. For many inmates, being incarcerated is only the latest in a long line of difficulties and many experience that they are not in control of their life. Oftentimes creating a sense of control drives antisocial behaviors in a correctional setting.

Besides RFT, Functional Contextualism and the six processes, the therapeutic stance in ACT is of importance in delivering the format. A nontechnical anchor to think of is Compassion and to model the processes in the model. Compassion has been described as both being aware of the suffering of others and to wish to do something about that suffering (Gilbert, 2009). If we apply it to an ACT model, it means that the therapist values the clients' suffering and wants to act in ways that reduce that suffering (Values and committed action). In order to do that the therapist needs to be in contact with the present moment in order to detect the signs of suffering, to be accepting the emotions that show up when someone is experiencing suffering, and to hold thoughts about what might be possible to affect judgments of the client and of the world lightly (defuse) that might otherwise be barriers to acting in a compassionate way. Self-as-context helps therapists take perspective on the situation. This also models the desired processes to the clients, for example, a client who tests the therapist by detailing violence, a therapist can then model acceptance by staying on the subject and showing compassion to the client.

Empirical support: applications of ACT model in forensic and criminal settings

While ACT has a solid outcome base at the moment (e.g., A-Tjak et al., 2015; Hooper & Larsson, 2015), the outcome data of ACT in forensic and correctional settings are scarce, in fact only three papers have so far applied ACT to an inmate population (González-Menéndez, Fernández, Rodríguez, & Villagrá, 2014; Lanza, García, Lamelas, & González-Menéndez, 2014; Lanza & Menéndez, 2013) in treating substance abuse, but with good results, which is encouraging. In addition, the longer-term effects in the Gonzáles-Menéndez (2014) study indicate a more stable improvement compared to traditional CBT rhymes with the idea of using the values component in ACT.

Recently a group delivered format of ACT-informed training called "Achieving your potential" has been implemented to "facilitate prosocial behavior change, improving interactions with staff, compliance with unit rules, reducing harm to self and others, realistic decision-making, and development of short and long-term goals" with so far promising preliminary results of increasing psychological flexibility (Rainey-Gibson & Davis, 2016) showing that a group-format delivery of ACT is feasible in this setting.

Staff

There is another group in the area of forensic and correctional psychology that is often overlooked, the group that is employed to work with inmates. Correction officers offer PTSD at twice the rate of American war veterans (Spinaris, Denhof, & Kellaway, 2012). This surely impacts on the quality of rehabilitation or psychological treatment they can help inmates get through. One of the basic principles of ACT is that suffering is universal, and based on processes that are not inherently pathological, it lends itself to nonclinical applications for both prevention and actually enhancement. There are applications of ACT in the area of organizational management and occupational psychology (Hooper & Larsson, 2015); this has been used effectively to help social workers cope with work-related stress (Brinkborg, Michanek, Hesser, & Berglund, 2011), increased productivity in call-center staff and also indications that interventions at management levels trickle down in the organization to increase mental health at other levels letting companies make more money (Hooper & Larsson, 2015). There is also a new marriage of ACT and evolution science (Wilson, Hayes, Biglan, & Embry, 2014) in the shape of prosocial wherein groups and organizations are made to function better through a combination approach.

PROSOCIAL

The prosocial (www.prosocialgroups.org) model starts off in the work of Ostrom, who won the Nobel Memorial Prize in Economic Sciences for research on what makes groups able to successfully handle common-pool resources, one of the toughest challenges of any group. Ostrom and colleagues developed a set of principles for how groups worked who managed to solve these types of problems, regardless of geographical location or cultural specifics. The rules laid out by Ostrom (1990) are:

- Group identity/purpose
  - Ostrom found that a strong group identity is vital for the group to work, not by arbitrarily forming a group but knowing who else are in the group, what it takes to become and stay a member and what the purpose of being a member is.
- Proportionate benefits and costs
  - It is important for groups that everyone puts in their share of work over time, that if someone goes above and beyond to facilitate the group's purpose they read benefits accordingly and that if someone is given extra benefit from the group, that is mirrored in doing more than the rest. For example, a shift leader has more pay than her colleagues because of her added responsibility.
- Consensus decision-making
  - This makes everyone invested in the decision, as we are more likely to work for something that is an individual choice. In addition, good decisions are often based on local knowledge that individuals in the group have and that higher up leadership might lack in certain aspects.
- Monitoring
  - Groups decide how to monitor that people comply with decisions and proportionate benefit/cost ratio. Even if groups have a high level of trust, it is good practice to decide on a monitoring system, if people are able to lapse or actively play the group then it is less likely to succeed.
- Graded sanctioning
  - If transgressions are identified the sanctioning needs to be graded. Most transgressors respond quickly to a friendly notice, if it is delivered quickly and reliably. However, more severe or repeated transgressions need to have a graded ladder of sanctions to finally being expelled from the group.
- Fast and fair conflict resolution
  - When conflicts arise, they need to be resolved quickly and fairly, possibly through an elected member of the group that can be trusted to be impartial. It can be one chosen on the spot by both or more parties.
- Local autonomy
  - Groups nested within larger organizations, such as correctional staff it needs enough authority to organize themselves in accordance with the above in order to work effectively.
- Polycentric governance
  - Groups nested within a larger society have multiple dependencies: federal and state regulations, unions etc. and are required to relate to all of these in a manner consistent with the above rules.

Most of these will be recognized from organizational literature or management studies. What Act brings to these eight rules is a "how" in that for almost each of these steps, internal events will show up to limit them. So merging these rules with the ACT model helps deal with barriers to these rules and become more effective groups. Readers might get all the way to the last two bullet points before thinking "this can't be done within a correctional setting" and then defusion might be useful. What if it could be adapted?

In the prosocial approach the Matrix ACT tool is used, it is a simple 2 × 2 matrix that helps people sort out internal events from external behavior vertically and experiential avoidance from values guided behavior laterally. This enables a quick guide to the kinds of behaviors that the group wants to see and what stops people from performing. ACT interventions may then help to deal with the barriers. This model was used to deal with the Ebola crisis in Sierra Leone; it was so popular that it is now being used to help with domestic problems in the same country. In North America, it has been used in dealing with the environmental problems associated with logging. Business applications of prosocial have been implemented in Australia, North America, and Europe (Styles, 2015). There is an internet platform that is set up to help groups get started using the model and that is free to use, as long as the data can be used in improving prosocial.

Do you want to know more?

If the reader is intrigued to know more about ACT and how to apply it to their own practice, consider joining the Association for Contextual Behavior Science (www.contextualscience.org) where practitioners and both basic and applied researchers comingle and drive this work ahead. It has a free depository for clinical protocols and an active community where generosity and sharing is used strategically to disseminate ACT and RFT. For an example on how to use a group ACT format in correctional settings, see Chapter 3 in *Forensic CBT: A Handbook for Clinical Practice* (Amrod & Hayes, 2013). The core ACT book is *Acceptance and Commitment Therapy: The process and practice of mindful change* (Hayes, Strosahl, & Wilson, 2012) and is a good place to get started. The two greatest ways of learning ACT is to attend experiential workshops and courses, and to start applying it in your own practice.

References

Amrod, J., & Hayes, S. C. (2013). ACT for the Incarcerated. In R. C. Tafrate & D. Mitchell (Eds.), *Forensic CBT: A Handbook for Clinical Practice* (pp. 43–65). Oxford: Wiley. doi:10.1002/9781118589878.ch3

A-Tjak, J. G. L., Davis, M. L., Morina, N., Powers, M. B., Smits, J. A. J., & Emmelkamp, P. M. G. (2015). A Meta-Analysis of the Efficacy of Acceptance and Commitment Therapy for Clinically Relevant Mental and Physical Health Problems. *Psychotherapy and Psychosomatics*, *84*(1), 30–36. doi:10.1159/000365764

Brinkborg, H., Michanek, J., Hesser, H., & Berglund, G. (2011). Acceptance and Commitment Therapy for the Treatment of Stress among Social Workers: A Randomized Controlled Trial. *Behaviour Research and Therapy*, *49*(6–7), 389–398. doi:10.1016/j.brat.2011.03.009

Dymond, S. & Roche, B. (2013). *Advances in Relational Frame Theory: Research & Application.* Oakland, CA: New Harbinger Publications.

Foody, M., Barnes-Holmes, Y., & Barnes-Holmes, D. (2013). An Empirical Investigation of Hierarchical versus Distinction Relations in a Self-based ACT Exercise. *International Journal of Psychology and Psychological Therapy 13, 3, 373–388.*

Gilbert, P. (2009). Introducing Compassion-Focused Therapy. *Advances in Psychiatric Treatment, 15*(3), 199–208. doi:10.1192/apt.bp.107.005264

González-Menéndez, A., Fernández, P., Rodríguez, F., & Villagrá, P. (2014). Long-Term Outcomes of Acceptance and Commitment Therapy in Drug-Dependent Female Inmates: A Randomized Controlled Trial. *International Journal of Clinical and Health Psychology, 14*(1), 18–27. doi:10.1016/S1697-2600(14)70033-X

Hayes, S. C., & Smith, S. (2005). *Get out of your mind and into your life: The new Acceptance and Commitment Therapy.* Oakland, CA: New Harbinger Publications.

Hayes, S. C., Barnes-Holmes, D., & Roche, B. (2001). *Relational Frame Theory. A Post-Skinnerian Account of Human Language and Cognition.* New York: Kluwer Academic/Plenum Publishers.

Hayes, S. C., Strosahl, K. D., & Wilson, K. G. (1999). *Acceptance and Commitment Therapy: An Experiential Approach to Behavior Change.* New York: Guilford Press.

Hayes, S. C., Strosahl, K. D., & Wilson, K. G. (2012). *Acceptance and Commitment Therapy: The Process and Practice of Mindful Change.* New York: Guilford Press.

Hooper, N., & Larsson, A. (2015). *The Research Journey of Acceptance and Commitment Therapy (ACT).* London: Palgrave Macmillan.

Hooper, N., Saunders, J., & McHugh, L. A. (2010). The Derived Generalization of Thought Suppression. *Learning & Behavior, 38*(2), 160–168.

Hooper, N., Stewart, I., Duffy, C., Freegard, G., & McHugh, L. A. (2012). Modelling the Direct and Indirect Effects of Thought Suppression on Personal Choice. *Journal of Contextual Behavioral Science*, 1–10. 1, 73–82.

Lanza, P. V., & Menéndez, A. G. (2013). Acceptance and Commitment Therapy for Drug Abuse in Incarcerated Women. *Psicothema, 25*(3), 307–312. doi:10.7334/psicothema2012.292

Lanza, P. V., García, P. F., Lamelas, F. R., & González-Menéndez, A. (2014). Acceptance and Commitment Therapy versus Cognitive Behavioral Therapy in the Treatment of Substance Use Disorder with Incarcerated Women. *Journal of Clinical Psychology, 70*(7), 644–657. doi:10.1002/jclp.22060

Larsson, A., Hooper, N., Osborne, L. A., Bennett, P., & McHugh, L. (2015). Using Brief Cognitive Restructuring and Cognitive Defusion Techniques to Cope with Negative Thoughts. *Behavior Modification*, 1–44. doi:10.1177/0145445515621488

McHugh, L. (2011). A New Approach in Psychotherapy: ACT (Acceptance and Commitment Therapy). *The World Journal of Biological Psychiatry, 12*(Suppl 1), 76–79. doi:10.3109/15622975.2011.603225

Moran, D. J. (2008). The Three Waves of Behavior Therapy: Course Corrections or Navigation Errors? [Special Issue]. *The Behavior Therapist, 31*(8), 147–157.

Ostrom, E. (1990). *Governing the Commons: The Evolution of Institutions for Collective Action.* Cambridge University Press. ISBN 0-521-40599-8.

Polk, K. L., & Schoendorff, B. (2014). *The ACT Matrix: A New Approach to Building Psychological Flexibility across Settings and Populations.* Oakland, CA: Context press.

Rainey-Gibson, E., & Davis, J. (2016). Acceptance and Commitment Therapy in a Maximum Security Prison. Presented at: *World Congress of the Association for Contextual Behavior Science*. Seattle, USA.

Ramnerö, J., & Törneke, N. (2011). *The ABCs of Human Behavior: Behavioral Principles for the Practicing Clinician*. Oakland, CA: New Harbinger Publications.

Skinner, B. F. (1971). *Beyond Freedom and Dignity*. New York: Alfred A. Knopf.

Spinaris, C. G., Denhof, M. D., & Kellaway, J. A. (2012). Posttraumatic Stress Disorder in United States Corrections Professionals: Prevalence and Impact on Health and Functioning. Retrieved from http://nicic.gov/library/026910

Styles, R. (2016). Outstanding Evidence for PROSOCIAL in a Government Agency Setting. *PROSOCIAL Magazine*. Retrieved from http://magazine.prosocialgroups.org/outstanding-evidence-for-prosocial-in-a-government-agency-setting/

Törneke, N. (2010). *Learning RFT: An Introduction to Relational Frame Theory*. Oakland, CA: Context press.

Villatte, M., Villatte, J. L., & Hayes, S. C. (2016). *Mastering the Clinical Conversation*. New York: The Guilford Press.

Wegner, D. M., & Gold, D. B. (1995). Fanning Old Flames: Emotional and Cognitive Effects of Suppressing Thoughts of a Past Relationship. *Journal of Personality and Social Psychology, 68*(5), 782.

Wenzlaff, R. M., & Wegner, D. M. (2000). Thought Suppression. *Annual Review of Psychology, 51*(1), 59–91.

Wilson, D. S., Hayes, S. C., Biglan, A., & Embry, D. D. (2014). Evolving the Future: Toward a Science of Intentional Change. *Behavioral and Brain Sciences, 37*(4), 395–460. doi:10.1017/S0140525X13001593

Wilson, K. G., & DuFrene, T. (2009). *Mindfulness for Two. An Acceptance and Commitment Therapy Approach to Mindfulness in Psychotherapy*. New York: Guilford Press.

3 Advances and emerging clinical forensic psychological trends with Juvenile Fire Setting and Bomb Making behavior

Ronn Johnson and Tanna M. Jacob

Juvenile Fire Setting and Bomb Making (JFSB) functions as a major public safety risk (Johnson, 2016). As an area of practice, JFSB poses a challenge from both a clinical and forensic psychological perspectives. Forensically, a significant part of the empirical work has focused on identifying risk factors for recidivism (Peters and Freeman, 2016) or to designate an assessment category that may be used to identify what type of offender status could be assigned (Johnson, Beckenbach, and Killbourne, 2013). Globally, government entities and researchers have evaluated the prevalence of JFSB with the intent being to highlight characteristics among those youth who have a proclivity for this type of unwanted behavior. However, the only consistent finding among all the research is there are multiple, complex variables working together within each individual that may or may not culminate into engaging in juvenile fire-setting behavior. Nonetheless, one fact does remain consistent, JFSB behavior is an international phenomenon that coincides with diverse costs (e.g., public safety, property damage, and human injury or death) (Johnson, 2016).

The well-documented public safety issues and associated costs associated with JSFB has also fueled an emerging interest in this area as a clinical forensic psychological research topic (Johnson, 2016). For example, investigations conducted in Australia (Tanner, Hasking, and Martin, 2015; Watt, Geritz, Hasan, Harden, and Doley, 2014), Canada (Hanson, Mackey, Atkinson, Staley, and Pignatiello, 1995), Sweden (Ekbrand and Uhnoo, 2015), and the United States (Federal Bureau of Investigation, 2012; Kolko and Vernberg, 2017; Putnam and Kirkpatrick, 2005), among others; demonstrate an ongoing trend in examining the prevalence and impact of JFSB. Still, there remain several challenges for the diverse group of professionals who are charged with responding to these cases. For example, a representative number of mental health professionals who come in contact with these juveniles often lack the relevant expertise required to identify the central underlying public safety issues beyond the initial incident, or properly assess for risk, create an individual and community safety plan, and utilize appropriate interventions or introduce evidenced-based treatment approaches (Johnson, 2015).

In terms of JSFB prevalence, roughly between 1993 and 1997, the US Bureau of Alcohol, Tobacco, and Firearms found that juveniles were involved in approximately 13,500 cases of actual or attempted explosive incidents. In 2005, 54% of the arson arrests in the United States were juveniles (Putnam and Kirkpatrick, 2005). Since 2007, the National Fire Protection Association noted that approximately 40% of those arrested for arson were under the age of 18. Additionally, of the roughly 113,000 annual fires intentionally set by those juveniles, the property damage has exceeded one billion dollars. Furthermore, the estimated personal injury or loss of life as a result of those fires is over 1,700 (Peters and Freeman, 2016). In 2012, the US Federal Emergency Management Agency (FEMA), who tracks federally reported juvenile arson cases, reported that for that year, JFSB resulted in an average of 1,000 injuries or death and caused over $280 million in property damage (Johnson, 2015). That same report stated the number of child-related fires occurring between 2005 and 2009 was over 56,000. These statistics are not unique to the United States. In Australia, adolescent offenders accounted for 38% of arson cases (Watt et al., 2014). In 2009, Sweden reported 230 documented juvenile fire-setting incidents that occurred specifically on school property resulted in annual costs of approximately 50 million euro (Ekbrand and Uhnoo, 2015). Research conducted in 2013 on 1,698 juvenile arson cases in San Diego County, California found that in 550 of these cases the juvenile involved was age 10 or younger. Of those, approximately 216 were elementary school between first and sixth grade. Furthermore, of the total cases referred, 31% indicated the JFSB took place on the school bus, at the bus stop, or on campus (Johnson, Jones, and Said, 2013). It is not uncommon for these juveniles to share certain risk factors associated with the expression of JFSB. These factors include physical, emotional, and/or sexual abuse, unsettled family environments, a history of familial criminal behavior, or medical and/or psychiatric conditions, which can contribute to the engagement in maladaptive behaviors such as JFSB (Del Bove and Mackey, 2011; Dickens et al., 2009; van Goozen, Fairchild, and Harold, 2008; Kolko and Kazdin, 1994; Trickett, Negriff, Juye, and Peckins, 2011).

The negative effects of experiencing community or familial violence, abuse of any nature, and/or interpersonal trauma such as sexual or physical assault during childhood and/or adolescence have the potential to severely impact multiple life domains as the individual attempts to find ways to cope. The reactions to trauma may be gauged on a continuum, which depends upon many factors in juveniles (e.g., developmental history, trait resiliency, and level of adversity). The role these factors play on the expression of behavior may be explained using a particular theoretical approach used by the assessing clinician. For example, the Medical Model, Diathesis Stress Model, and the Behavioral/Environmental Model all cite developmental and environmental influences in one's ability or inability to manage the impact of trauma (Inaba and Cohen, 2014; Overbeek, Vollebergh, Meeus, Engels, and Luijpers, 2001). While much of the research focuses on JFSB as a criterion of a yet-determined mental diagnosis (Johnson et al., 2013), specifically antisocial personality disorder (Vaughn et al., 2010; Watt et al., 2014) or the much-needed development of a more complete assessment tool (Johnson et al., 2013), this advanced and emerging trends chapter examines the need for mental health professionals to examine JFSB behavior not as a symptom of mental illness, but rather as an attempt to cope and relieve the stressors occurring from within their environment. From this point of view, JFSB must be reviewed from a more systemic biopsychosocial perspective in order to properly assess, psychoeducate, and treat a diverse group of clients. So how can professionals working with children and adolescents recognize the potential risk factors associated with JFSB in order to implement psychoeducation programs and prevention strategies aimed at lowering rates of recidivism?

Forensic risk factors and predictors of psychopathological behavior

Gender assumes a role in the likelihood of engaging in JFSB for the purposes of understanding prevalence, such that males are more likely than females to set fires (Del Bove, Caprara, Pastorelli, and Paciello, 2008; Kolko, 1985). Research continues to examine the exact role gender assumes as a determinate among those who engage in JFSB. For juveniles who engage in JFSB, perhaps the central question to be asked is what purpose is the behavior serving to their current experience or crisis (i.e., avoidance or attention seeking) (Johnson et al., 2013)? To date, multiple findings have identified a nexus between JFSB to a history of abuse, familial/life stressors, low socioeconomic status, and interpersonal conflicts, further adding to the narrative that negative systemic vulnerabilities correlate with an inability to respond adaptively to stress (Alder, Nunn, Northan, Lebnan and Ross, 1994; Johnson et al., 2013; Kolko and Kazdin, 1991; Kolko and Kazdin, 1994; McCarty and McMahon, 2005; Root, MacKay, Henderson, Del Bove and Warling, 2008; Tanner et al., 2015; Watt et al. 2014). It is not atypical for adolescents to struggle with ways in which to successfully manage life and/or environmental stressors by engaging in questionable or dangerous behaviors. As a result of their attempt to control their negative internal emotions, they utilize maladaptive or psychopathological coping strategies, which only exacerbates the problem. Therefore, the juvenile's perceived or actual lack of control and/or low cognitive reappraisal when interpreting or evaluating various aspects of their lives may be a predictor the use of maladaptive strategies such as JFSB (Elzy et al., 2013; Gottfredson and Hirschi, 1990).

Research conducted by Ferdinand and Verhulst (1996) identified a correlation between internalizing and externalizing problems in adolescents, specifically the presence of anxiety/depression and engaging in delinquent behavior. Results of the 6-year longitudinal study determined the need for increased emphasis on the correlation of environmental and social influences, such as familial history of mental illness and discord, experience of trauma and/or abuse, and adverse living conditions when determining risk level of developing or engaging in internal and/or external behaviors (Overbeek et al., 2001).

Tanner et al. (2015) utilized the BIS/BAS scale (Carver and White, 1994), which measures positive and negative personality dispositions such as extraversion, positive affectivity, positive and negative temperament, reward seeking, harm voidance, and susceptibility to punishment. Findings indicate personality variables predicted JFSB such that adolescents scoring high on areas of behavioral activation (fun seeking and drive) but low on behavioral inhibition (impulse control) were predictors of JFSB. Further findings identified engagement in nonsuicidal self-injury coupled with poor sociodemographics and/or experiencing adverse life events may result in an increased inability to successfully respond to stressors that then amplified the likelihood of using fire to cope (Tanner et al., 2015).

Of course, not all individuals who experience biopsychosocial adversities engage in JFSB. Individual levels of resiliency and other characteristics, such as quality of support system, act as protective factors. These finding underscore the importance of using multiple assessments to identify protective factors and individual strengths as well as potential underlying psychological processes. In doing a more thorough and ongoing assessment, examiners reduce the risk of misdiagnosis, over diagnosis, or missed diagnosis when interviewing and/or assessing children and adolescents identified as exhibiting JFSB.

Myths, misdiagnoses, and missed psychiatric diagnoses with JFSBs

From a human growth perspective, engaging in some form of norm-violating behavior is part of natural manifestation of development and does not always signal emotional disturbance (Overbeek et al., 2001). In this case, the issue may be differentiating between what is "normal/typical" JFSB from that which is determined to be problematic (Watt et al., 2014). One of the diagnostic myths is the uninformed notion that JFSB is indicative of meeting a criteria corresponding to a personality disorder. This type of myth in practice is expected to interfere with treatment plans and unduly places limits on us as professionals working collaboratively in an advisable restorative justice framework. Other clinical JFSB myths include that "playing with matches" or a general curiosity of fire is part of normal development (Gaynor, 1996). In practice, this myth can function as a precursor to the development of a personality disorder, which also perpetuates misdiagnoses, overdiagnoses, or all together missed diagnosis (Johnson, 2015). Moreover, providers must avoid overlooking the possible presence of adverse environmental, developmental, and other psychocultural factors that also influence the behavior. Another myth-building issue is noted in inexperienced professionals or first responders who have initial contact with juvenile fire setters or bomb makers and perpetuate, through their actions or lack thereof, a "boys will be boys" or "girls don't play with fire" approach to these JFSB cases. As a result, these gender-stereotypical myths in practice actually jeopardize the quality of the risk assessments and interfere with culturally responsive treatment interventions.

In terms of missed diagnoses, the work of Johnson and Jones (2014) and Johnson et al. (2013) demonstrated that symptoms associated with JFSB might also be characteristic of other mental disorders. An examiner that is less experienced with JFSB cases may be perplexed during the diagnostic phase of assessment. This unfortunate circumstance would be especially problematic if said examiner is under the mistaken impression that there must only be one all-encompassing diagnosis (Duff and Kinderman, 2008; Hopwood and Sellborn, 2013). To mitigate against these unwanted clinical forensic psychological practices, providers are encouraged to gather detailed biopsychosocial information, properly assessing and accurately interpreting results, consult with experts familiar with JFSB, and remain up-to-date with the most current research pertaining to this clinical forensic population. As a result, an examiner by default creates a more thorough opportunity to appropriately evaluate and render a more culturally responsive diagnose JFSB.

There are recently developed culturally responsive treatment options that have yielded significant results. In addition, the available assessment measures will augment the clinical professionals' ability to generate an inaccurate diagnosis or prevent overdiagnoses that may unwittingly impede treatment (Johnson, 2015). JFSB may or may not meet the threshold of one or more DSM-5 diagnostic criteria for mental disorders (Johnson, 2015; Kolko and Vernberg, 2017). Therefore, the importance of accurate, ongoing assessment is of significant clinical relevance. The use of valid assessment measures to assess for risk level among the JFSB population is key to creating an evidence-based treatment plan.

Development of new assessment tools

Research on JFSB reveals an increased use of assessment instruments with uniform psychometric properties to specifically target adolescent misuse of fire (Johnson et al., 2013; Johnson, 2015). This trend is prompting advancements within the field to identify, modify, or create measures aimed at providing this clinical forensic population with the assessment of adaptive coping skills required in order to develop protective factors to reduce likelihood of recidivism. The use of valid and reliable assessment tools help accurately assesses motivational risk level aids.

In 1999, Schwartzman and colleagues developed a JFSB typology based on the juvenile's level of motivation for engaging in the behavior. This typology consists of two main levels; Accidental, which tend to be children, aged 5–10, who exhibit a general curiosity about fire, have no known psychiatric issues, and are not aware of the potential consequences that could result from engaging in fire play; and Pathological, which may include those who are A: Using JFBS as a means of coping with adverse life stressors and can be any age, B: Adolescents aged 11–15 who exhibit symptoms associated with aggression such as conduct disorder, C: Those who engage in JFSB to self-harm or are reinforced by the sight and smell of fire, D: Individuals who are cognitively impaired, E: Those who are enabled by other groups or to deliberately entice others, and F: Intentional wild land fire types whose goal is destruction to inhabited areas.

Non-licensed paraprofessionals working with JFSB tend to utilize the Juvenile Fire Setter Intervention Handbook created by FEMA (FEMA, 2002, as cited in Johnson, 2016). This handbook is a training tool designed to instruct law and fire safety officials and helps with knowing how to create education-based early prevention fire-safety programs. The FEMA approach is not intended for qualified mental health professionals. However, the FEMA model is commonly used as a screening tool for JFSB (Johnson, 2016). Research conducted by Slavkin (2000) found that the theoretical foundation for the development of the FEMA model used to assess JFSB by fire departments was based on data that was over 35 years old. Additionally, many of the empirical studies on which validity is based are now 20 years old. Research on the methodology, statistical correlations, and advancements in the field of JFSB have surpassed data collected 20-plus years ago, making the widespread use of any model aimed at evaluating that which impacts matters of public safety and are lacking validation, reliability, and strong empirical evidence of efficacy is concerning (Johnson, 2016). Furthermore, the lack of training in the ability to accurately interpret assessment scores by the fire official administering the tool has resulted in the failure of a significant number of individuals who endorsed clinical levels of symptomology to be referred to mental health professionals (Slavkin, 2000). In fact, research conducted by Andrews, Bonta, and Wormith (2011), found that much of the interpretation by fire officials is based on personal and/or professional experience, gut feelings about the juvenile, and anecdotal knowledge, all of which has been shown to be an inaccurate gauge in the identification of JFSB typology. According to Johnson (2016), who reviewed the FEMA manuals from 1979 to the current 2014 printing, "[The FEMA manual] fails to make substantial connections between the information and the applicability in the use of the FEMA model in the categorization of juvenile fire setting typology." It is for these reasons, professionals working with JFSB modify or create a consistent psychometric measure aimed as properly identifying factors contributing to the motivation behind the behavior in order to address public safety concerns, promote awareness, and reduce recidivism.

Many juveniles referred for fire setting fall into the pathological typology, yet they do not meet all the diagnostic criteria of any one particular

DSM-5 diagnosis, Johnson (2014) has developed the DSM-5 Quadrant which is an evidenced-based clinical portrait that is unique to JFSB. Moreover, the DSM-5 Quadrant is an extremely valuable tool when developing an evidenced-based treatment plan and interventions. In practice, the Quadrant is a framework incorporating the four most research-based disorders that are associated with JFSB; Attention Deficit/Hyperactivity Disorder (AD/HD), Autism Spectrum Disorder (ASD), Conduct Disorder/Oppositional Defiant Disorder (CD/ODD), and Posttraumatic Stress Disorder (PTSD) (see Figure 3.1). In utilizing this framework, qualified and licensed mental health examiners reduce the risk of misdiagnosis, overdiagnosis, or missed diagnosis. Additionally, since comorbidity within this population is common, once the juvenile has been referred to a mental health professional, evidenced-based practices can be implemented which properly align to the individual's needs and severity of clinical symptom criteria identified within each quadrant (Johnson, 2015).

Johnson's DSM-5 Quadrant emerged out of a need to improve clinical forensic research and practice with JFSB behavior by blending discrete variables (e.g., gender, birth order, familial history) and continuous variables (age, socioeconomic status, family functioning, etc.), which aid in the recognition of overlapping symptoms associated with the common mental disorders (AD/HD, ASD, CD/ODD, and PTSD) potentially contributing to JFSB. The four disorders that make up the Quadrant's framework are not sole clinical indicators of JFSB, just as individuals displaying some of the symptoms associated with any disorder meet the criteria for a clinical diagnosis. Yet, the Quadrant offers a graphic representation of relevant and exploratory correlations that allow the examiner to separate clinical factors beyond their initial referring behaviors. For example, some juveniles involved in JFSB exhibit clinical symptoms that are consistent with AD/HD; however, a diagnosis of AD/HD is not indicative of engaging in JFSB (Johnson, 2015). Similarly, ASD is often associated with a restricted set of social skills. This deficit has been noted as a factor in some who engage in fire-setting behavior (Radley and Shaherbano, 2011). This finding links with AD/HD, as it is a common co-occurring disorder among those diagnosed with ASD (Antshel, Zhang-James, and Faraone, 2013; Johnson, 2015; Matson and Cervantes, 2014), but not consistently indicative of JFSB. Alternatively, the predominant diagnosis among juveniles exhibiting JFSB when compared to non-fire-setters is CD or ODD (Becker, Luebbe, Stoppelbein, Greening, and Fite, 2012; Kazdin and Kolko, 1986), but again, the behavior and the diagnosis may not go hand-and-hand. Research conducted by Cruise and Ford (2011) found that adjudicated juveniles tend to present with any number of symptoms associated with repeated exposure to trauma (PTSD). As a result, these juveniles frequently use maladaptive coping strategies (JFSB) in an attempt to manage emotional dysregulation. However, not everyone exposed to trauma engages in JFSB.

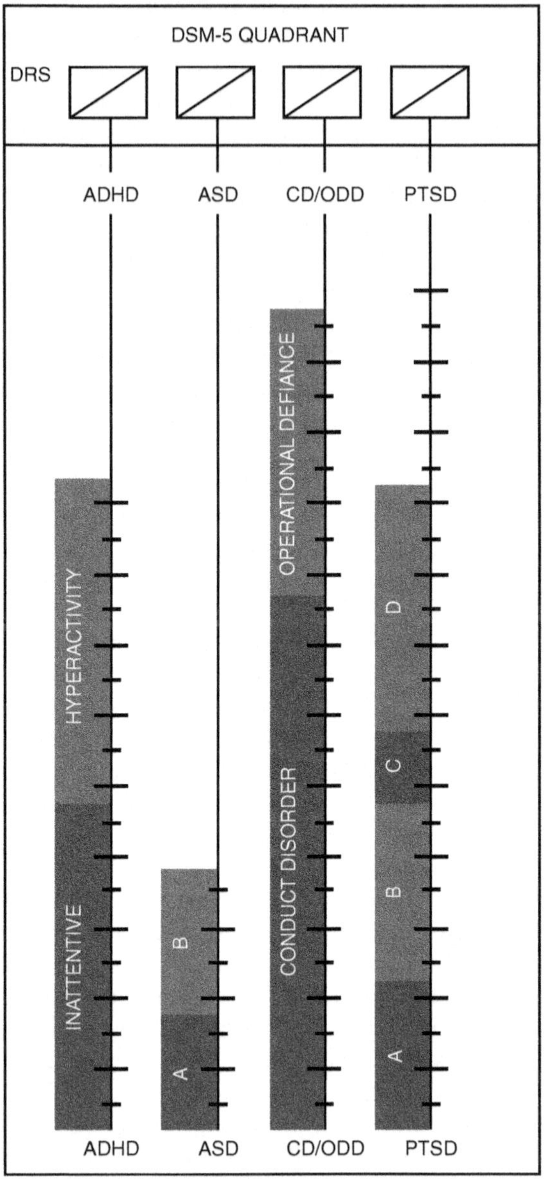

Figure 3.1 DSM-5 Quadrant.

The Quadrant functions also functions as a research-based "diagnostic decision aid and visual communication tool" to reduce misdiagnosis, over diagnosis, or missed diagnosis which hinders treatment, public safety, and referrals to appropriate resources (Johnson, 2015). The diagnostic research and common symptomology displayed by JFSB has established relationships. This alignment makes the use of the DSM-5 Quadrant valuable when evaluating youth involved in fire-setting or bomb-making behaviors, as it goes beyond the clinical or forensic psychological issues that may never be considered or addressed by merely relying on the FEMA manual or the use of a simple risk-assessment level based on the gut feelings of an inexperienced examiner. Rather, there is a recommended practice of using the ongoing gathering of clinical forensic information, such as interviewing and assessments, in order to expand the diagnostic process among this clinical population. In doing so, clinical professionals are expected to secure a more exhaustive understanding of the needs of the juvenile and the capacity to craft more informed treatment choices based on relevant clinical forensic psychological findings (Johnson, 2015).

The ability of mental health practitioners working with JFSB to identify potential risk factors facilitates the use of appropriate measures and intervention strategies that are projected to mitigate recidivism. One of the current psychometric measures for assessing individual JFSB behaviors across multiple domains is the Fire Risk Assessment Tool for Youth (FRAT-Y) (Stadolnik, 2010). This third-generation comprehensive assessment tool detects the presence and severity of risk factors associated with fire-setting behavior among juveniles (Stadolnik, 2016). The FRAT-Y is applicable for use with children and adolescents aged 5–17. The tool examines 17 risk factors associated with fire setting, including primary and secondary motivations using structured assessment methods, risk profiles, and intervention/treatment planning worksheets (Johnson et al., 2013). Furthermore, the FRAT-Y aims to utilize statistical research evidence to reduce the tendency to engage in subjective clinical judgments when determining individual risk factors among the JFSB population (FirePsych, Inc., 2010).

While only two potential assessments, specifically geared for use with juvenile fire setting and bomb makers, are mentioned here, there are a number of assessments available to identify a myriad of risk factors which guide clinicians in development treatment plans with adjudicated youth (Stadolnik, 2016), some of these include Achenbach System of Empirically Based Assessment (Achenbach and Rescorla, 2013), Bell Relationship Inventory for Adolescents (Bell, 2005), Jesness Inventory-Revised (Jesness, 2003), Millon Pre-Adolescent Clinical Inventory (Millon, Tringone, Millon, and Grossman, 2005), and the Trauma Symptoms Checklist-Children (Briere, 1996). In many cases, the individual assessing the juvenile is an examiner representing the US fire service. Once level of risk has been determined and initial public safety risk has been lowered, referral to a licensed mental health professional assures continued treatment and clinically informed support for the juvenile and their family to help reduce recidivism risks by exploring rooted biopsychosocial issues contributing to the JFSB.

Sample of interventions and treatment options

The licensed mental health professional administering appropriate interventions should maintain communication with initial fire service examiner as the JFSB case could result in court appearances. Therefore, the clinician should be knowledgeable in forensic adaptations pertaining to the criminal justice system and comfortable being called as a witness of fact. The introduction of an empirically based treatment plan for JFSB bridges the gap between further public safety risks and enduring family issues (Johnson, 2016).

As research suggested, it is not uncommon for those exhibiting JFSB to have co-occurring mental disorders. This, along with interpersonal/familial challenges, adds to the risks and highlights the importance of taking a multisystemic, culturally responsive approach when building a treatment plan with the entire family. Research discussed earlier in this chapter has illustrated that JFSB can be triggered by multiple factors. Mental health professionals need to adopt culturally responsive interventions into their treatment plans, which function as both aggravators and mitigators to challenge worldviews and increase emotional awareness thus reducing recidivism (Johnson, 2013; Johnson, 2016).

There are promising interventions and treatments options that research has demonstrated significant results among adjudicated youth that combine elements of psychoeducation and ethnoracial contexts to provide individuals with an understanding of how intergenerational and interpersonal relationships, communication, and current stressors impact their life (Sikkema et al., 2013). Interventions and treatment options for professionals who work with adjudicated youth involved in JFSB should incorporate a variety of culturally responsive methods (Johnson, 2013) in order to identify and target the range of coping styles unique to the client in hopes of increasing resiliency and decreasing cognitive distortions that may be driving the dysfunctional behaviors. In doing so, professionals are mindful of the ethnoracial traumatizing challenges these juveniles have to overcome. One desired culturally responsive area to address would be able to assist the client to develop an increased sense of self-worth, recognize areas of control they have over aspects of their life, and acquire adaptive coping strategies that can be adjusted for every phase of life or achieving restorative healing.

Aiding in the development of these areas is essential in making concerted efforts to reduce the psychological and psychosocial impact that often coincides with adjudication and/or incarceration as a result of JFSB. Furthermore, as the client makes positive gains in coping abilities, the qualified clinician experiences increased awareness that not all emotionally reactive behaviors, such as JFSB, are indicative of an underlying mental disease, but rather an attempt to avoid adverse emotional states (Johnson et al., 2013).

Interventions vary depending on risk levels and age of client. For those falling in the low risk classification typology, the evidenced-based interventions take more of a psychoeducation approach and focus on safety and prevention (Johnson et al., 2013). Examples of such empirically supported interventions are PLAY SAFE! BE SAFE! (BIC Corporation, 1994/2017) and "Sparky.org" (National Fire Protection Association, 2015). The multimedia kit PLAY SAFE! BE SAFE! is designed to teach preschool children about fire safety and the dangers of playing with fire. Sparky.org is a website which uses a multi-tiered educational curriculum (*Learn Not to Burn*) to raise awareness and aid in the building of appropriate behaviors and strategies related to fire safety. Each lesson is tailored for specific age groups (Pre-school, Kindergarten, 1st Grade, 2nd Grade, 3rd Grade, 4th Grade, and 5th Grade) and addresses firefighting history and science, fire detection and safety, and first aid topics (National Fire Protection Association, 2015).

For children and adolescents who are categorized as medium risk, psychoeducation is secondary to clinical forensic psychological matters. Interventions need to concentrate on the aggravating effects of existing psychopathology and the acquisition of adaptive coping strategies in order to prevent behavior reoccurrence (Johnson et al., 2013). Cognitive Behavioral Therapies (CBTs) have been heavily researched with juvenile populations including JFSB (Peters and Freeman, 2016; Sharp, Blaakman, Cole, and Cole, 2006). CBTs are effective due in part to the fact that the model helps the individual understand how a maladaptive thought process fuels emotions, which results in a display of dysfunctional behaviors. CBTs work by challenging the client's dysfunctional beliefs in order to promote more realistic, problem-solving abilities.

Lanktree et al. (2012) examined a promising treatment model, the Integrative Treatment of Complex Trauma (ITCT). ITCT was normed using English-speaking, racially diverse traumatized innercity children, and adolescents, aged 8–17, with a low socioeconomic status. The ITCT is a therapy that utilizes attachment theory, the self-trauma model, and trauma-focused CBT. Therapy focuses on social and cultural issues, coping skills development, support systems, family/caretaker relationships, and attachment issues. The interventions are structured, but can be customized to each client based on specific areas of need (Lanktree et al., 2012). This is not a short-term treatment. The average course of treatment takes approximately eight months of weekly sessions. Upon which research findings indicated that participants showed the greatest reduction of symptoms relating to anxiety, depression, and PTSD, all of which are related to the onset of JFSB, noted as the most commonly occurring disorders among those who exhibit JFSB, and are featured in Johnson's DSM-5 Quadrant (Lanktree et al., 2012; Johnson, 2015).

In terms of those identified as high risk and due to the increased risk to self and others, immediate safety protocols need to be taken such as admission to an inpatient care facility or detention (Johnson et al., 2013). However,

FEMA reported in 2002 that less than 1% of juveniles meet the criteria for this classification.

While there are certainly more therapeutic modalities available to treat JFSB, the aforementioned evidenced-based interventions and treatment methods demonstrate an increased awareness of the need to develop or modify evidence-based practices for youth displaying JFSB to recognize the underlying factors contributing to the behavior. As with any at-risk population, safety needs to be in the forefront as the identified population is still developing mentally (Gerson and Rappaport, 2013). Adaptive coping strategies, such as positive reappraisal, problem-solving, exercising self-control, and seeking support, plus the development of various skills, such as adaptive communication styles, building healthy relationships, and recognizing triggers all need to be assessed throughout the course of treatment in order to ensure positive growth, reduction of symptoms, and attainment of therapeutic goals (Gerson and Rappaport, 2013; Hetzel-Riggins and Meads, 2011). It is clear that advanced and emerging trends in forensic practice require the incorporation of the information contained in this chapter.

Conclusions: implications for clinical forensic psychological practice and research

In this chapter, we examined the confluence of clinical and forensic psychological issues that are involved in working with JFSB cases. The primacy of public safety concerns in the work with this patient population are advanced and emerging trends for forensic practice. For example, contemporary juvenile forensic practice that often includes the court. Professionals working with JFSB cases must work within a multidisciplinary framework that often involves targeted programs, such as mental health courts and culturally responsive case management approaches to assist these juveniles.

The chapter offered two overarching recommendations for service providers, clinicians, and administrators at juvenile justice agencies. First, even though psychological treatment resources are limited at agencies, allocating a portion of specialty mental health services to JFSB may help agencies combat the issue of recidivism by offering evidenced-based interventions proactively to indirectly prevent JFSB recidivism rather than acting reactively. Second, with the number of ethnic minorities involved in JFSB, service providers may benefit from incorporating small culturally responsive techniques into clinical diagnostic interviews to facilitate clinically and forensically useful answers. For example, culturally responsive clinical interviewing is expected to result in better identification of mental health difficulties. Still, there are identified legal and public policy considerations that obligate juvenile justice agencies across all jurisdictions to evaluate and provide treatment for JFSB cases. To effectively allocate clinical psychological services, agencies must be able to appropriately determine which JFSBs have a diagnosable mental disorder, meaning those that meet clinical diagnostic criteria. Future research must add to the prevalence estimates for JFSBs because currently there is limited research on prevalence of mental disorders for JFSB who are usually at a lower risk to reoffend when they receive evidenced-based treatments (e.g., FATJAM) (Johnson, 2010; Johnson et al., 2013).

Future research must also focus on aiding a diverse group of providers in a way that assists them and administrators in allocating mental health resources. Such efforts would aid JFSB chiefly by contributing to the knowledge base for culturally responsive clinical assessment techniques. This type of research is projected to demonstrate a promising combination of methodological strengths including seeking out more difficult JFSB cases to take full advantage of the advanced and emerging forensic trends.

References

Achenbach, T. M., & Rescorla, L. A. (2013). The Achenbach system of empirically based assessment (ASEBA): Applications in forensic contexts. In R. P. Archer, E. A. Wheeler, R. P. Archer, and E. A. Wheeler (Eds.), *Forensic uses of clinical assessment instruments* (pp. 311–345). New York: Routledge and Taylor & Francis Group.

Andrews, D. A., Bonta, J., & Wormith, J. S. (2011). The risk-need-responsivity (RNR) model: Does adding the good lives model contribute to effective crime prevention? *Criminal Justice and Behavior, 38*(7), 735–755.

Antshel, K. M., Zhang-James, Y., & Faraone, S. V. (2013). The comorbidity of ADHD and autism spectrum disorder. *Expert Review of Neurotherapeutics, 13*(10), 1117–1128.

BIC Corporation. (1994/2017). *Play safe! Be safe!* Retrieved from www.playsafebesafe.com

Becker, S. P., Luebbe, A. M., Stoppelbein, L., Greening, L., & Fite, P. J. (2012). Aggression among children with ADHD, anxiety, or co-occurring symptoms: Competing exacerbation and attenuation hypotheses. *Journal of Abnormal Child Psychology, 40*, 527–542.

Bell, M. D. (2005). *Bell relationship inventory for adolescents.* Los Angeles, CA: Western Psychological Services.

Briere, J. (2005). *Trauma symptom checklist for young children (TSCYC).* Odessa, FL: Professional Manual. Psychological Assessment Resources, Inc.

Carver, C. S., & White, T. L. (1994). Behavioral inhibition, behavioral activation, and affective responses to impending reward and punishment: The BIS/BAS scales. *Journal of Personality and Social Psychology, 67*(2), 319–333.

Cruise, K. R., & Ford, J. D. (2011). Trauma exposure and PTSD in justice-involved youth. *Child Youth Care Forum, 40*, 337–343. doi:10.1007/s10566-011-9149-3

Del Bove, G., & Mackey, S. (2011). An empirically derived classification system for juvenile firesetters. *Criminal Justice and Behavior, 38*(8), 796–817.

Del Bove, G., Caprara, G. V., Pastorelli, C., & Paciello, M. (2008). Juvenile firesetting in Italy: Relationship to aggression, psychopathology, personality, self-efficacy, and school functioning. *European Child and Adolescent Psychology, 17*, 235–244. doi:10.1007/s00787-007-0664-6

Dickens, G., Sugarman, P., Edgar, S., Hofberg, K., Tewari, S., & Ahmed, F. (2009). Recidivism and dangerousness in arsonists. *Journal of Forensic Psychiatry and Psychology, 20*(5), 621–39.

Ekbrand, H., & Uhnoo, S. (2015). Juvenile firesetting in schools. *Journal of Youth Studies, 18*(10), 1291–1308. doi:10.1080/13676261.2015.1039970

Elzy, M., Clark, C., Dollard, N., & Hummer, V. (2013). Adolescent girls' use of avoidant and approach coping as moderators between trauma exposure and trauma symptoms. *Journal of Family Violence, 28*, 763–770. doi:10.1007/s10896-013-9546-5

Federal Bureau of Investigation. (2012). *Uniform crime report: Crime in the United States, 2012*. Retrieved from www.fbi.gov

Federal Emergency Management Agency. (2012). *Understanding youth firesetting behaviors*. Retrieved from www.usfa.fema.gov/fireservice/prevention_education/strategies/arson/aaw12/understanding.shtm

Ferdinand, R. F., & Verhulst, F. C. (1996). The prevalence of self-reported problems in young adults from the general population. *Journal of Social Psychiatry and Psychiatric Epidemiology, 31*, 10–20. doi:10.1007/BF00789117

FirePsych, Inc. (2010) *FirePsych*. Retrieved from www.firepsych.com/index.php?option1/4 808 com_content&view1/4article&id1/469&Itemid1/428

Gaynor, J. (1996). Firesetting. In M. Lewis (Ed.), *Child and adolescent psychiatry: A comprehensive textbook* (2nd ed., pp. 591–603). Baltimore, MD: Williams & Wilkins.

Gerson, R., & Rappaport, N. (2013). Traumatic stress and posttraumatic stress disorder in youth: Recent research findings on clinical impact, assessment, and treatment. *Journal of Adolescent Health, 52*, 137–143. doi:10.106/j/jadolhealth.2012.06.018

Hanson, M., Mackey, S., Atkinson, L., Staley, S., & Pignatiello, A. (1995). Firesetting during the preschool period: Assessment and intervention issues. *Canadian Journal of Psychiatry, 40*(6), 299–303. doi:10.1177/070674379504000604

Hetzel-Riggin, M. D., & Meads, C. L. (2011). Childhood violence and adult partner maltreatment: The roles of coping style and psychological distress. *Journal of Family Violence, 26*, 585–593. doi:10.1007/s10896-11-9395-z

Inaba, D. S., & Cohen, W. E. (2014). *Uppers, downers, all arounders* (8th ed.). Ashland, OR: CNS Publications.

Jesness, C. F. (2003). *Jesness inventory-revised: Technical manual*. North Tonawanda, NY: Multi-Health Systems.

Johnson. R. (2013). Forensic and cross-culturally responsive assessment using the DSM-5: Just the F.A.C.T.S. *Journal of Theory Construction and Testing, 3*, 18–22.

Johnson, R. (2014). Towards a forensic psychological evaluation of juvenile fire setters: Parent power. *Journal of Forensic Research, 5*, 214. doi:10.4172.21577145.10000214

Johnson, R. (2015). Towards an evidence-based clinical forensic diagnostic assessment framework for juvenile fire setting and bomb making: DSM-5 quadrant. *Journal of Forensic Psychology Practice, 15*, 275–293. doi:10.1080/15228932.2015.1022479

Johnson, R. (2016). Culturally responsive family therapy with post-risk assessment juvenile fire setting and bomb making: A forensic psychological paradigm. *Journal of Psychology and Psychotherapy, 6*(3), 270–280. doi:10.4172/2161-0487.1000270.

Johnson, R., & Jones, P. (2014). Identification of parental endorsement patterns: An example of the importance of professional attunement to the clinical-forensic risk markers in juvenile fire-setting and bomb-making. *American Journal of Forensic Psychology, 32*(2), 25–42.

Johnson, R., Beckenbach, H., & Killbourne, S. (2013). Forensic psychological public safety risk assessment integrated with culturally responsive treatment for juvenile fire setters: DSM-5 implications. *Journal of Criminal Psychology, 3*(1), 49–65. doi:10.1108/20093821211307767

Johnson, R., Jones, P., & Said, C. (2013). *Database of demographic and clinical data of referred cases 1998–2013*. Juvenile Arson and Explosive Research and Intervention Center. San Diego, CA: Burn Institute.

Kazdin, A. E., & Kolko, D. J. (1986). Parent psychopathology and family functioning among childhood firesetters. *Journal of Abnormal Child Psychology, 14*, 315–329. doi:10.1007/BF00915449

Kolko, D. J. (1985). Juvenile fire setting: A review and methodological critique. *Clinical Psychology Review, 5*, 345–376. doi:10.106/0272-7358(85)90012-1

Kolko, D. J., & Kazdin, A. E. (1991). Motives of childhood firesetting: Firesetting characteristics and psychological correlates. *Journal of Psychology and Psychiatry, 32*(3), 535–549. doi:10.1111/j.1469-7610.1991.tb00330.x

Kolko, D. J., & Kazdin, A. E. (1994). Children's descriptions of their firesetting incidents: Characteristics and relationship to recidivism. *Journal of the American Academy of Child and Adolescent Psychiatry, 33*, 114–122. doi:10.1097/00004583-199401000-00015

Kolko, D. J., & Vernberg, E. M. (2017). *Assessment and intervention with children and adolescents who misuse fire: Practitioners guide*. New York. Oxford Press.

Lanktree, C. B., Briere, J., Godbout, N., Hodges, M., Chen, K., Trimm, L., ... Freed, W. (2012). Treating multitraumatized, socially marginalized children: Results of a naturalistic treatment outcome study. *Journal of Aggression, Maltreatment, and Trauma, 21*(8). 813–828. doi:10.1080/10926771.2012.722588

Matson, J. L., & Cervantes, P. E. (2014). Commonly studied comorbid psychopathologies among persons with autism spectrum disorder. *Research in Developmental Disabilities, 35*(5), 952–962.

Millon, T., Tringone, R., Millon, C., & Grossman, S. (2005). *Millon Pre-Adolescent Clinical Inventory manual*. Minneapolis, MN: Pearson.

National Fire Protection Agency. (2015). Retrieved from www.Sparky.org

Overbeek, G., Vollebergh, W., Meeus, W., Engels, R., & Luijpers, E. (2001). Course, co-occurrence, and longitudinal associations of emotional disturbance and delinquency from adolescence to young adulthood: A six-year three-wave study. *Journal of Youth and Adolescence, 30*, 401–427. doi:10.1023/a:1010441131941

Peters, B., & Freeman, B. (2016). Juvenile firesetting. In L. Kraus (Ed.), *Adjudicated youth, an issue of child and adolescent psychiatric clinics* (pp. 99–106). Philadelphia, PA: Elsevier.

Putnam, C. T., & Kirkpatrick, J. T. (2005). Juvenile firesetting: A research overview. *Juvenile Justice Bulletin*. Retrieved from www.ncjrs.gov

Radley, J., & Shaherbano, Z. (2011). Asperger syndrome and arson: A case study. *Advances in Mental Health and Intellectual Disabilities, 5*(6), 32–36.

Schwartzman, P., Fineman, K., Slavkin, M., Mieszala, P., Thomas, J., ... Baer, B. (1999). *Juvenile fire setter mental health intervention: A comprehensive discussion of treatment, service delivery, and training of providers*. Fairport, NY: Office of Juvenile Justice and Delinquency Prevention.

Sikkema, K. J., Ranby, K. W., Meade, C. S., Hansen, N. B., Wilson, P. A., & Kochman, A. (2013). Reductions in traumatic stress following a coping intervention were mediated by decreases in avoidance coping for people living with HIV/AIDS and childhood sexual abuse. *Journal of Consulting and Clinical Psychology, 81*(2). 274–283. doi:10.1037/a0030144

Sharp, D. L., Blaakman, S. W., Cole, E. C., & Cole, R. E. (2006). Evidence-based multidisciplinary strategies for working with children who set fires. *American Psychiatric Nurses Association, 11*(6). 329–337.

Slavkin M. L. (2000) *Juvenile firesetting: An exploratory analysis.* Unpublished doctoral dissertation: Indiana Universit, Bloomington

Stadolnik, R. (2010). *Firesetting Risk Assessment Tool for Youth (FRAT-Y): Professional manual.* Norfolk, MA: FirePsych.

Stadolnik, R. (2016). Promising practice in the development of assessment and treatment models for juvenile firesetting/arson. In R. M. Doley, G. L. Dickens, and T. A. Gannon (Eds.), *The psychology of arson: A practical guide to understanding and managing deliberate firesetters* (pp. 243–259). London and New York: Routledge and Taylor & Francis Group.

Tanner, A. K., Hasking, P., & Martin, G. (2015). Non-suicidal self-injury and firesetting: Shared and unique correlates among school-based adolescents. *Journal of Youth Adolescences, 44*, 964–978. doi:10.1007/s10964-014-0119-6

Trickett, P. K., Negriff, S., Juye, J., & Peckins, M. (2011). Child maltreatment and adolescent development. *Journal of Research on Adolescence, 21*(1), 3–20.

Van Goozen, S. H. M., Fairchild, G., & Harold, G. T. (2008). The role of neurobiological deficits in childhood antisocial behavior. *Current Directions in Psychological Science, 17*(3), 224–228. doi:10.1111/j.1467-8721.2008.00579.x

Vaughn, M. G., Fu, Q., Delisi, M., Wright, J. P., Beaver, K. M., Perron, B. E., & Howard, M. O. L. (2010). Prevalence and correlates of fire-setting in the United States: Results from the National Epidemiological Survey on Alcohol and Related Conditions. *Comprehensive Psychiatry, 51*, 217–223

Watt, B. D., Geritz, K., Hasan, T., Harden, S., & Doley, R. (2014). Prevalence and correlates of firesetting behaviours among offending and non-offending youth. *Legal and Criminology Psychology, 20,* 19–36. doi:10.1111/lcrp.12062

4 Preventing false confessions during interrogations

Phillip R. Neely, Jr.

Introduction

Hritz (2010) described false confessions as "an admission to a criminal act—usually accompanied by a narrative of how and why the crime occurred—that the confessor did not commit" (Hritz, Blau, & Tomezsko, 2010, p. 3). False confessions may have many different causes, but the result is usually convictions for crimes that the confessor knew nothing about. Dr. Richard Leo (2004) has identified three causes for false confessions: voluntary, compliant, and persuaded (Drizin & Leo, 2004, p. 924). Whenever a case of exoneration is featured in the media questions arise as to this can of thing can happen in a post-DNA Criminal Justice System, but according to information presented by the Innocence Project (2015) 23% of wrongful convictions were based on false confessions or admissions ("False confessions," 2015, para. 1).

Background

More than 25 years ago, the phenomenon of false confessions began to be researched as to its causes. Kassin and Wrightsman (1985) studied false confessions and concluded that the main factors of voluntary, coerced, and coerced-internalized are the three major categories of false confessions (Kassin & Kiechel, 1996, p. 1). But much more research was needed on this subject to determine if an innocent person could really be convinced that committed a crime they were not involved in, and under what conditions does it happen under. In the 1990s, simple research was conducted with volunteers that were asked to take a typing test. The participants thought that the test was a simple word per minute test, but before starting, were shown a specific button and were told not to push the button, or it would ruin the test. During the test, the controller would switch off the machine then accuse the typist of pushing the button. The experiment reveled that some of the typists could be made to believe that they pushed the forbidden button (Kassin & Kiechel, 1996, p. 2).

Some would believe that an innocent person would find it relatively easy to resist any temptation to falsely confess to a crime, but Guyll (2013) differ in this opinion and suggest, "innocent suspects underestimate the threat of interrogation and that resisting pressure to confess can diminish suspect's physiologic resources and lead to false confession" (Guyll et al., 2013, p. 370). Incredibly under lab conditions people can come to visualize and recall details of a false memory of criminal behavior... not only juveniles, but adults as well (Shaw and Porter, 2014, para. 1). Police are not allowed to physically use high-stress techniques such as waterboarding, but an interrogation of a suspect for twenty hours could illicit the same physiological effect according to Harmon (2009), prolonged stress and subsequent changes in the body's hormone levels can have a negative effect on memory and learning...information presented by the captor can become part of the suspects memory (Harmon, 2009, para. 4).

Voluntary false confessions

Drizin (2004) describes voluntary false confession as knowingly giving an affirmative response with little or no police pressure. These can be people that come forward and falsely confess to crimes for a variety of reasons including limited mental ability, a need for notoriety, to protect someone else, guilt of the way they treated the victim; even Munchausen Syndrome. According to the Wall Street Journal (2013) "young people are more prone to admitting to guilt for crimes that they did not commit" (Zusha, 2013, para. 3). The case of *State v. Lamonica* (2010) Lamonica was arrested and convicted of multiple counts of aggravated rape of his two sons after coming forth and confessing to the crimes despite the boys insisting that the crimes never happened. Lamonica was suffering from clinical depression, some anxiety and guilt, and feelings of worthlessness. Lamonica, at trial claimed this his confession was false but none the less, was convicted (State v. Lamonica, 2010).

The voluntary false confession is the first category of false confessions in which a person, that is usually not even a suspect comes forward and places themselves in the spotlight of a case as the guilty party. Cases of particularly high profile, or news worthiness stand a greater chance of a false confessor. Gudjonsson (1992, 2003) placed those who suffer feelings of guilt, need to protect another, self-esteem issues, mental illness, or even to take revenge on another as the most popular reasons (Gudjonsson, 2003, p. 1). More than 200 people came forward in 1932 and claimed to be the kidnapper and murderer of famous aviator Charles Lindbergh's baby, more than thirty confessors admitted to the murder of Elizabeth Short (Ramsland, 2010, para. 2).

Compliant false confessions

Compliant false confessions differ from voluntary false confessions because the person making the false confession tends to initially deny an accusation, but at some point during the interview, or interrogation becomes compliant with the accusation and the police theory of how the crime occurred; usually to make the session stop or to gain a reward. Leo (2004) believes that compliant confessors eventually succumb to police interrogation pressure (Drizin & Leo, 2004, p. 3). Compliant confessors are more likely when the suspect is a juvenile, mentally handicapped, grieving the death of a loved one, or because he has not slept in days (Rosen, 2014, para. 3).

Persuaded false confession

Persuaded false confessions most often occur when the confessor is usually someone of low intelligence, or persuadable confessors knows that they are innocent but has uncertainties concerning an underlying feeling of guilt. Even though they cannot remember the crime, these persons can be easily convinced that they played a part in the crime, and will associate what they are being told during the interview to integrate those facts into their own memories. Persuadable confessors will learn the details of a crime either through police interview, media, or another source and rehearse that information while applying themselves to the scenario until the story makes sense to them. Once this occurs an interviewer, and even the confessor himself may not be able to distinguish their story from the truth. According to Leo (2008), persuadable confessors are especially vulnerable to interview techniques in which the interview gives details of a crime then ask the confessor to fill in the blanks (Leo, 2008, p. 26). In the case of *Crowe v. City of San Diego* (2010) in which the United States Court of Appeals, Ninth Circuit, ruled that Michael Crowe was persuaded to confess to the murder of his 12-year-old sister after hours of denying in a police interrogation (Crowe v. County of San Diego, 2010).

Coercion false confession

In the case of *Fox v. Hayes* (2010), The United States Court of Appeals, Seventh Circuit, ruled that Kevin Fox was coerced to confess to the rape and murder of his 3-year-old daughter after 14 hours of interrogation in a small, windowless room, Kevin Fox simply gave up (Fox v. Hayes, 2010). John E. Reid began his career as a polygrapher in 1947, and decades later has become regarded as one of the most preeminently experts in the field of police interrogation techniques with the advent of The Reid Technique; a popular interview method taught and used all over the nation. The Reid Technique involves three different components – factual analysis, interviewing, and interrogation ("The Reid technique," 2015, para. 5). This technique has been roundly criticized by false confession experts as a major source of the problem concerning false confessions. In an article about the Reid Technique, Novella (2014) wrote that Reid's most famous case, and the one that established his fame, was later found to be in error – the 1955 case of Darrel Parker, who was convicted of killing his wife based upon a confession obtained by Reid himself (Novella, 2014, para. 9), despite the claim from Reid that when properly followed his method is specifically designed to weed out innocent persons as suspects.

Coerced-reactive false confessions

The definition of a coerced-reactive false confession according to McCann (1998) is the individual is pressured or induced to confess by some other person other than law enforcement (McCann, 1998, p. 450). Coerced-reactive false confessions are unique in that law enforcement is the secondary party involved in the genesis of the false confession. Spouses, relatives, gangs, and friends can be sources of promises, threats, and pleas that pressure the coerced-false confessor to come forward and introduce themselves to law enforcement as the guilty party.

Solutions and recommendations

Law enforcement has an ethical responsibility to prevent false confessions; there are few other miscarriages of justice that damages the fabric of jurisprudence more than the conviction, incarcerating or worse, the execution of an innocent person. Despite an apparent knowledge of details of a crime, motive, and opportunity of the confessor to have committed the crime, every good detective knows that a confession to a crime is only as good as its corroboration with tangible facts. Detectives have the legal and ethical obligation to know every detail of their case. Detectives are admonished to know every detail and nuance of a case with such familiarity that they are well prepared to recognize and debunk an instance of false confession.

Recognizing what a false confession is primary in addressing this phenomenon. Officers should be educated on false confession personality profiles and limit interrogations to persons who pose a high probability of being a viable suspect in the crime (Leo, 2006). Training of law enforcement personnel that conduct criminal interrogations to be sensitive to the psychological causes and vulnerabilities that may influence suspects to confess to a crime that they did not commit.

Ethical issues associated with this forensic area

The ethical issues focus on the major developments in research addressing the link between **corruption** and growth, the multifaceted character of **corruption**, and the potential for **corruption** to counterbalance strides toward more significant trade openness. The foundation of the chapter is formed by an examination of the relevance and extent of the impact of **corruption** on growth, the influence of **corruption** on financial performance, and the connection between governance quality and **corruption**. This chapter contributes to the literature by providing evidence on the causal mechanisms and transmission channels in the **corruption**-growth connection, the function that **corruption** has with regard to economic growth, and the interaction between **corruption** and democracy. An overview of the conception of **corruption** within U.S. democratic liberalism, including the consisting of bribery. The relationship between markets and **corruption**, including in regard to the impact that markets have on government, is discussed.

Corruption, commonly defined as the abuse of public power for private gain, is an endemic feature of political life around the globe. Corrupt practices are widely condemned and a consistent target of laws and investigations, even in regions where such behavior is common. There have been few comprehensive assessments of what research has learned about the impact of corrupt officials on ordinary citizens, the costs of corruption and poor governance, the underlying causes of corruption and weak governance, and the role of income and wealth as both a cause and a consequence of corruption (Ionescu, 2014).

In this chapter, the goal is to explore the impact of democracy on corruption, the underlying conditions that create corrupt incentives, the propensity of the public employee to engage in corrupt behavior, and the relationship between regulations, institutions, and corruption. The purpose of this chapter is to gain a profound understanding of the interconnection among institutional beliefs as an operation of the interaction of educational accomplishment and corruption, community perceptions of amplified corruption, corruption as a vital aspect hindering socioeconomic strategies, and the destructive effects of corruption (Ionescu, 2014).

Applications in diverse forensic settings

There is a huge amount of study on the grounds and penalties of corruption, educational interest is relatively fresh. In fact, most of this writing points in time from the last 50 years. A majority of the scholars charged to take on the concerns of the relationship joining corruption and fiscal advance thought that it could essentially play a constructive role under undeniable conditions. There were theoretical creators of the grease-the-wheel school of thinking. For instance, recommendations that corruption can advance economic development in countries with privileged antagonistic toward growth by incentivizing elected officials to assist entrepreneurs, lower investor doubt on upcoming government intervention, and damaging bad economic strategies (Johnson, Ruger, Sorens, & Yamarik, 2014).

Some critics thought that perhaps corruption in certain occurrences may deliver a solution to a vital hurdle to expansion. Specifically, it would assist in promoting wealth creation, reduce the red tape, and permit entrepreneurs and minorities from antagonistic prejudice against them. Most notoriously, it was argued that during reconstruction, corruption perhaps could be the unique way of surrounding customary laws or governmental regulations which impede financial development. As a matter of fact, it was believed to be good or proficient governance in such a case that may essentially be worse for financial development than an inflexible, overcentralized, corrupt bureaucracy is one with an inflexible, overcentralized, truthful bureaucracy (Johnson et al., 2014).

Future research directions

Homicide or felony cases are often reliant on eyewitnesses who give faulty or false testimony or evidence because of mistaken identity between the time the crime or infraction was committed and the actual court date. In addition, the witness can be propositioned, intimidated by lawyers and police officer and detectives, or offered a fee to compromise their testimony. Thereby altering the facts as they actually were to what they are desired to be by the authority's false representation or lazy police work.

The Innocence Project was cofounded by Barry Scheck who is one of several individuals that started this unique crime solving scientific approach to assisting the criminal justice system. Crime solving and interrogation tactics in the policing system has taken a drastic back seat to actually utilizing sound old-fashion police work. Nowadays, the detectives are catching solid citizens, interrogating them without reading them their Miranda rights, coercing them into confessions; sending them through the court system knowing that they have the wrong individual, destroying innocent citizen's lives as they are sentenced to life in prison or worse. There have been many documentaries that have exposed the criminal justice system and their tactics concerning their cruel methods of misrepresenting the legal system. The authorities predominately focus their attention on the minority communities, because they are mostly uneducated in the area of the legal system or knowing their basic Constitution rights as listed in the United States Constitution. They are least to be able to afford an experienced attorney, are easily intimidated by the authorities, and they don't understand the language of the legal lingo used by investigating officers. In other words, the authorities prey on easy targets to fill their jails or quotas and the minorities are often their primary targets.

Knowing all this to be fact, the Innocence Project was put into operation, which is an nonprofit legal organization located around the world just to represent the inmates that have become a victim of our powerful American criminal justice system. It has been responsible for exonerating over three hundred inmates that were wrongfully convicted of crimes they actually didn't commit; but have become a product of the legal misrepresentation and malpractice stated earlier. Joining forces with Barry Scheck is Peter Nuefld who got the project up and running in the Yeshiva University in New, York City in 1992. In 2014 alone, in the first quarter, the Innocence Project exonerated 25 inmates that were on death row or facing life in prison. This nonprofit legal organization is committed to exonerating wrongfully convicted individuals through the use of DNA testing primarily and they are also reforming the criminal justice methods to prevent tentative injustices. Currently, the Innocence Project has managed to, with scientific research, using DNA and forensic, freed 329 wrongfully convicted inmates, plus 18 that have spent prison time on death row.

There was more influence to initiate this endeavor; the United States Department of Criminal Justice performed a study with the interest of the

United States Senate and several others, in concern about the incorrect identification of suspects by the so-called eyewitnesses. The percentage of incorrect witnessing incidents was greater than seventy percent across the entire study. Deciding that science was a positive alternative to perfect nonbiased objective fact finding in police work, the Innocence Project was consider a positive first start attempt to reform the current way the justice system was getting their facts. So, Nuefld and Scheck moved forward in conjunction with the Cardozo School of Law in New York City.

The National Innocence Project Organization, however, only take on cases that can be solved by using DNA post testing to prove innocence. DNA exonerations in the United States have been steady increasing; these stories have been becoming more familiar as more individuals become released after DNA post testing has been performed. This may mean that the criminal justice system has not fully embraced the DNA posting method whole-heartedly. Instead, it has continued to utilize the antiquated system of lazy justice. Police officers and judges that are advancing their personal agenda to alter the outcome of the verdict to fit their personal agenda, this is unacceptable. These adverse behaviors cannot be ignored or over looked and allowed to continue and plague our criminal justice system. In the south as well as the north, the power and influence of the judges over the sentencing process is grossly misrepresented, their throwing the book at innocent citizens that have been victimized by the bureau. It is even evident that the judges know this but hesitate to change the system by insisting that the Constitution is strictly followed in their court of law. For example, 18 inmates were sentenced to death prior to the Innocence Project in 1992; the project took their case, established their innocence, and gained their release. The average sentence lived out by prisoners on death row, etc. before being exonerated is thirteen and a half years.

DNA exonerates almost 50% of people of color. There have been successful exonerations in over 35 different states including Washington, DC. The Innocence Project has been involved in 172 of the three hundred and fourteen DNA exonerations in the United States. There have been numerous others that have been helped by Innocence Project Network Organizations, attorneys that are linked to and assist in the mission to reform the criminal justice system collectively. Also, plus or minus 50% of the inmates that have been exonerated have been financially compensated for the amount of years they have suffered unjustly.

Mainly, the problem is, when dealing with cases that take months and years to make it to court, the witness, by this time has difficulty accurately recalling the critical facts that would have been fresh in the mind when the event occurred. The human brain has is not capable of accurately recalling acute details in regards to remembering detailed instructions, scene in a movie, or events as an eye witness's account. The same incident witnessed by several different individuals will have little similarities as they explain details. Thereby, inducing inaccurate details and conclusions in delicate cases that are determining the faith of a would-be innocent citizen caught up in a misrepresentation of justice, which will cost them their liberty or their life. In the meantime as the inmates spend time in prison, the system eventually will assimilate him or her and change their way of thinking and their rational concept of who they are. The prison system is a hard life, the fear of being taken advantage of by harden criminals, the prison guards are not looking out for the welfare of the inmates, and your family is left to fend for themselves in a society that has not learn to recognize the difficult situation that has been dealt to the love ones. The authorities are gradually accepting the merits and the good works of the Innocent Project here in the United States and around the world.

Conclusion

The leading section argues the preceding writings on corruption and salary disparity, and the next section assess the past literature about economic growth and salary inequality.

1 Corruption—Pay inequity nexus
    - Corruption is the misuse of trusted authority for private benefit. It is by and large recognized as the misuse of government office to remove payment in the delivery of public services. It is one of the foremost problems to partisan, fiscal, and social growth. It undercuts the rule of law and chips away at the institutional foundations of good governance upon which sustained development and progress hinge on. The indigent in the community are often the toughest hit by the consequences of corruption, remaining the most dependent on public amenities and the least capable of compensating the elevated expense related with deceit, enticement and other forms of corrupt actions, to achieve those amenities (Batabyal & Chowdhury, 2015).
2 Economic growth—Income disparity nexus
    - Economic growth provides contradictory predictions concerning the effects of economic growth on the dissemination of pay and the salaries of the poverty-stricken. Some examples display that economic growth improves development and lowers discrimination. Economic growth might affect the indigent via two means: comprehensive development and modifications in the dissemination of pay. They discover that economic growth unreasonably boosts income of the low-income quintile and lowers income disparity. Approximately 40% of the extended effect of economic expansion on the income increase of the poorest quintile is the consequence of lessening in income inequality. Approximately 60% is owed to the influence of an economic advancement on cumulative expansion (Batabyal & Chowdhury, 2015).

The majority of studies on the connection between dishonesty and financial development concentrate on emerging countries. However, is the influence of corruption in the industrialized world also harmful and vital? The question is raised whether corruption is beneficial in completely restructured financial systems; however perhaps not given the illustrations of corruption remain in all locations of reconstruction. Nevertheless, corruption may enact a similar part in an industrialized country if authority displays the quality so cold-heartedness, overcentralization and sclerosis due to extreme rule. It is believed the state government size and expenditure levels manipulate the amount of corruption taking place (Johnson et al., 2014).

Additional models demonstrate economic deficiencies, for example, information irregularities and businesses costs, might be particularly binding on the indigent who do not possess collateral and adequate credit histories. As a result, rigorous enforcement of these credit limitations will excessively affect the indigent. Additionally, these credit restraints diminish the effectiveness of capital distribution and intensify income disproportion by hindering the flow of money to indigent persons. Since this view, economic growth assists the indigent by both increasing the effectiveness of wealth allocation and by lessening credit restrictions that more comprehensively confine the poor (Batabyal & Chowdhury, 2015).

The theoretical forecasts of the results of economic sector growth on pay inequity are not undisputed. The lessons on the dissemination of pay assumes that there might be an upturned u-shaped association among income inequity and financial growth. As individuals transfer from lesser income agronomic sector, income inequity grows. Nevertheless, as the agronomic sector minimizes combines with the agronomic pay increasing, this movement reverses and income inequity lowers. Furthermore, the sectorial model is vital for the correlation between financial growth and pay inequity implies economic growth might lower inequity to a less-significant extent in countries having greater modern sectors (Batabyal & Chowdhury, 2015).

References

Batabyal, S., & Chowdhury, A. (2015). Curbing corruption, financial development and income inequality. *Progress in Development Studies, 15*(1), 49–72. doi: 10.1177/1464993414546980. www.pidsjournal.com

Crowe v. County of San Diego, Nos. 05-55467, 05-55542, 05-56311, 05-56364. Find Law 3 (United States Court of Appeals, Ninth Circuit 2010).

Drizin, S. A., & Leo, R. A. (2004). The problem of false confessions in the post-DNA World [supplemental material]. *N.C.L., REV., 891*(924), 924. Retrieved from http://courses2cit.cornell.edu/sociallaw/student_project/FalseConfessions.html

Fox v. Hayes, No. 08-3736 Find Law: 36 (United States Court of Appeals, Seventh Circuit 2010).

Gudjonsson, G. (2003). *The psychology of interrogations and confessions* [supplemental material]. Compass Port, UK. doi:523ed20e

Guyll, M., Madon, S., Yang, Y., Lanin, D. G., Scherr, K., & Greathouse, S. (2013). Innocence and resisting confession during interrogation: Effects on physiologic activity [supplemental material]. *Law and Human Behavior, 37*(5), 366–375. doi: 10.1037/ihb000044

Harmon, K. (2009, September 21). *How torture may inhibit accurate confessions* [Article]. Scientific American. Retrieved from www.scientificamerican.com/blog/post/how-torture-may-inhibit-accurate-co-2009-09

How prevalent are false confessions? (2015). Retrieved from www.innocenccproject.org

Hritz, A., Blau, M., & Tomezsko, S. (2010). *False confessions.* Retrieved from http://course2cit.cornell.edu/sociallaw/student_projects/FalseConfessions.html

Ionescu, L. (2014). The role of government auditing in curbing corruption. *Economics, Management & Financial Markets, 9*(3), 122–127. Retrieved from www.addletonacademicpublishers.com

Johnson, N., Ruger, W., Sorens, J., & Yamarik, S. (2014). Corruption, regulation, and growth: An empirical study of the United States. *Economics of Governance, 15*(1), 51–69. doi:10.1007/s10101-013-0132-3. Retrieved from www.springerlink.com/home/main.mpx

Kassin, S. M., & Kiechel, K. L. (1996). The social psychology of false confessions: Compliance, Internalization, and confabulation [supplemental]. *Psychological Science*, 7, 125–128. Retrieved from www.forensicpsych.umwblogs.org/issue and debates/false confessions

Leo, R. (2008, May). *Persuaded false confessions.* Paper presented at the Law and Society Association, Montreal, Quebec. Retrieved from http://citation.com/metal/p235858-index.html

McCann, J. (1998). A conceptual framework for identifying various types of false confessions. *Behavioral Sciences*, 441–453. Retrieved from www.google.com/?gws_rd=ssl#safe=off&q=maccann+coerced-reactive+false+confessions

Novella, S. (2014). *The Reid technique of investigation.* Retrieved from htttp://theness.com/neurologicablog/index.php/the-reid-technique-of-investigation

Ramsland, K. (2010). *Buddhist temple massacre.* Retrieved January 26, 2015, from www.crimelibrary.com/criminalmind/forensic/buddhist_temple/9.html

Rauch, J. (2014). The case for corruption. *Atlantic, 313*(2), 19–22. Retrieved from www.atlanticmedia.com

Rosen, M. (2014). *Where the truth lies: The phenomenon of false confessions.* Retrieved from http://jjay.cuny.edu/docket/4263.php

Shaw, J., & Porter, S. (2014, November 14). Constructing rich false memories of committing crime [supplemental material]. *Psychological Science.* doi: 10.1177/0956797614562862

State v. Lamonica, KA 1366 Court reporter 89 (Court of Appeals of Louisiana First Circuit 2010).

The Reid technique. (2015). Retrieved from www.reid.com/educational_info/critic technique.html

Zusha, E. (2013, September 8). False confessions dogs teens. *Wall Street Journal.* Retrieved from www.wsj.com/articles/SB10001424127887324906304579014930133O2

5 The dynamic role of the forensic psychologist in emerging issues in correctional mental health

Kori Ryan and Heather McMahon

As prisons continue to be one of the largest providers of mental health services in the United States, the intersection between psychology, law, and the criminal justice system provides a need for an interdisciplinary approach to assessment and treatment. This intersection requires savvy clinical psychologists who are well trained to treat mental health issues in correctional environments. In addition, as the overlap between mental health and the correctional system expands, new areas for forensic psychology will continue to emerge. The United States has one of the highest incarcerations rates in the world. Within the U.S. federal and state prisons, 1,574,700 persons were incarcerated at the close of 2013 (Bureau of Justice Statistics, 2014) with more than 6,899,000 (i.e., 1 of every 34 U.S. adults; Carson & Sabol, 2012) persons under correctional oversight (e.g., probation, parole; Bureau of Justice Statistics, 2014). The United States has the highest rate of incarceration in the world with 707 inmates per 100,000 persons (Lamb & Weinberger, 2014).

With an increase in incarceration, concurrent with a decrease in community resources for individuals with mental health issues, prisons have become de facto treatment environments as people are arrested for various crimes. In addition, the stressful nature of the correctional environment, including lengthy sentences, solitary confinement, and dangerous conditions enhance the possibility of developing mental health issues, even where they did not exist before. Not only does this increase indicate a need for well-trained clinicians and clinically oriented practitioners in the correctional environment, but it also raises significant legal questions that forensic psychologists are uniquely qualified to address. This chapter addresses some of these evolving and emerging issues, such as addressing areas of competency within the correctional environment, diversity issues, issues of aging inmates, risk assessment, immigration, and parole decisions as well as issues of community supervision. This chapter is not intended to be exhaustive, rather, it is meant to highlight some of the recent issues that are arising within the broad issue of increased mental health concerns in correctional environments.

Prevalence of mental health issues in prison

The average length of all sentences for U.S. prisoners is 63 months (Bonczar, Hughes, Wilson, & Ditton, 2011). Inmates in the United States are spending more time in prison than other industrialized nations (Bonczar, et al., 2011). Many of those inmates are suffering from mental health problems. James and Glaze (2005), for the Bureau of Justice Statistics, found that 56% of federal inmates and 64% of jail inmates were identified as having a mental health problem. These percentages added up to over 700,000 state inmates, over 78,000 federal inmates, and over 479,000 jail inmates being identified as having issues with mental health. Serious mental illnesses, such as affective disorders (i.e., bipolar disorder and major depression) and psychotic disorders (e.g., Schizophrenia), are overrepresented in the criminal justice system (Steadman, Osher, Robbins, Case, & Samuels, 2009). In addition to the large percentage of inmates being identified as having a mental illness, those in state prisons with a mental illness were disproportionately housed in administrative segregation or supermax units (Fellner, 2007; Haney, 2003; Lovell, Cloyes, Allen, & Rhodes, 2000). Additionally, inmates with mental illnesses often have co-occurring substance abuse disorders, which can create a complex clinical picture. In the general prison population, it is estimated that 18%–30% of males and 10%–24% of females have alcohol use disorders. Drug-use disorders occur in approximately 10%–24% of male inmates and in 30%–60% of female inmates (Fazel, Bains, & Doll, 2006). In jail detainees and state inmates who have mental illnesses, approximately 75% also had a co-occurring drug and alcohol dependence disorder (Abram & Teplin, 1991; James & Glaze, 2006).

Assessment issues

The forensic psychologist faces some obstacles when working within the correctional setting. The psychologist's code of ethics is often hampered by the needs and desires of the correctional facilities and the courts. Privacy is an expectation in all doctor/patient, evaluator/patient, therapist/patient relationships; however, the realities of the correctional setting does not always allow this to happen. At times, forensic psychologists are called upon to conduct assessments without that privacy. There are times when an evaluator is not allowed to be alone with an inmate, as in times where the inmate poses a risk of harm to themselves, others, or the institution. There are times when the facility might require that a correctional officer sit in on the evaluation. There are situations of lockdowns in correctional settings, or when an inmate is placed in administrative segregation, but an evaluation has to be conducted all the same.

The National Commission on Correctional Health Care created guidelines for the mental health assessment of inmates. The guidelines include the need to screen inmates within 2 hours of admission by a certified professional or technical worker. They recommend that a mental health evaluation be completed by a qualified mental health professional within 14 days of admission. They also recommend that staff is trained in identifying the symptoms of mental illness in case an offender is not identified by the initial screenings (NIC, 2004). The guidelines identify the following elements that should be present in a mental health assessment: psychiatric history, current psychotropic medications, current suicide risk assessment, history of suicidal behavior, current and past drug and alcohol use, history of violent behavior, history of sexual offenses, history of victimization, educational history, history of head trauma, history of seizures, emotional response to incarceration, and intelligence testing for intellectual disabilities (NIC, 2004).

Forensic mental health assessments are used for a variety of purposes. The criminal justice system frequently asks for complex questions to be answered by qualified mental health professionals. Those issues can range from answering criminal competency questions, to questions about treatment, release planning, and suicide and risk assessment. Psychological assessments can be used at all points in the criminal justice system. Psychological assessments can begin during pretrial detention, including assessments of competency to stand trial and risk assessments to determine if an individual is safe to release pending the rest of their court proceedings. During the sentencing phase, competency to be sentenced, risk assessments, and assessments addressing aggravating or mitigating factors might be needed. After sentencing, risk assessments, needs assessments, or other psychological evaluations might be completed to help make recommendations for probation/ parole supervision, or to help assign a security level within a prison. While an individual is in prison, they may require any number of psychological assessments to be completed, including risk of suicide, diagnostic clarification, and competency to make treatment decisions. Psychological assessments can be used to determine who is appropriate to be released from incarceration, and upon release, psychological assessments can be used to help determine risk for recidivism, or needs while out in the community.

Forensic psychologists can also help to distinguish between genuine mental health symptoms and malingering. They can help to educate staff that the presence of malingering does not necessarily preclude the presence of a genuine mental disorder, and help correctional staff to be able to distinguish between what is genuine and what is malingered. With a large prison population, there is a great need for forensic psychologists to help navigate these difficult judicial or corrections questions.

Forensic psychologists are also in a unique position for assessing and treating individuals with complex psychological presentations and presentations of comorbid conditions. Forensic clinicians are uniquely qualified to help meet the needs of a growing number of individuals who have mental health needs and also present with challenges due to being a special, or growing population within the correctional setting.

Issues of competency

The U.S. Constitution states that an individual cannot go to trial or be convicted if that individual is incapable of defending himself against legal charges (Pate v. Robinson, 1966). Competency evaluations are the most common forensic issue evaluated with approximately 60,000 defendants evaluated annually (Morris & DeYoung, 2012; Pirelli, Gottdiener, & Zapf, 2011). Research suggests that number has been steadily increasing (Zapf, Roesch, & Pirelli, 2014). Competency to stand trial evaluations, treatment, and adjudication take up more financial resources than any other forensic psychology venture (Zapf, Skeem, & Golding, 2005). With the increasing number of competency evaluations, so likely is the financial burden that they pose increasing.

In addition to the financial burden, there is also a burden with regard to space required to house these individuals, particularly if they are being charged with a felony. Traditionally, individuals awaiting their competency evaluations and hearings are housed in jail facilities. They can contribute to overcrowding in jails as their cases are placed on hold once the issue of competency is raised in their proceedings. After an individual has been deemed incompetent, it can be difficult finding placement. The most common placement depends on the jurisdiction, but many states rely on state hospitals to house these individuals. Hospital space is limited, which often creates backlogs of incompetent individuals waiting to get into the hospital.

As forensic psychiatric hospital beds fill up, the necessity to provide these types of evaluations in the jails may increase. Once an individual has been deemed incompetent, they will likely require another evaluation to indicate that they are ready to proceed. As problems for bed space continue, it is likely that forensic psychologists may be asked to help make determinations for placement at a facility or within the community. Some states have developed community-based restoration programs. Forensic psychologists can be tasked with determining who is appropriate for the different types of programs, as either part of the competency evaluation or in addition to it.

Other types of competency may need to be assessed by a qualified mental health professional that will likely increase as the number of prisoners with special needs increases. Competence to be sentenced, competence to waive counsel, competence to testify, and competence to be executed are additional evaluations that may need to happen at a higher rate with an increase in the inmate population. While some of the abilities assessed may overlap to other areas of competence, forensic psychologists should make it clear that competence in one specific area does not presume competence in another area. For example, competence to be sentenced goes beyond competency to stand trial. It specifically applies to the period of time between the moment the adjudication process ends and the moment the sentence is rendered. Competence to testify and waive counsel are similar in that they apply to specific areas of the trial process that go beyond the general competency to stand trial.

Ford v. Wainright (1986) held that the Eighth Amendment, which discusses Cruel and Unusual Punishment, bars the execution of individuals who are deemed incompetent at the time of execution. The field of forensic psychology has debated whether psychologists should be involved in these types of evaluations. No clear answers have emerged. There are no set guidelines to follow to help professionals with this complex question, but nevertheless, forensic psychologists are often called upon to help with these determinations.

Though this chapter discusses criminal justice settings, forensic psychologists might be called upon to assess areas of civil competence in correctional settings. Competency to make treatment decisions is another type of competency evaluation that may need to be conducted. An individual receiving treatment must provide informed consent before the treatment begins. Informed consent requires three conditions. Those conditions include having adequate information to make a decision, being free of coercion, and competence, or rather capacity to make the decision (Appelbaum & Grisso, 1995). The concept of coercion, especially legal coercion, is particularly relevant with the forensic population facing charges. Competency to consent to treatment goes beyond informed consent and is typically conducted for a specific reason. It focuses on the reasoning that leads up to the treatment decision, not necessarily the decision itself (Buchanan & Brock, 2001). Treatment in correctional facilities has become increasingly inmate/patient focused. Requests for this type of assessment usually come after an inmate/patient has refused a specific recommended treatment. Providers who do not obtain informed consent from a patient can be charged with battery or negligence if they treat someone against their will. As with criminal competency, if there is a suspicion of impaired decision-making, an assessment needs to be requested.

Special populations

The prison population is arguably representative of policies enacted to manage emerging issues in criminal justice. While an exhaustive review of these policies, including arrest policies, drug laws, systemic racism, amongst others is beyond the scope of this chapter (and done in thorough fashion elsewhere), there are several issues of relevance to forensic psychologists. Aging and the related aging issues, increased understanding of gender diversity, issues of immigration, and the increasing use of community supervision (probation and parole) in corrections are a few of the increasingly complex and under attended areas in forensic psychology. As these populations increase, forensic psychologists need to be aware of, and attuned to, present and future issues in assessment and intervention.

Older inmates and older criminals

The population of individuals aged 65 and up is increasing exponentially in the general public. In 2006, there were 37.3 million people over the age of 65 and by 2030, that number is expected to increase to approximately 71.5 million, or around 20% of the general population (Leatherman & Goethe, 2009). In most correctional research, elderly or geriatric inmates are defined as being age 40 and up. The most recently available data from the Federal Bureau of Prisons (USDOJ, 2016) indicates that the number of federal inmates aged 50 and older grew 25% from 2009 to 2013, while the younger populations decreased. The number of state inmates aged 40 and older has doubled over the course of three decades. In 1974, inmates 40 years or older made up 16% of the state prison population, and by 2004, they made up 33% of the overall prison population (Porter, Bushway, Tsao, & Smith, 2016). With a growing number of older inmates, coupled with lengthy prison sentences and high incarceration rates, it is likely that the percentage of older inmates will continue to rise. Psychological and neuropsychological assessments will likely be key in being able to lead to better psychiatric and medical care of geriatric inmates, and help facilities with the growing population.

Geriatric inmates differ from the general inmate/patient population in many ways. They have an increase in medical problems, tend to have multiple medical or mental health care needs, tend to rely on polypharmacy to treat those needs, and may suffer from irreversible conditions such as dementia or Alzheimer's. Because of the comorbidity present in many geriatric patients, they may present with a complex clinical picture. This is coupled with a correctional environment that may be hostile to the growing needs of elderly inmates. Correctional settings are often stressful. Elderly inmates may have access to fewer medications, or have less access to adequate nutrition, proper medical care, or have physical limitations that are difficult to navigate in the correctional setting. Inmates in general have a higher prevalence of history of substance ingestion, which may further complicate clinical presentations regarding neurocognitive functioning. Forensic psychologists are equipped to provide assessments to help clear up that clinical picture and can provide a cohesive history and psychological assessments that can help to pinpoint the problems and ultimately get the inmates the treatment that they need. This can help to relieve the burden of the cost of unneeded medical or mental health interventions.

Evaluations within the criminal setting are likely to increase as the older population increases. Competency to stand trial evaluations can be difficult when dealing with the aging correctional population. Issues dealing with irreversible conditions (e.g., dementia) may require additional psychological and/or neuropsychological testing. The evaluator may have to render an opinion on the issue of whether the individual has the possibility of being restored to competence due to their limitations. This can be tricky to assess as some neurocognitive disorders can be reversed (e.g., vitamin deficiency) but may present the same as an irreversible dementia. Forensic evaluators may have to render an opinion based on little outside data and may not have access to neuropsychological or medical testing that could prove or disprove these theories. Evaluations of criminal responsibility may be on the rise in the older population. Issues of delirium or other medical conditions may be grounds for an insanity plea, in addition to the more traditional mental health diagnoses that are associated with this type of plea.

Civil issues may also come up while older individuals are incarcerated. These can include the capacity to make legal decisions after sentencing (e.g., file an appeal), the capacity to make treatment decisions, and the capacity to make a living will or an advance directive. Assessments of elder abuse may arise in correctional settings as geriatric inmates may be vulnerable to victimization by other inmates or staff. These issues need to be assessed by a qualified professional.

Geriatric detainees and inmates cost more money to house, and as their numbers grow, so does the burden of the cost on taxpayers. There have been debates across the country in considering whether older inmates should be released through early release or medical programs. Forensic psychologists can assist with these determinations by providing the necessary assessments needed to help make a decision that weighs the needs of the institution, the needs of society, and the needs of the inmate. Forensic psychologists can provide risk assessments to determine who is most likely to recidivate. They can also provide adaptive functioning assessments, amongst other things, to help officials make these difficult release determinations and help to determine placement within the community or correctional system.

In addition to criminal facilities, older offenders are also increasing in the civil commitment areas, including in states who have Sexually Violent Predator laws and civil commitment facilities. Forensic psychologists are able to provide an individualized approach to risk assessment with older sexual offenders in which they take into account treatment, future stability, mental health issues, including dementia, and can make recommendations for this population that is likely to require special consideration when released back into the community.

Gender issues in the correctional system

While the majority of individuals under correctional supervision are males, female representation is increasing for both adult and juvenile females, and at a faster rate than males (Puzzanchera, 2009; Glaze & Kaeble, 2014). Despite their increasing representation in the criminal justice system, gender-based differences are slow to be recognized. As Shepherd and colleagues summarize (2013), early research in gender and offending suggested that men and women had similar paths to offending, thus the assessment process would be similar across genders.

The pathway to prison for many women disproportionately for black women) is through, directly or indirectly, domestic violence and/or childhood abuse (Bloom, Owen, & Covington, 2003). For example, Loring and Beaudoin (2000), in a study evaluating 251 victim-perpetrators (women arrested for crimes who are also victims of domestic violence), found that threats against the women (the victim-perpetrators) included threats of isolation, harming of victim property, harm or kill a nonfamily member, to harm or kill the women's child, to harm or kill pets, to prevent medical care, or to harm or kill the victim themselves led to involvement in the legal system. Other crimes included bank robbery, theft or fraud, murder or attempted murder of a third party, witnessing child abuse, etc. (Loring & Beaudoin, 2000). In another study, Loring and Bolden Hines (2004) found that in their study of 107 women who were involved in a family violence center and committed an illegal behavior, 75% of these women had their pet threatened or hurt by their abuser, and 24% of these women reported committing their crime due to coercion by their partner. Additionally, while rehabilitation may be a stated goal of the correctional system, the availability of trauma-informed treatment in jails and prisons are sporadically available.

Studies support that while areas of need may overlap for males and females, they may not have the same level of impact and may need different consideration (Hollin & Palmer, 2006). Females may also have different needs than males. Issues such as trauma, relationships, access to medical treatment, drug use, and others appear to be not necessarily unique to female offenders, but in different levels and with different effect, with the understanding that overall, women have varying needs (Bartlett & Somers, 2016; Kreis, Gillings, Svanberg, & Schwanner, 2016; Van Voorhis & Presser, 2001). Recognizing the high rates of trauma and the varying criminogenic needs of the female prison population may provide forensic psychologists with additional information about atypical symptoms, appropriate recommendations, and the information to provide guidance in the form of legal opinions.

Many tools derived for clinical use and utilized or adapted for forensic purposes, such as the Minnesota Multiphasic Inventory-2 (MMPI-2) have separate scoring guidelines for males and females to recognize differences between genders. Researchers have increasingly highlighted the needs for considering risk assessment instruments and other assessment tools that are designed specifically for females, and utilized for forensic purposes. For example, the Level of Service Inventory-Revised (LSI-R) is utilized in the United Kingdom to assess domains of static and dynamic criminogenic needs. Hollin and Palmer (2006) identified that there is a lack of examination of the LSI-R in terms of use with females, and in their study found some differences across the domains; Smith et al. (2009) found similar predictive validity with the LSI-R between males and females. Other risk-assessment tools, such as the Historical Clinical Risk Management-20 (HCR-20) has had mixed results in its ability to assess for risk in women (e.g., Nicholls, Ogloff, & Douglas, 2004; de Vogel & de Ruiter, 2005). Davidson, Sorenson, and Reedy (2016) found that certain scales were better predictors of general and assaultive disciplinary actions by females on the Personality Assessment

Inventory (PAI), a commonly used assessment in correctional settings. For female juveniles, Shepherd, Luebbers, and Dolan (2013) examined the Structured Assessment of Violence Risk in Youth (SAVRY), the Youth Level of Service/Case Management Inventory (YLS/CMI) and the Psychopathy Checklist: Youth Version (PCL:YV) and found limitations in predictive validity. Gender-specific risk assessment instruments or scales are being developed in response to the identified need to be gender responsive; research has suggested that validity is increased when gender responsiveness was incorporated into the assessment process (Wright, Van Voorhis, Salisbury, & Bauman, 2012).

Much like other medical areas, psychological assessments have been adapted for women after being normed on males. Given the high rates of trauma, substance abuse, and differences in assessment responses, understanding their unique pathways and behaviors and how gender may impact assessment is imperative for forensic psychologists.

Trans inmates*

The needs of biological males tend to dominate the conversation in corrections, but corrections cannot escape the need to consider gender fluidity. Individuals who identify as transgender are an increasing population in the correctional system, and issues of medical necessity and sex reassignment are becoming increasingly relevant in psychological assessment. Exact numbers seem to be difficult to determine; Brown and McDuffy, in 2009 estimated approximately 750 trans* individuals were incarcerated in the United States. The Bureau of Justice Statistics (2013) estimated that there were 3,209 transgender adult inmates in state and federal prison and 1,079 in local jails in 2011. A recent report regarding transgender individuals in the Texas Department of Corrections increased from 67 to 333 from 2014 to 2016 (McGaughy, 2016). It is even more difficult to determine the number of transgender juveniles in correctional facilities; the Office of Juvenile Justice and Delinquency Prevention (Developmental Services Group, 2014) reported that few jurisdictions record gender identity. Reports suggest that historically, male to female transgender inmates were not allowed access to commissary items that were accessible for inmates housed in female facilities, nor were individuals provided gender-appropriate undergarments, allowed to wear their hair in particular ways, were referred to, purposefully, as the wrongly identified gender, not provided appropriate medical treatment (including substance abuse and mental health treatment), amongst other issues (Transgender Law Center, 2005). Additionally, questions of whether or not gender confirmation surgery is considered "medically necessary" resulted in several lawsuits against the California Department of Corrections and Rehabilitation (CDCR) who initially denied the surgery as being medically unnecessary. The American Medical Association has stated that gender confirmation surgery is a medically necessary intervention in the treatment of gender dysphoria (AMA, 2008). The first inmate to sue CDCR for gender reassignment surgery was Michelle Norsworthy, who was released on parole before a higher court could rule on her case and later settled with CDCR, although issues of transgender inmates are not new (see, e.g., *Farmer v Brennan,* 1994. The U.S. Supreme Court ruled that prisons have a duty to protect inmates from the violent actions of other inmates under the Eighth Amendment).

In 2015, *Quine v Beard* established a settlement in which the California Department of Corrections provided clothing and other commissary items (such as makeup) that were traditionally only for female inmates. In addition, and of significant relevance for forensic psychology, is that the California Department of Corrections also agreed to refer Quine for sex reassignment surgery, and be rehoused in a female CDCR facility. Quine has since has sex reassignment surgery and as of this writing, is housed at the Central California Women's Facility (Associated Press, 2017; Reed, 2017). Other states' Department of Corrections have also had to respond to issues of transgender inmate placement and appropriate needs, such as Washington D.C., New Hampshire, and Georgia (Patterson, 2016). While the transgender population remains a relatively low percentage of all inmates, studies suggest that the transgender-incarcerated population is growing. The population is over represented in the penal system (Glezer, McNiel, & Binder, 2013) and transgendered individuals may be arrested at higher rates than the general population (Minster & Daley, 2003) meaning the issues surrounding appropriate treatment of transgender inmates will continue to develop.

Despite the growing legal and policy support for considering issues of gender in the forensic area, the issue of transgender individuals in the criminal justice system has been relatively unaddressed in the forensic psychological literature. Webb, Heyne, Holmes, and Peyta (2016) call psychologists to action to adapt or rethink assessment tools as the tools refer to gender. Webb and colleagues highlight the very significant issue that many tests require to "check the box" along the gender binary (i.e., male or female) that are normed on cisgender individuals. By choosing "male" or "female," the psychologist may miss the true scope of understanding the complexity of gender identity, co-occurring mental health issues, or minority stress. Keo-Meier, Herman, Reisner, Pardo, Sharp, and Babcock (2015), for example, found that testosterone treatment impacted scores on the MMPI-2, for transgender men one of the most widely used and studied psychological assessments, demonstrated a reduction in endorsed psychological distress. Additionally, Webb and colleagues described these changes as being more normative to the gender the test-taker identified with, rather than biological sex. Significantly, more research is needed to even suggest that transgender individuals might be assessed based on their gender identity. In another article, Keo-Meier and Fitzgerald (2017) emphasize the need for further competence in assessment with transgender clients "above and beyond the general population" (p. 51) to better assess issues of psychopathology and gender identity dysphoria. As Webb et al. (2016) state, "the potential for harm to transgender and nonbinary individuals is obvious when examining problems that arise when norms for these populations to do not exist" (n.p.). There are few answers pertaining to those who do not identify on the gender binary, or the impact of transition on assessment. While the American Psychological Association has published the Guideline for Psychological Practice with Transgender and Gender Nonconforming People (2015), the guidelines provide little guidance in terms of psychological testing with trans* and gender nonconforming persons. Additionally, advocates disagree that gender dysophoria be considered a psychiatric condition, but rather a medical one. It is likely that forensic psychologists may be limited in providing findings that meet the standards for testimony in regard to broad issues of transgender and non-binary individuals. Broadly speaking, understanding gender beyond biological sex is becoming an increasing area of need for forensic psychologists.

Immigration

The United States has long been a melting pot, with roots in diversity and immigrant populations. However, due to a confusing and lengthy process for legal immigration as well as issues of poverty and lack of access to social services, immigrants may find themselves on the wrong side of the law. Psychological assessment has been conducted with immigrants in areas of hardship, asylum, domestic violence, cognitive impairments, or as victims of a crime, as a means to stay in the United States. Asylum may be granted to individuals who have entered the United States and are seeking protection to stay (compared to a refugee, who is outside of the United States and is petitioning to enter). Asylum can be provided to individuals "who can demonstrate that they have been persecuted or fear being persecuted because of their race, religion, nationality, membership in a particular social group, or political opinion" (De Jesus-Rentas, Boehnlein, & Sparr, 2010, p. 491). A lengthy explanation of the asylum process is beyond the scope of this chapter; De Jesus-Rentas and colleagues present a thorough summary of the difficulties presented for immigrants seeking asylum. Two typical questions asked of the forensic psychologist in an asylum evaluation include, "to ascertain, to the extent possible, the validity of persecution claims made by the asylum seeker," and "to justify the veracity of the asylum petitioner's claims based on a professional assessment of the psychology of the asylum seeker" (Vaisman-Tzachor, 2014, n.p.) and the role of the forensic psychologist is to provide information about the hardship it would cause the petitioner if they were to be deported. In other immigration hardship evaluations, the hardship of others (spouses, children, etc.) would be evaluated. In other evaluations, the impact of the victimization is considered for the individual's case to remain in the United States.

In addition to the forensic psychological assessment needs mentioned, immigrant men and women are finding themselves incarcerated, either in the prison system or in detention centers. Individuals (often mothers and their children) can be held in "family detention centers" which are prison-like holding centers for immigrant families. At least one family detention center was previously used as a prison: the T. Donn Hutto Residential Center in Taylor, Texas. Corrections Corporation of America (CCA), a private prison corporation owns the facility. CCA contracts with Immigration and Customs Enforcement (ICE) for the facility and operations (CCA, 2018). There are facilities in Texas, New Mexico, and Pennsylvania. The American Civil Liberties Union (ACLU) settled a lawsuit with ICE in 2007 regarding living conditions at the T. Donn Hutto Residential Center (ACLU, 2007). The lawsuit included allegations that children were forced to wear prison uniforms, were not receiving adequate educational opportunities, and were being threatened with separation from their parents, amongst others. In 2010 and 2011, the ACLU was investigating allegations of rampant sexual assault against female detainees at the same facility (ACLU, 2010). The provision of mental health care in detention facilities is believed to be poor (Antonius & Martin, 2015), and immigrants who are ordered to psychiatric treatment may find themselves placed in detention centers instead (Venters & Keller, 2009).

Recent changes in administration have enhanced involvement of law enforcement officials in detaining, arresting, and deporting immigrants, including those who have been engaging in the process to become lawful citizens. This increased involvement with law enforcement by otherwise law-abiding individuals may complicate the assessment process. Additionally, it may increase the number of immigrants seeking psychological evaluations to assist with their cases. Given immigrants seeking legal protection in the United States are not flush with excess funds, seeking services for their cases is likely to be a difficult process.

De Las Fuentes, Duffer, and Vazquez (2013) emphasized the complex nature of these assessments for forensic psychologists; not only must they be well attuned to immigration law, they must also consider issues of diversity and be culturally competent, which is another area of study where more research is needed. Touching on the gender issues addressed in the section prior, immigrants are often women traveling alone or dependent on male family members (United Nations Population Fund, 2013). Considering these complex issues facing immigrants, particularly ones with histories of trauma, revictimization and retraumatization, and poor interactions with the U.S. government as well as complicated criminal histories, is likely to be an increased area of need for forensic psychologists, particularly those who speak the same languages as immigrants seeking an opportunity to stay in the United States legally. Finally, of note, Antonius and Martin (2015) described how immigrants may find themselves in a "gray area" when it comes to areas of competency evaluations. Immigrants may find themselves addressing issues of competency, which is complicated by the different legal classifications (civil issues versus criminal issues). The picture is also complicated by the fact that immigrants are not necessarily afforded the same rights as citizens in terms of legal protections and due process (Korngold, Ochoa, Inlender, McNeil, & Binder, 2015). The authors contend that recent changes in immigration law to address some of these issues is likely to increase the need for competency evaluations in immigrants.

Parole decisions

In addition to having more individuals incarcerated and having longer sentences than other nations, the United States also uses "control of freedom" more often than other nations (Justice Policy Institute, 2011). This includes any supervision or placement in the community or elsewhere that is not incarceration. This can includes things such as probation, parole, community service, and halfway houses. With a growing jail and prison population, and a lack of bed space available to house offenders, community supervision or incarceration alternatives may be on the rise. Forensic psychologists will likely be called upon to help make decisions regarding an individual's appropriateness for community supervision, their needs within the community, or other decisions that can help with placement.

As mentioned earlier, geriatric inmates are growing in number. Parole decisions may need to include evaluations to see if the inmates require guardianship or conservatorship upon leaving the correctional facilities. Parole decisions may need to rely on adaptive functioning assessments to determine the most appropriate setting for an individual (i.e., nursing home vs. group home vs. independently living).

Other relevant assessments for supervision outside of correctional facilities might include civil commitment evaluations. Upon release from a facility, if an individual is deemed a threat to themselves or others, they may require commitment to a hospital. If an individual is deemed a possible threat to society in another manner, they might also be evaluated for civil commitment. One such example would be sex offenders in states with sexually violent predator laws.

Sex offenders made up about 9% of the state prison population in 1999, with approximately 1.2 million inmates in state prisons who had committed sex-related offenses (Burdon, Kilian, Koutsenok, & Prendergast, 2001). According to the Bureau of Justice, the number of prisoners sentenced for violent sexual assault had increased by an annual average of nearly 15% between the years of 1980 and 1997 (Greenfeld, 1997). This was a faster rate of growth than any other category of violent crime, except for drug trafficking. The use of community supervision or shorter jail sentences has increased over the years as well (Drake & Barnoski, 2006). After the U.S. Supreme Court decision of Kansas v. Hendricks (1997), Sexually Violent Predator (SVP) laws have multiplied. Twenty U.S. states now have SVP laws.

With the rise of states who have enacted sexually violent predator laws and programs, the number of evaluations required by qualified mental health professionals has also risen. Complicated, and controversial, assessments are often required that include the risk of future sexual offending. When an individual is placed in a SVP program, they may need reevaluations at regular intervals to determine if they are appropriate for release. This also requires specialized skills and a competent mental health clinician to complete the evaluation.

Officer and staff screening, hiring, and training

Public safety service is inherently stressful and dangerous; the stakes are high for both the officer and the agency employing them. Given the liability and safety concerns in these positions, psychologists have been tasked with providing pre-employment psychological screenings with recommendations for hire. These comprehensive assessments, though individually targeted, can assess for stress resilience, ability and strategies for coping, problematic behavioral patterns, judgment and decision-making, behavioral control, etc. (Johnson, 2013). The ability to control one's behavior in stressful situations is paramount to public safety roles, and making the appropriate initial hires through comprehensive prescreening processes can prevent lengthy, costly, and unpleasant consequences for the hiring agency.

To continue, correctional officers are one example of service roles in public safety that require an assessment component for hire. The Bureau of Labor Statistics (2014) describe the role of a correctional officer as those who oversee those who have been arrested; these officers are placed in a high-stress, dangerous environments and are expected to keep order and handle crises. A study by Rogers (2001) found that correctional staff reported high rates of depression, feelings of hopelessness, and suicidal ideation. Approximately 50% reported excessive tiredness, 44% had frequent headaches, and 12% had monthly migraines. Correctional officers are frequently hypervigilent due to the potential dangerous environment that they work in. This can have serious repercussions on both the physical and mental health of correctional staff.

Forensic psychologists are able to help in the realm of hiring officers who are the best fit for correctional environments and can help to train officers to better function in the difficult environment. Psychologists are able to administer and interpret psychological assessments that will enable departments to hire officers who are suited to corrections and law enforcement. In California, Government Code 1031 requires that officers meet "psychological suitability." The California Department of Corrections and Rehabilitation (CDCR) requires that applicants are required to participate in the Peace Officer Psychological Examination (POPE) which consists of the Minnesota Multiphasic Personality Inventory Second Edition (MMPI-2), the Sixteen Personality Factor Questionnaire (16PF), a psychological history questionnaire, and an interview with a psychologist.

This is established through the Peace Officer Standards of Training (POST). This aligns with the International Association of Chiefs of Police Guidelines (IACP) (2009) which states, "A written test battery, including objective, job-related psychological assessment *instruments*, should be administered to every candidate" (p. 3, emphasis added). The IACP guidelines (2009) also encourage the use of a semistructured interview and instruments validated on public safety applicants.

The exact guidelines for preemployment screenings vary from state to state, but many will utilize the POST standards as a basis for their process (Johnson, 2013) and 90% use testing in some capacity (Cochrane et al, 2003). One test, in particular, the Matrix-Predictive Uniform Law Enforcement Selection Inventory (M-PULSE), allows psychologists to identify officers who are likely to practice in a way that creates liability problems in the future. The M-PULSE can identify issues that are directly linked to officer misconduct. If misconduct occurs, psychologists can also conduct fitness for duty evaluations. The IACP indicates that the purpose of referring an employee for a fitness for duty evaluation is based on the following criteria: (1) that there is an objective and reasonable basis that the employee may not be able to safely and/or effectively perform their duties and (2) that this inability is due to a psychological condition or impairment.

There are several reasons why a correctional officer might be referred for this type of evaluation. If an officer is using excessive force, is involved in multiple questionable incidences, or has an incident outside of work (e.g., is arrested or involved in fighting) they might be referred for an evaluation. They might also be referred if they are displaying unusual behavior such as exhibiting excessive absenteeism, displaying anger toward their coworkers, or if there is suspected use of drugs or alcohol. The IACP guidelines require that the evaluator for fitness for duty evaluations be a licensed psychologist or board certified psychiatrist with law-enforcement experience.

Forensic psychologists can also help with training officers to identify mental health issues comorbidity issues, crisis communication, etc. Several incidents have been presented in the media where correctional officers either ignored individuals with mental health issues or purposefully contributed to inmates with mental health conditions being seriously injured or even dying. Whether it has been occurring more often, or if the presence of media and social media allows for the general population to become more aware of it, it is present and is a problem. This indicates that there is a need for additional training and hiring of officers who understand the needs of incarcerated individuals with mental health condition. This will help to decrease violence against inmates, violence against staff, and other dangerous situations. Lavoie and colleagues (2006) indicated that while most officers who worked at a maximum-security prison had some form of mental health training, approximately 80% did not feel that the training had adequately prepared them to work with mentally disordered offenders.

Expanding the role of forensic psychologists in corrections

Leadership in forensic mental health

In addition to forensic assessment, forensic psychologists can provide a range of expertise to address other significant issues in corrections. For example, several court case decisions (Ruiz v. Estelle 1980, and Madrid v. Gomez 1995) established the framework and criteria for adequate mental health services. Some states' correctional facilities are monitored by court appointed monitors due to a failure to provide adequate mental health care, such as California (see *Coleman v. Brown*), and other states are recognizing that correctional facilities are failing to meet the needs of mentally ill inmates, for example, New York State has recommended the closure of Rikers Island due to persistent issues, some related to mental health (Goodman, 2017).

Forensic psychologists can play a large role in helping to ameliorate some of these identified problems. Forensic psychologists have developed clinical and forensic skills in assessment and can offer their expertise in assessment in many ways. Forensic psychologists are in a unique position and can lend their expertise to lead multidisciplinary treatment teams. Traditionally, psychiatrists have led treatment teams; however, psychologists can offer expertise in assessment, diagnosing, and treatment that other professionals do not possess and are necessary for the purpose of identifying the needs of patient/inmates. Psychologists have gained prescribing privileges in Louisiana, New Mexico, Illinois, and Iowa. With a national shortage of psychiatrists, particularly in correctional settings, if prescribing privileges for psychologists continues to gain traction, forensic psychologists may be in a unique position to fill a national gap in service. Forensic psychologists can prescribe psychiatric medication as a total approach to mental health care in correctional settings; with unique training with criminal behavior, psychopathology, and assessment, forensic psychologists may provide a necessary service in correctional or forensic facilities.

Forensic psychologists may also provide a much-needed service in terms of mortality assessments. Suicide is the leading cause of death in jails, accounting for approximately one-third of jail deaths (Department of Justice, Office of Justice Programs, 2015). The mortality rate for drug and alcohol intoxication is also significant in correctional settings; OJP (2015) has indicated that the mortality rate for drug and alcohol intoxication rose 69% between 2012 and 2013. Further, in state prisons, the mortality rates for individuals over the age of 55 has increased at an average rate of 5% every year since 2001. Specialized training in suicide assessment, comorbidity, and psychological autopsy provides the opportunity to address issues of suicide and preventable mortality in corrections. Suicide risk assessments tend to focus on the individual endorsing suicidal ideation or behaviors and are commonly practiced in correctional settings. Psychological autopsies are less common, but are a valuable tool of research. Psychological autopsies are conducted after a completed suicide and typically consist of multiple interviews with staff, peers, and family members as well as thorough record reviews. The psychological autopsy can help uncover systemic issues within the correctional facility or help to uncover the interactions between various risk factors or domains to help to decrease the amount of future completed suicides. Psychological autopsies require specialized knowledge in risk factors, underlying dynamics, and systemic issues which forensic psychologists can offer. Forensic psychologists may be able to address policy and system-level issues to improve the provision of care for mentally ill inmates. As described earlier, privacy is a considerable concern when assessing and treating inmates. Cell-side evaluations are not unusual, and safety of the institution is paramount, often to the detriment of patient confidentiality. Patient movement is often limited, which may prevent access to mental health treatment. As outlined earlier, this has become a significant issue in California.

Lastly, qualified psychologists are often not being recruited or paid adequately to work in a tough, serious, potentially dangerous setting. With large ratios, psychologists may act as crisis interventionists and little else. With the multitude of skills available to the forensic psychologist, significant emerging and existing issues may be able to be minimized. Expanding the role of the forensic psychologist provides significant and necessary contributions to the fields of psychology, criminal justice, and corrections.

The field of forensic psychology and the utility of forensic mental health assessments are relatively young in the field of psychology, and it is expected that their utility will continue to grow. With that, there are emerging special populations that make the field of forensic mental health assessment more complex. It is of the utmost importance that forensic psychologists not practice outside of their area of competency, however, with appropriate education and training, forensic psychologists are in a unique position to help with the intersection of emerging special populations and their intersection with a complex criminal justice system. The field of forensic psychology is dynamic; as laws change, so will the role of the forensic psychologist.

References

Abram, K.M., & Teplin, L.A. (1991). Co-occurring disorders among mentally ill jail detainees: Implications for public policy. *American Psychologist, 46*, 1036–1045.

American Civil Liberties Union (2007). *ACLU challenges prison-like conditions at Hutto Detention Center.* Accessed from: https://www.aclu.org/aclu-challenges-prison-conditions-hutto-detention-center

American Civil Liberties Union (2010). *Sexual abuse of female detainees at Hutto highlights ongoing failure of immigration detention center, says ACLU.* Accessed from: https://www.aclu.org/news/sexual-abuse-female-detainees-hutto-highlights-ongoing-failure-immigration-detention-system

American Medical Association, (2008). *American Medical Association House of Delegates Resolution 122: Removing financial barriers to care for transgender patients*. Chicago, IL: Author. Accessed from: http://www.tgender.net/taw/ama_resolutions.pdf

American Psychological Association. (2015). Guidelines for psychological practice with transgender and gender nonconforming people. *American Psychologist, 70*(9), 832–864. doi:10.1037/a0039906

Antonius, D., & Martin, P.S. (2015). Commentary: Mental health and immigrant detainees in the United States. *Journal of the American Academy of Psychiatry and the Law, 43*(3), 282–286.

Appelbaum, P., & Grisso, T. (1995). The MacArthur treatment competence study 1. Mental illness and competence to consent to treatment. *Law and Human Behavior, 19*(2), 105–126. doi:10.1007/BF01499321

Associated Press. (2017). First US inmate to receive state-funded sex-reassignment surgery moved to women's prison. *The Oregonian*. Accessed from www.oregonlive.com/today/index.ssf/2017/02/first_us_inmate_to_receive_sta.html.

Bartlett, A., & Somers, N. (2016). Are women really difficult? Challenges and solutions in the care of women in secure services. *Journal of Forensic Psychiatry and Psychology*, online. doi:10.1080/14789949.2016.1244281

Bloom, B, Owen ,B., & Covington, S. (2003). *Gender-responsive strategies: Research, practice, and guiding principles for women offenders.* Washington, D.C.: National Institute of Corrections

Bonczar, T., Hughes, T.A., Wilson, D., & Ditton, P.M. (2011). *State Prison Admissions: Sentence Length by Offense and Admission Type.* Bureau of Justice Statistics Published May 5, 2011. Washington, DC, Bureau of Justice Statistics: 2010. Accessed from http://bjs.ojp.usdoj.gov/index.cfm?ty=pbdetail&iid=2174.

Buchanan, A.E., & Brock, D.W. (2001). Determinations of competence. In T. A. Mappes & D. DeGrazia (EDs.), *Biomedical ethics* (5th ed., 109–114). New York, NY: McGraw-Hill.

Burdon, W.M., Kilian, T.C., Koutsenok, I., & Prendergast, M.L. (2001). Treating substance abusing sex offenders in a correctional environment: Lessons from the California experience. *Offender Substance Abuse Report* (January/February), 3, 4, 11, 12.

Bureau of Justice Statistics. (2013). *Sexual Victimization in Prisons and Jails Reported by Inmates, 2011–2012: National Inmate Survey, 2011–2012.* Washington, DC, Office of Justice Programs.

Bureau of Labor Statistics, U.S. Department of Labor. (2014). *Occupational Outlook Handbook, Correctional Officers and Bailiffs.* Retrieved from https://www.bls.gov/ooh/protective-service/correctional-officers.htm.

California Peace Officer Standards and Training Commission. (POST) Regulation 1955.

Carson, E.A., Sabol, W.J. (2012). *Prisoners in 2011.* Washington, D.C.: Bureau of Justice Statistics.

Carson, E.A. (2014). *Prisoners in 2013.* Washington D.C.; Bureau of Justice Statistics.

Cochrane, R. E., Tett, R. P., & Vandecreek, L. (2003). Psychological testing and the selection of police officers: A national survey. *Criminal Justice and Behavior, 30*(5), 511–537.

Corrections Corporation of America, (2018). T. Don Hutto Residential Center. Accessed from: http://www.correctionscorp.com/facilities/t-don-hutto-residential-center

Davidson, M., Sorenson, J.R., & Reidy, T. (2016). Gender responsiveness in corrections: Estimating female inmate misconduct risk using the Personality

Assessment Inventory (PAI). *Law and Human Behavior, 40*(1), 72–81. doi:10.1037/lhb0000157

Department of Justice, Office of Justice Programs. (2015, August 4). *Deaths in local jails and state prisons increased for the third consecutive year* [Press Release]. Retrieved from https://ojp.gov/newsroom/pressreleases/2015/ojp08042015.pdf.

Developmental Services Group, Inc. (2014). *LGBTQ Youths in the Juvenile Justice System. Literature Review.* Washington, DC, Office of Juvenile Justice and Delinquency Prevention. Accessed from www.ojjdp.gov/mpg/litreviews/LGBTQYouthsintheJuvenileJusticeSystem.pdf.

De Jesus-Rentas, G., Boehlein, J., & Sparr, L. (2010). Central American victims of gang violence as asylum seekers: The role of the forensic expert. *Journal of the American Academy of Psychiatry and Law, 38*(4), 490–498.

de las Fuentes, C., Duffer, M.R., & Vasquez, M.J. (2013). Gendered borders: Forensic evaluations of immigrant women. *Women & Therapy, 36*(3–4), 302–318. doi:10.1080/02703149.2013.79778

de Vogel, V., & de Ruiter, C. (2005). The HCR-20 in personality disordered female offenders: A comparison with a matched sample of males. *Clinical Psychology and Psychotherapy, 12*, 226–240. doi:10.1002/cpp.452

Drake, E., & Barnoski, R. (2006). *Sex Offenders in Washington State: Key Findings and Trends.* Olympia, Washington State Institute for Public Policy, Document No. 06 03–1201.

Farmer v. Brennan, 511 U.S. 825 (1994).

Fazel, S., Bains, P., & Doll, H. (2006). Substance abuse and dependence in prisoners: A systematic review. *Addiction, 101*(2). 181–191. doi:10.1111/j.13600443.2006.01316.x

Fellner, J. (2007, July). Keep mentally ill out of solitary confinement. *Huffington Post.* Accessed from www.hrw.org/news/2007/07/19/keep-mentally-ill-out-solitary-confinement.

Ford v. Wainwright, 477 U.S. 399 (1986).

Glaze, LE. & Kaeble, D. (2014) *Correctional populations in the United States, 2013.* Washington, D.C.: U.S. Department of Justice, Bureau of Justice Statistics

Glezer, A., McNiel, D.E., & Binder, R.L. (2013). Transgendered and incarcerated: A review of the literature, current policies and laws, and ethics. *The Journal of the American Academy of Psychiatry and the Law, 41*(4), 551–559.

Goodman, J.D. (2017, March). Mayor backs plan to close Riker's and open jails elsewhere. *New York Times.* Accessed from www.nytimes.com/2017/03/31/nyregion/mayor-de-blasio-is-said-to-back-plan-to-close-jails-on-rikers-island.html?_r=0.

Greenfeld, L.A. (1997). Sexual offenses and offenders: An analysis of data on rape and sexual assault. *Bureau of Justice Statistics.* February 1997, NCJ-163392.

Haney, C. (2003). Mental health issues in long-term solitary and "supermax" confinement. *Crime and Delinquency, 49*(1), 124–156. doi:10.1177/0011128702239239

Hollin, C.R. & Palmer, E.J. (2006). Criminogenic need and women offenders: A critique of the literature. *Legal and Criminological Psychology, 11*, 179–195. doi:10.1348/135532505X57991

International Association of Chiefs of Police, Police Psychological Services Section. (2009a). *Guidelines for Psychological Fitness-for-Duty Evaluations.* Arlington, VA.

International Association of Chiefs of Police, Police Psychological Services Section. (2009b). *Guidelines for Preemployment Psychological Evaluations.* Arlington, VA: Author.

James, D.J., & Glaze, L.E. (2006). *Mental Health Problems of Prison and Jail Inmates: Bureau of Justice Special Report.* Washington, DC, U.S. Department of Justice.

Johnson, R. (2013). Forensic psychological evaluations for behavioral disorders in police officers: Reducing negligent hire and retention risks. In J.B. Helfgott (Ed.), *Criminal Psychology* (pp. 253–278). Santa Barbara, CA: Praeger.

Justice Policy Institute. (2011). *Finding Direction: Expanding Criminal Justice Options by Considering Policies of Other Nations*, April 2011.

Kansas v. Hendricks, 521 U.S. 346 (1997).

Keo-Meier, C., & Fitzgerald, K.M. (2017). Affirmative psychological testing and neurocognitive assessment with transgender adults. *Psychiatry Clinics of North America*, 40, 51–64. doi:10.1016/j.psc.2016.10.10.011

Keo-Meier, C., Herman, L., Resiner, S.L., Pardo, S.T., & Babcock, J.C. (2015). Testosterone treatment and MMPI-2 improvement in transgender men: A prospective controlled study. *Journal of Consulting and Clinical Psychology, 83*(1), 143–56. doi:10.1037/a0037599

Kreis, M.F., Gillings, K., Svanberg, J., & Schwanner, M. (2016). Relational pathways to substance misuse and drug-related offending in women: The role of trauma, insecure attachment, and shame. *International Journal of Forensic Mental Health, 1*, 35–47. doi:10.1080/14999013.2015.1134725

Korngold, C., Ochoa, K., Inlender, T., McNiel, D., & Binder, R. (2015). Mental health and immigrant detainees in the United States: Competency and self-representation. *Journal of the American Academy of Psychiatry and Law, 43*(3), 277–281.

Lamb, H., & Weinberger, L. (2014). Decarceration of U.S. Jails and Prisons: Where Will Persons With Serious Mental Illness Go?. *Journal of the American Academy of Psychiatry and the Law*. 42. 489–494.

Lavoie, J.A., Connolly, D.A., & Roesch, R. (2006). Correctional officers' perceptions of inmates with mental illness: The role of training and burnout syndrome. *International Journal of Forensic Mental Health, 5*(2), 151–166.

Leatherman, M.E., & Goethe, K.E. (2009). Substituted decision making: Elder guardianship, *Journal of Psychiatric Practice, 15*(6).

Loring, M.T. and Beaudoin, P. (2000). Battered women as coerced victim-perpetrators. *Journal of Emotional Abuse, 2*. 3–14.

Loring, M.T. and Bolden-Hines, T.A. (2004). Pet abuse by batterers as a means of coercing women into committing illegal behavior. *Journal of Emotional Abuse, 4*, 27–37.

Lovell, D., Cloyes, K., Allen, D., & Rhodes, L. (2000). Who lives in super-maximum custody? Washington state study. *Federal Probation, 64*(2), 33–38.

McGaughy, L. (2016). Number of Texas prison inmates coming out as transgender at all-time high. *Dallas News*. Accessed from www.dallasnews.com/news/lgbt/2016/09/27/number-texas-prison-inmates-coming-transgenderatall-time-high.

Minster, S., & Daley, C. (2003). *Trans Realities: A Legal Needs Assessment of San Francisco's Transgender Communities*. San Francisco, CA: The National Center for Lesbian Rights and the Transgender Law Center.

Morris, D.R., & DeYoung, N.J. (2012). Psycholegal abilities and restoration of competence to stand trial. *Behavioral Sciences and the Law*, 30, 710–728.

National Institute of Corrections. (2004). *Effective Prison Mental Health Services*. U.S. Department of Corrections. Washington, DC: Author.

Nicholls, T.L., Ogloff, J., & Douglas, K. (2004). Assessing risk for violence among male and female civil psychiatric patients: The HCR-20, PCL:SV, and McNiel & Binder's screening measure. *Behavioral Sciences and the Law*, 22, 127–158. doi:10.1002/bsl.579

Pate v. Robinson (1966) 383 U.S. 375, 378.

Patterson, B. (2016). Justice department takes steps to protect transgender prisoners. *Mother Jones*. Accessed from www.motherjones.com/politics/2016/03/doj-trans-inmate-guidelines.

Pirelli, G., Gottdiener, W.H., & Zapf, P.A. (2011). A meta-analytic review of competency to stand trial research. *Psychology, Public Policy, and Law, 17*(1), 1–53. doi:10.1037/a002

Porter, L.C., Bushway, S.D., Tsao, H.S., & Smith, H.L. (2016). How the U.S. prison boom has changed the age distribution of the prison population. *Criminology*, 54, 30–55. doi:10.1111/1745–9125.12094

Puzzanchera, C. (2009) *Juvenile Arrests 2008*. Washington, DC: U.S. Department of Justice, Office of Juvenile Justice and Delinquency Prevention

Quine v. Beard, C 14–02726 JST (N.D. Cal. 2015).

Reed, E. (2017, February). N.H correctional facilities tackle the complex issue of transgender inmates. *Concord Monitor*. Accessed from Transgender Law Center (2005). *Safety Inside: Testimony before the National Prison Rape Elimination Commission*. San Franscico, CA: Transgender Law Center.

Rogers, J.B. (2001). *FOCUS I Survey and Final Report: A Summary of the Findings: Families Officers and Corrections Understanding Stress*. Washington, DC: U.S. Department of Justice.

Shepherd, S.M., Luebbers, S., & Dolan, M. (2013). Gender and ethnicity in juvenile risk assessment. *Criminal Justice and Behavior, 40*(4), 388–408. doi: 10.1177/0093854812456776

Smith, P., Cullen, F.T., & Latessa, E.J. (2009). Can 14,737 women be wrong? A meta-analysis of the LSI-R and recidivism for female offenders. *Criminology and Public Policy, 8,* (1). 183–208

Steadman, H.J., Osher, F., Robbins, P.C., Case, B., & Samuels, S. (2009). Prevalence of serious mental illness among jail inmates. *Psychiatric Services*, 60, 761–765.

Transgender Law Center. (2005). *Safety Inside: Testimony before the National Prison Rape Elimination Commission*. San Franscico, CA: Transgender Law Center.

United Nations Population Fund. (2013). *New Trends in Migration. 46th Session*. Accessed from www.unfpa.org/pds/migration.html.

U.S. Department of Justice, Office of the Inspector General (USDOJ). (2016). *The Impact of an Aging Inmate Population on the Federal Bureau of Prisons*. Evaluation and Inspections Division, May 2015.

Vaisman-Tzachor, R. (2014). Psychological assessment protocols for asylum applications in federal immigration courts. *The Annals of Psychotherapy*, December. Accessed from www.annalsofpsychotherapy.com/articles/fall14.php.

Van Voorhis, P., & Presser, L. (2001). *Classification of Women Offenders: A National Assessment of Current Practices*. Washington, DC: National Institute of Corrections.

Venters, H., & Keller, A.S. (2009). The health of immigrant detainees. *The Boston Globe*. Accessed from http://archive.boston.com/bostonglobe/editorial_opinion/oped/articles/2009/04/11/the_health_of_immigrant_detainees/.

Webb, A., Heyne, G., Homes, J., & Peta, J.L. (2016). Which box to check: Assessment normal for gender and the impact for transgender and nonbinary populations. *Division 44 Newsletter*, 44. Accessed 7 April, 2017, www.apadivisions.org/division-44/publications/newsletters/division/2016/04/nonbinary-populations.aspx.

Wright, E.M, Van Voorhis, P., Salisbury, E.J, & Bauman, A. (2012). Gender-responsive lessons learned and policy implications for women: A review. *Criminal Justice and Behavior, 39.* 1612–1632. doi: 10.1177/0093854812451088

Zapf, P.A., Roesch, R., & Pirelli, G. (2014). Assessing competency to stand trial. In I.B. Weiner & R.K. Otto (Eds.), *The Handbook of Forensic Psychology* (4th ed., pp. 281–314). New York: Wiley.

Part II

Emerging technological advancements within forensic sciences

6 Comparative perspectives on digital forensic technology

Hollianne Marshall

Introduction

With the advent of technology and the rapid sharing of digital information, there has also been a rapid development of the misuse of technology and fraudulent digital transactions. Combatting these issues has resulted in much confusion and difficulty surrounding the development and use of technologies to investigate digital and computer crime. Many scholars have discussed the constitutional implications of these types of investigations, and others have called into question the ethics of using certain software and technologies to enhance our investigative searches. Supreme Court cases have addressed the wrongful digital search and seizure in several criminal cases, but there is still a lack of clarity on how to interpret the 4th Amendment and other legislation when applied to the digital world. Controversy surrounds the fact that digital technology evolves at a much faster pace than legislation and creates problems within the criminal justice system. This chapter overviews the history of digital forensics, and explains various digital forensic methods and technologies as well as how they are applied. Next, ethical issues including boundaries in the search and seizure of digital information, privacy rights, and data storage, during and after investigations are discussed. Suggestions for improving ethical training of digital forensic investigators, guidelines for improving the appropriate collection of digital evidence and, creating clear boundaries for the use of this forensic information are discussed.

History of digital forensics

The history of digital forensic processes is quite short and has grown nearly as quickly as digital technology itself. Beginning in the early 1980s, digital forensics became a more commonly used tool (Carrier, 2003; Casey, 2011; Garfinkel, 2010; Nelson, Phillips, & Steuart, 2014; Pollitt, 2010). This was in response to the growing financial fraud through computers during the 1970s. During the late 1970s and 1980s, the idea of digital forensics related mostly to data recovery, and tools were designed with this specific goal in mind. The mid 1980s saw formal agencies at the FBI develop, to respond to growing computer crime. The Magnetic Media Program was the first agency tasked with handling electronic crime (this was later changed to the Computer Analysis and Response Team (CART) in the 1990s) (Casey, 2011; Goldfoot, 2011; Nelson, et al., 2014; Wegman, 2005). As technology has rapidly grown, so has the field of digital forensics, particularly in law enforcement investigation and throughout the judicial process (Casey, 2011; Reith, Carr, & Gunsch, 2002; Rogers, 2003). Toward the end of the 1980s, many digital forensic labs and digital forensic professional organizations were developing and specialized training in acquiring and recovering data became more widespread (Casey, 2011; Garfinkel, 2010; Nance & Ryan, 2011; Reith, et al., 2002). Independent labs started to contract with law-enforcement agency and digital forensic organizations began to develop certain standards for labs, analysts, and technology (Casey, 2011; Garfinkel, Farrell, Roussev, & Dinolt, 2009; Pollitt, 2010; Reith, et al., 2002). Digital forensic analysis quickly grew beyond data recovery into more invasive processes including decoding encryption, reimaging drives, to more contemporary "ethical hacking" and "cloud forensics" (Biggs & Vidalis, 2009; Casey, 2011; Garfinkel, 2010). The more contemporary digital forensic analyses involve several types of software and long periods of time to execute. A complicating issue is that not all types of analyses are executed in the same way and might not even use the same software (Beebe, 2009; Casey, 2011; Pollitt, 2010). Digital forensic analyses are consistently entangled in ethical decisions as their very nature provides access to privileged and/or private information. The ethics of acquisition and use of this private information during a criminal investigation receives challenges on more than one occasion (Broucek & Turner, 2013). Many of these ethical considerations arise from the lack of standardization within the field both in oversight and analytical processes (Garfinkel, 2010; Pollitt, 2010; Wegman, 2005). However, other issues surrounding the fair application of the 4th Amendment in the U.S. Constitution and the Electronic Communications Privacy Act (ECPA, 1986) with the use of digital forensic analysis in investigation continue to be prominent. The search and seizure of digital data are complex processes and legislation has not been able to keep the pace with technological advances and increasing digital crime (Bedi, 2014; Kerr, 2015; Salgado, 2005; Vaughn, 2014). When this is coupled with a lack of universal digital forensic analysis standards, questioning this type of investigation is critical. In order to fully understand the ethical issues, it is first important to understand the basics of digital forensics.

Digital forensics: processes and technology

Digital forensics is the act of searching through seized electronics and data to find information that can be used and interpreted in a criminal court case

(Wegman, 2005; Wolf, Shafer, & Grendon, 2006). Within digital forensics, there are two major processes: *Acquisition* and *Extraction* (Beebe & Clark, 2005; Nelson, et al., 2014; Reith, et al., 2002). Acquisition deals with acquiring data in some capacity through copying or imaging data onto a new drive or into a new format (Casey, 2011; Carrier, 2003; Carrier & Grand, 2004). Data acquisition, quite simply, is the acquiring of data through selected software. Data acquisition format is the format that data are in (binary code, encrypted, etc.). The literature is unclear on best practice for software or process in acquiring data, and it is apparent that there are no current standard software programs or methods for acquiring data (Adams, 2008; Rogers, 2003). Several common processes for acquisition include the following:

- Physical data copy are data in physical space on a disk
- Logical data copy are data in logical format—sequential or partitioned
- Remote, live, and memory acquisition—a procedure for acquiring a memory image when a system has been infected or data are fractured

(Carrier & Grand, 2004; Casey, 2011)

The many different routes to acquire data can create ethical dilemmas because data are not acquired in the same method by any agency, or for each individual case (Garfinkel, 2010; Nelson et al., 2014). This choice of process depends on the specific program used for data acquisition and the time available for this procedure (Nolan, O'Sullivan, Branson, & Waits, 2005). Many tools to acquire data are considered untrustworthy, particularly if data are stored in memory (Beebe & Clark, 2005; Garfinkel, 2010; Nelson, et al., 2014; Reith, et al., 2002). This creates many discrepancies across jurisdiction for data acquisition, but there are also issues with how much and which type of data can or cannot be seized during the course of an investigation (Garfinkel, 2010; Goldfoot, 2011; Nance & Ryan, 2011). Much of this will be discussed later in the chapter in conjunction with search and seizure policies. There are growing issues with the acquisition of data remotely and also problems with data that are stored on third-party servers and cloud storage (Biggs & Vidalis, 2009; Garfinkel, 2010). Data acquisition creates both a complexity problem and a quantity problem during extraction (Beebe & Clark, 2005; Garfinkel, 2010; Pollitt, 2011). The complexity issue is one that acquiring readable and/or worthwhile data requires extensive training and a skillset that is rather rare for individuals within the field. The quantity problem is such that in any situation, the amount of data acquired from an individual computer is so large that it is impossible to search for evidence item by item (Beebe & Clark, 2005; Garfinkel, 2010; Pollitt, 2011).

Extraction refers to searching, viewing, or analyzing the data that is already acquired (Garfinkel, 2010; Nelson et al., 2009; Pollitt, 2011). Similar to data acquisition, the extraction process depends on the program used, but also depends on the state of the data at the time of acquisition. Some common extraction processes include the following:

- Data viewing: This is the process of viewing data that has been copied and ready for search. Viewing data becomes problematic if it is a "general" search and even at times if it is not.
- Keyword searching: This combats the quantity issues that data acquisition creates. Programs allow analysts to search for key words rather than going through each piece of data individually. Keyword searches arguably lead to bias in the search and possibly less exculpatory evidence.
- Decompressing: A process to extract data that has been compressed for disk space; this process can lead to broken or unreadable data.
- Carving: Data carving is part of data recovery; users can identify data that has no file allocation by searching for clusters belonging to the file extension.
- Decrypting: Clarifies unclear (encrypted) data. Similar to decompression, this process can create data that is unusable or inaccurate to first inception of data.

(Carrier & Spafford, 2004; Garfinkel, 2010; Nance & Ryan, 2011; Walls, Levine, Liberatore, & Shields, 2011; Wolf, et al., 2006)

Digital forensic technologies

There are many different digital forensic technologies. Much of the literature discusses the abundance in availability of several types of technology (Baryamureeba & Tushabe, 2004; Bedi, 2014; Garfinkel, 2010; Nelson, et al., 2014; Walls, et al., 2011). There would be too many to discuss within the confines of this chapter. However, these are a few of the very common technologies used in digital forensic analysis and it is important to keep in mind that many different types of software can execute these functions.

Digital replication technology is used often during acquisition. This process can share completely different physical databases, mirror back up data, and make images. Many compare digital replication technology to taking photos of a crime scene in the physical world. Others suggest that digital replication is what violates the search of actual property (Cole, Gupta, Gurugubelli, & Rogers, 2015; Kessler, 2004; Ma, Shen, Chen, & Zhang, 2015; Walls, et al., 2011; Wang, Cannady, & Rosenbluth, 2005).

Information encryption technology is based on mathematical algorithms and the encryption key created by the initial user. This technology can scramble information or make previously unreadable information clear. This process relies heavily on having a key or being able to crack a key. Critics of this technology assert there are issues with accuracy with when using decryption both with interpreting new data and also with being certain it is restored to its original form (Cole et al., 2015; Garfinkel, 2010; Kessler, 2004; Ma, et al., 2015; Walls, et al., 2011).

Data Recovery Technology restores data that are deleted, in invisible space, or that are damaged. After recovery, data are usually viewable in their original state. The same standpoint that this process may not be entirely accurate applies to this type of technology (Cole, et al., 2015; Garfinkel, 2010; Kessler, 2004; Ma, et al., 2015; Pollitt, 2011; Wang, et al., 2005).

Data interception technology is similar to surveillance this technology collects data transactions that are intercepted before they meet the end user. Essentially this is real-time interception of conversations and data transfers. Many take issue with this type of technology because while it searches to intercept certain digital transactions, it also views other unrelated digital transactions, thus creating a true lack of individual privacy (Cole, et al., 2015; Pollitt, 2011; Walls, et al., 2011).

Data spoofing is used mostly to detect and understand network attacks; some refer to this as "ethical hacking." Hacking a system to record the methods of network attack in order to discover threats and create new protections (Garfinkel, 2010; Ma, et al., 2015; Nolan, et al., 2005).

Digital time-stamping is a technology that detects the time that digital transactions occurred and can also detect if digital time stamps have been edited or tampered with (Ma, et al., 2015).

Cloud methods are newer proposed methods of digital forensic analyses. These methods can accomplish intrusion detection, network monitoring, and support vector machine (Beebe, 2009; Garfinkel, 2010; Ma, et al., 2015; Ruan, 2013). Some suggest that cloud computing is a better option for current forensics issues. More can be done in real time, many of the processes can reduce the time it takes to run a search and much of the trouble of extracting cloud data is greatly reduced by using cloud technology as part of the digital forensic process (Ma, et al., 2015; Wang, et al., 2005).

The inconsistencies across industries with data evaluation and extraction continue to raise ethical questions. Particularly because there are no industry standards or "preferred" methods; the processing of data can be entirely different from one case to the next (Garfinkel, 2010; Nelson, et al., 2014; Pollitt, 2011). The complexity problem created by data acquisition is one where data are acquired but are in need of very high skill levels to process, this makes those question investigations if someone did not have the proper training to identify all data, and more importantly any data that could have served as exculpatory evidence; evidence to show a crime may not have been committed (Garfinkel, 2010; Jekot, 2007; Kessler, 2004; Nelson, et al., 2014; Rogers, 2003). There are also many cited cases, some of which discussed later in the chapter, explaining how methods of extraction and reading data can also fall outside the realm of the constitution (Goldfoot, 2011; Vaughn, 2014).

Digital forensic processes

The lack of standardization across digital forensic training and digital forensic labs calls into question the ethics of applying these procedures in criminal investigation (Carrier, 2003; Carrier & Spafford, 2004; Elyas, Maynard, Ahmad, & Lonie, 2014; Garfinkel, 2010; Nelson, et al., 2014; Wolf, et al., 2006). Collecting and saving data varies with each investigation, and there is no oversight of the search through data (Casey, 2011; Nelson, et al., 2014). There are also ethical considerations with how much data to collect and how long to save it. With no governing guidelines, data can be collected and saved in any capacity for any length of time. Issues with software reliability and validity also plague the field of digital forensics (Garfinkel, 2010; Nelson, et al., 2014; Pollitti, 2011). Many argue that software is constantly evolving and often cannot guarantee accurate extraction. Beyond this, there are always questions with how much data to collect and if it is a violation to copy an entire drive if this is not where key data are located. This brings focus back to unclear legislation and an industry that lacks universal analysis standards.

Searching data

A search through data occurs after seizure and acquisition, and is often a time consuming process (Casey, 2011; Garfinkel, 2010; Pollitti, 2011). Searching data can have inherent bias because it is impossible to go through data piece by piece. Therefore, key terms and other search mechanisms are required to reduce the time and effort needed for the search and for the investigation (Beebe & Clark, 2005; Reith, et al., 2002; Salgado, 2005). Since data are so cumbersome, and it is nearly impossible to read every last piece of data, often keyword searches are relied upon to find necessary supporting information. However, this keyword search is often the only information searched for, rather than also searching for other exculpatory evidence (Adams, 2008; Bedi, 2014; Ruan, 2013; Vaughn, 2014).

Describing data

Often describing and interpreting data is subjective. Each digital forensic analyst can have a different standpoint on the meaning of a piece of data. Data can also have different meaning to investigators, prosecutors, juries, and judges. There are no clear guidelines for what finding certain pieces of data mean and often data are only of importance given their relationship to a specific type of crime (Garfinkel, 2010; Nelson, et al., 2014; Pollitti, 2011). This becomes even more complicated when describing data that have been decrypted or accessed from a cloud or third-party server. In these cases, interpretation and description of data are limited based on what types of complete data are eventually acquired (Carrier, 2003; Nelson, et al., 2014; Ruan, 2013). Some challenge acquiring partial data, especially partial data from searches, because there would be no way to infer that a given piece of data meant that an individual intended to commit a crime. This is the other significant problem in data description, finding data does not always mean criminal intent (Garfinkel et al., 2009; Pollitti, 2011; Walls, et al., 2011).

Limitations in application

Investigations are limited to available technology. This technology varies by socioeconomics in both the law enforcement agency and jurisdiction (Garfinkel, 2010; Pollitti, 2011). Technologies also vary by contracted digital forensics labs and the capability of the lab is dependent upon the tools it has. These characteristics all limit the application of digital forensics in a fair and just way (Broucek & Turner, 2013). Some argue for a cohesive set of procedures and a cohesive set of technologies and software among digital forensics labs in order to create fairness within process and practice (Wolf, et al., 2006; Garfinkel, et al., 2009; Ma, et al., 2015). Others argue that since digital forensic labs are not always needed for criminal cases, calling for these kinds of regulations are unreasonable (Pollitti, 2011). In any case, the ethics of not using the exact same methods for each case and different technology among cases can potentially be biasing the result of the criminal proceedings.

Digital foresncis analysis standards and application

Digital forensic investigations are different from physical forensic investigations in many ways, but because there are no federal or state standards governing digital forensic analysis and investigation, this complicates the application of this technology even further (Garfinkel, 2010; Nelson, et al., 2014). In order to understand this, the nature of the technology used for these forensic investigations should be overviewed.

Private corporations serve as the developers for digital forensic technology and software used for analysis. These corporations often contract their software with digital forensic labs. Law enforcement then contract with the forensic labs or individuals who execute the forensic investigations (Adams, 2008; Nelson, et al., 2014; Pollitt, 2011). As such, those who are conducting a digital forensic analysis are likely not a part of law-enforcement and lack training in criminal investigation (Garfinkel, 2010; Nelson, et al., 2014; Wang, et al., 2005).

Most digital forensic examiners are trained in ethical considerations; however, this training is often cited to be incomplete or not emphasized appropriately throughout training (Carrier, 2003; Carrier & Spafford, 2004; Wolf, et al., 2006). Because of the lack of clear standards, many digital forensic analysts adhere to the ethics of professional organizations that they belong to. However, with a lack of state and national standards, many digital forensic labs are not operating under consistent standards. The disconnect between ethical standards and training in digital forensics arguably leads to many unforeseen issues during in an investigation (Carrier, 2003; Carrier & Spafford, 2004; Garfinkel, 2010; Nelson, et al., 2014; Wolf, et al., 2006). Because digital forensic training is focused primarily on technical competency, and not criminal investigation, analysts are equipped to search data but do not have a full understanding of the implications of data within a criminal investigation (Garfinkel, et al., 2009; Nelson, et al., 2014; Pollitti, 2011). The lack of industry guidelines and gray areas in training create ethical considerations in the application of digital forensics and the application of these procedures during investigations.

U.S. Constitution and digital forensic analysis

The 4th Amendment in the United States Constitution affords the liberty of privacy and protection against the forcible search and seizure of property by government, particularly for incrimination, but not limited to this (Goodman, Murphy, Streetman, & Sweet, 2001; Kerr, 2015; Ruan, 2013; Vaughn, 2014). Focused on search and seizure, the Courts generally seek to answer four questions when applying the 4th Amendment to all cases both digitally and physically (Bedi, 2014; Kerr, 2015; Vaughn, 2014). First, there is a consideration of society's perception of reasonableness. Does society have a custom to expect reasonable privacy in the instance? Second, the content of the information is considered, is there something about the information that *should* be private (i.e., medical records, psychological evaluations, etc.)? Third, were any laws and policies violated to obtain the evidence or information; did the government violate any laws or public policy (Slobogin, 2008; Solove, 2002; Wegman, 2005)? Finally, should society be protected from this type of behavior, are the consequences of this behavior by law enforcement dangerous to society (Bedi, 2014; Kerr, 2015; Vaughn, 2014; Wegman, 2005)?

Physical facts are the concern of search and seizure laws, namely what can be seized and what spaces can be entered. However, there are no limits on information one can learn from these physical facts. Because the Supreme Court holds that reasonableness is part of the foundation of the 4th Amendment, general searches are deemed unreasonable (Garfinkel, et al., 2009; Nelson, et al., 2014; Slobogin, 2008; Vaughn, 2014; Wegman, 2005). This is why warrants are necessary in most searches by law enforcement. Aside from the reasonableness standard, particularity is a requirement. Search warrants must specify and be "particular" about areas to be searched and items to be seized. The Plain View doctrine is the final part of the foundation of the 4th Amendment (Bedi, 2014; Kerr, 2015; Vaughn, 2014). Items in plain view seen during searches are admitted as evidence during a trial. In traditional searches, a warrant is obtained and a search begins once the property is entered, this voids any reasonable expectation of privacy by the residents of the home being searched (Bedi, 2014; Kerr, 2015; Nelson, et al., 2014; Solove, 2002; Vaughn, 2014; Wegman, 2005). Walking from room to room does not constitute different searches although the warrant typically specifies which rooms and which property needs to be searched and/or seized. In other words, when opening a cabinet, anything in plain view can be seized, but the cabinet may not be searched unless specified by the warrant. There are vast differences in a computer search. Weeks or months go by before a trained analyst even begins the search. A range of software programs are used to perform the search (many of which change their features constantly) and purposively find evidence to support the original warrant (Bedi, 2014; Garfinkel, et al., 2009; Jekot, 2007; Kerr, 2015; Nelson, et al., 2014; Vaughn, 2014; Wegman, 2005).

Electronic Communications Privacy Act

While the 4th Amendment applies to government searches, the ECPA applies to everything else (Burnside, 1987; Martin & Cendrowski, 2014). The ECPA (1986) consists of three general parts: the Wiretap Act, The Pen Register Statute, and the Stored Communications Act. The Wiretap Act and the Pen Register Statute both serve to regulate prospective surveillance, namely, communications in transit (Martin & Cendrowski, 2014; Mulligan, 2003). These are communications that are intercepted prior to being received by the end user. The Stored Communications Act governs what would be considered retrospective surveillance; this attempts to protect the privacy of communications that are stored, particularly long term (Burnside, 1987; Martin & Cendrowski, 2014; Mulligan, 2003). All parts of the ECPA are designed to keep communications private, but an exception within this Act exists allowing the government to compel a third party to turn over the information (Mulligan, 2003; Steere, 1998). In a very general way, the ECPA invokes protection depending on whether an email has been opened, how long it is stored, and where it is stored; these factors all play a part in whether or not this information can be seized and used during an investigation (Burnside, 1987; Martin & Cendrowski, 2014; Mulligan, 2003). The ECPA intended to protect individuals from certain seizures of communications records by law enforcement. At first, this act seemingly served as a protection, but development of the internet and the evolution of email have changed how this protection is viewed and whether or not certain communications are protected by either the ECPA or the 4th Amendment (Burnside, 1987; Martin & Cendrowski, 2014; Mulligan, 2003; Steere, 1998).

Interpretation of legislation and digital forensics

Many different court opinions on the application of the 4th Amendment applied to both physical and digital spaces are written. Milestone cases are frequently discussed within the literature to give a better understanding of how this legislation is interpreted and what precedents are set.

Initially, in *Weeks v. United States* (1914), it was determined that law enforcement or government cannot enter an individual's private space without a warrant and seize property. This is where a precedent is set that property seized improperly is inadmissible as evidence, affording protection against law enforcement that choose not to follow proper procedures (Bedi, 2014; Garfinkel, 2010; Kerr, 2015; Ma, et al., 2015; Vaughn, 2014).

Next, *Katz v. United States* (1967) expanded the 4th Amendment where rulings assert that the focus of the 4th Amendment is the individual not the space. Because of the focus on individual persons, there is a call to keep up with technological innovations and courts maintain the spirit of the 4th Amendment to include areas where an individual has a justifiable expectation of privacy. This expectation has two parts: First, that the individual exhibits an expectation of privacy and second, that society also finds this expectation reasonable. After this ruling in *Katz v. United States*, a physical intrusion is not necessary for privacy to be violated; privacy can be violated if there is a reasonable expectation to have privacy at the time of the search (Bedi, 2014; Garfinkel, 2010; Kerr, 2015; Ma, et al., 2015; Vaughn, 2014).

In the *United States v. Miller* (1976), the court found that individuals do not have the expectation of privacy for records turned over to the bank. Having turned over records to a third party creates an unreasonable expectation for privacy. The knowledge that individuals at the bank can access this information at any time also limits the expectation of privacy. This precedent changes much in the idea of third-party storage of information relating to digital storage and digital records (Bedi, 2014; Garfinkel, 2010; Kerr, 2015; Ma, et al., 2015; Vaughn, 2014).

Smith v. Maryland (1979) is referred to as the first "pen register" case. Pen registers record numbers dialed by telephone. The decision in this case took the approach that because only phone numbers are recorded by the pen register and because we receive a phone bill with all numbers listed each month, this is not private information, and the expectation of such would be unreasonable (Bedi, 2014; Garfinkel, 2010; Kerr, 2015; Ma, et al., 2015; Vaughn, 2014).

In *Kyllo v. United States* (2001), the court recognized that defining "search," changes as technology develops. Because prior rulings and precedents specify the lack of need to physically intrude to violate privacy that this precedent also applies to surveillance searches. This case is cited as discussing that the government may not use their own device to search a home for "details that would be previously unknowable," unless a warrant is given directing that type of surveillance to occur. This decision makes it possible to argue for more detailed warrants in a digital search and seizure (Bedi, 2014; Garfinkel, 2010; Kerr, 2015; Ma, et al., 2015; Vaughn, 2014).

In a contradiction to this ideal, the Communications Assistance for Law Enforcement Act (CALEA) 1994, requires cooperation from communications providers by making them employ technologies to assist law enforcement in authorized searches. However, it is specified that this does not apply to ISP's or email. Despite this specification, many suggest there are loopholes that extend this requirement to assist law enforcement beyond the scope that the CALEA intended (Bedi, 2014; Garfinkel, 2010; Kerr, 2015; Ma, et al., 2015; Vaughn, 2014).

Contemporary issues

More recently, courts have been specifically addressing electronic seizures. In the *United States v. Jones* (2012) a hearing took place of a case involving the installation of a GPS device on the car of a wife of a suspected drug trafficker. There was a warrant for this installation but the device was installed outside the time frame specified by the warrant and outside the jurisdiction of the warrant. (The warrant came from Washington, D.C., but the device was installed on the car in Maryland) While the mistake from law enforcement in the search ultimately favored the defendant in the court decision, it is also discussed that the length of time that data are collected in these ways is of particular concern. Individuals cannot reasonably expect that their driving patterns are observed each day for a month, perhaps in one specific event or time frame, but not daily for an extended period of time. This indicates that individuals do retain privacy even if there are warrants to track our movements (Bedi, 2014; Garfinkel, 2010; Kerr, 2015; Ma, et al., 2015; Vaughn, 2014).

In *Riley v. California* (2013), the issue of privacy and cellphones is addressed. A cellphone was taken off of a person during an arrest and the contents of the phone were searched. The search resulted in implicating the suspect in many criminal activities. Typically, a search of the immediate area of person is considered acceptable to search for weapons and other items that can put law enforcement in immediate danger. The courts ruled that a cellphone is not an item that needs to be searched in this capacity during an arrest, as it is not linked to the safety of law enforcement or the suspect. The court decided that the search of seized cellphones during an arrest is an unreasonable search and that there is an expectation of privacy with the cellphone so a warrant to search the phone would be necessary (Bedi, 2014; Garfinkel, 2010; Kerr, 2015; Ma, et al., 2015; Vaughn, 2014).

The recent case (2013) regarding the NSA collecting phone records with cooperation from Verizon combines earlier court rulings in order to give an injunction to the NSA. In this case, the courts ruled that while individuals have no reasonable expectation that the listing of calls made is private, the length of time they are collecting data violates that reasonableness clause. It is unreasonable to expect that data like phone records over the course of years would not be private in some capacity (Bedi, 2014; Garfinkel, 2010; Kerr, 2015; Ma, et al., 2015; Vaughn, 2014).

The courts continue to apply the concept of "reasonable" expectation of privacy to cases of digital search and seizure. Rulings also pay close attention to the actions of law enforcement, indicating that particularity within a warrant is also important. The ideas of reasonableness and particularity are applied directly to digital forensic search and seizures; two particular methods were evaluated with these ideas in mind.

Ethical issues associated with digital forensic techniques

As discussed throughout this chapter there are countless ways in which to analyze digital data and information. Many of these technologies and methods are argued to have ethical issues associated with their use. Two examples of this will be discussed here using a popularly argued method and a consistently argued technological system.

Currently, "hashing" is a favored technique used to make digital forensic analysis "easier," and also considered by some to be effective for ruling out files that aren't necessary for an investigation (Casey, 2011; Roussev, 2009). Hashing makes it possible to distinguish between operating system files and user saved data. Hashing adds unique digits to data and is most often used to confirm that copied data remain unchanged from the original data (Roussev, 2009). Through this, it can also confirm that the original data did not change by the process of copying the data. Some scholars question if hashing constitutes an official search, since data are only being given a unique value and not actually being read individually (Casey, 2011; Roussev, 2009; Salgado, 2005). Others argue that the use of hashing to identify certain types of files must be part of a search, because it is not only finding information but also organizing it. This extends the question that if hashing identifies certain data in a search that are not specified by a warrant, would this void the search? These are some of the many issues that are called into question by the use of hashing and similar techniques. Because legislation is unclear on the specifics of digital information and property and there are no resolutions regarding appropriate techniques, much of the gray area surrounding reasonableness and particularity with hashing is ignored. Currently, there seems to be support for the use of techniques that are quick and efficient searching through data, regardless of the constitutional issues that surround the use of these methods (Salgado, 2005; Roussev, 2009). Hashing is considered to be highly reliable in confirming the protection and preservation of digital evidence and to some; this outweighs the potential risks to an investigation (Salgado, 2005).

Carnivore is a widely contested system launched by the FBI, which was changed to a system they called "DCS-1000," and then shut down so that the government could explore alternative "private" options (Eichenlaub, 2001; Gilman, 2001; Goodman, et al., 2001; Mocas, 2004; Slobogin, 2008; Solove, 2002). This system, and the newer variations, can attach to an internet service provider (ISP) and use surveillance to intercept and pick up pockets of data to use in a criminal investigation (Eichenlaub, 2001; Goodman, et al., 2001) Essentially Carnivore sifts through binary code to find certain data points, this system must sift through multitudes of data and communications in order to find the very specific information related to the case at hand (Goodman, et al., 2001; Mocas, 2004; Slobogin, 2008; Solove, 2002). Through this process, innocuous emails sent by colleagues or mundane Internet searches can also be collected and used as part of an investigation or to begin a new investigation (Eichenlaub, 2001; Gilman, 2001). The FBI touts these programs as better than the commercially marketed software that performs the same task. Reasons for this include the flexibility where the FBI can create specific searches for the system that match the court order, the data that are acquired are directly related to the search warrant and bypasses any other irrelevant data (Eichenlaub, 2001; Gilman, 2001; Goodman, et al., 2001; Mocas, 2004; Slobogin, 2008; Solove, 2002). Compared to commercial counterparts that do not have such a quality filter, many believe Carnivore and its successors to be more suited to investigation and more compliant with the 4th Amendment and ECPA (Eichenlaub, 2001). Critics of these systems argue that since transactional and substantive data end up in the same packet of data search with these surveillance systems, that this creates a blurry and ambiguous line of privacy violations (Eichenlaub, 2001; Gilman, 2001; Goodman, et al., 2001). Others discuss issues surrounding the FBI and the nearly infinite discretion they have with searches of this nature (Eichenlaub, 2001; Gilman, 2001). Another concern is that these types of searches look through 99% of information belonging to innocent users on an ISP in order to collect the data of relevance to the case. Basically, in order to find communication between two people, millions of other private communications of individuals, who are not under investigation, are also searched (Eichenlaub, 2001; Gilman, 2001; Goodman, et al., 2001; Mocas, 2004; Slobogin, 2008; Solove, 2002). Now that there is more secrecy surrounding the use of these programs, the ethics and legality of their use continues to be questioned.

Applications in diverse forensic settings

Several circuit courts recently wrote decisions regarding the application of the 4th Amendment to digital data and information. The 10th Circuit Courts argue that in certain digital search cases the plain view doctrine cannot be invoked if the search hits on evidence not specified in the warrant (Bedi, 2014; Kerr, 2005). The 9th Circuit Courts argue that there should be a protocol to follow when magistrates are issuing warrants for digital evidence. This discusses waving the plain view doctrine for digital evidence; however, the warrant and search must discuss risks involved with seizure and destruction of data, involve a third party to separate files, must have probable cause, and must destroy or return evidence outside the scope of the search (Bedi, 2014; Kerr, 2015). The 7th Circuit holds to the reasonableness standard and argues that using forensic tool kits (even though they do a general search) is reasonable given that not all data will be physically viewed. In other words, courts are still not in complete agreement as to how the 4th Amendment should be applied to digital data but all agree that reasonableness and particularity should at least be part of the consideration (Garfinkel, et al., 2009; Kerr, 2005; Nelson, et al, 2014; Vaughn, 2014).

Applying the 4th Amendment

Scholars have argued that there are three basic components of the 4th Amendment to be evaluated when dealing with digital information and property (Bedi, 2014; Kerr, 2015; Garfinkel, et al., 2009; Nelson, et al., 2014; Vaughn, 2014). Despite rapidly changing technologies, the courts generally agree that the foundations of the 4th Amendment interpretation are reasonableness and particularity. No matter what the technology, does the individual have a reasonable expectation to privacy, and how particular was the warrant for the search? This does not mean that law enforcement will not ever violate proper search or that individuals will not be unlawfully arrested for things based on evidence obtained. However, when the cases go to the higher courts, the assertion of reasonableness is the priority, as well as how to define a "search." The 4th Amendment is specified to have ties to the physical world and physical space as cited in the *United States v. Katz*, and has at times, been interpreted not to be inclusive of digital space (Bedi, 2014; Goodman, et al., 2001; Kerr, 2015; Ma, et al., 2015; Vaughn, 2014). First, the environment of the search needs to be addressed. The environment of disk storage is far different than the environment of home storage. Homes have visible items that can be searched while disk storage is essentially a series of illegible 1s and 0s (Elyas, et al., 2014). The analyst never searches the binary code directly, instead commands are given and then converted into information; this converted information is then searched (Bedi, 2014; Garfinkel, et al., 2009; Kerr, 2015; Ma, et al., 2015). The difference in the environment of the search and how evidence is collected is of particular concern.

Another component for evaluation in application of the 4th Amendment is ownership and control of the property or information being seized (Garfinkel, 2010; Ma, et al., 2015; Reith, et. al., 2002). There must be a legitimate relationship between property and the individual in order to use the 4th Amendment. This means that property must be owned and/or controlled by the individual. With digital data, this often may not be the case as data may be stored beyond the individuals control and data could be from a search that the individual never had ownership of (Kerr, 2015; Vaughn, 2014). Further, an analysis of the computer is not done on the original copy but on a bit stream copy made by the analyst. This complicates justification that this is a legitimate search of owned property. The issue surrounds how email communications are typically stored on a third-party server (Bedi, 2014; Garfinkel, 2010; Kerr, 2015; Ma, et al., 2015; Vaughn, 2014). Traditionally, the Supreme Court has ruled that third-party storage where the individual does not feel they retain ownership of the data does not have 4th Amendment protection. Because it is impossible to retain ownership or privacy rights to a communication that has already been received and opened, this only furthers the freedom of access of law enforcement and other private forensic analysis (Bedi, 2014; Garfinkel, 2010; Kerr, 2015; Ma, et al., 2015; Vaughn, 2014).

The issue of storage needs evaluation in these cases. Storage in a home relies upon the size of the home or storage space; one reasonably has control over items in the home and storage (Elyas, et al., 2014). If items in storage were destroyed, an individual would likely be aware of this. A home has a finite space in which items can be stored. Digital data storage can have infinite space, but it is unlikely that an individual retains control over the storage in this space. Further, if data are deleted from virtual storage, an individual would more than likely not be aware of this destruction. Computers also store a great deal of information that a typical user is unaware of, which is certainly different from physical space (Bedi, 2014; Garfinkel, 2010; Kerr, 2015; Ma, et al., 2015; Vaughn, 2014).

Finally, what is called the retrieval mechanism needs to be evaluated. In the physical world, once a search in physical space is complete, the search is finished and all the places where evidence might be were searched (Bedi, 2014; Garfinkel, 2010; Kerr, 2015; Ma, et al., 2015; Vaughn, 2014). In the digital world, the end of a search is not so easily reached. There is a lack of clarity when these searches officially "end," and they often take 9 months or more after electronic property are seized in order to complete a digital forensic search. In a standard search there is a time limit on how long the search can be conducted before it is forcibly terminated (Kerr, 2015; Nelson, et al., 2014; Vaughn, 2014). These rules are not equally applied to digital search and seizure; copies of data can be kept and searched indefinitely. Similarly, there are limitations to areas one can look in a physical search; however, in the digital realm, electronic data can be searched in all spaces (Bedi, 2014; Garfinkel, 2010; Kerr, 2015; Ma, et al., 2015; Vaughn, 2014). This becomes problematic because there is opportunity to discover commission of additional crimes in a digital search and this opportunity would not exist in a physical search (Nelson, et al., 2014). This differentiation between the search of physical space and digital space are incredibly important when deciding how to apply the 4th Amendment given its suggestion of physical space from *United States v. Katz* (Bedi, 2014; Garfinkel, 2010; Kerr, 2015; Ma, et al., 2015; Vaughn, 2014).

Future research directions

Emerging technologies in digital forensics aim to make data acquisition and extraction from clouds and decryption of data easier and more interpretable by the analyst. Other technologies are emerging to accommodate "ethical" hacking; hacking used to keep protections on networks and individual machines (Garfinkel, 2010; Nelson, et al., 2014). Newer technologies aid in an investigation but hinder legal interpretation of the guidelines for search and seizure. Legislation cannot develop as quickly as technology and this has enormous implications on investigation, judicial proceedings, and sentencing (Adams, 2008; Garfinkel, 2010; Wegman, 2005). With unclear standards for digital data acquisition and extraction during investigation, it is necessary to start researching into effective methods to combat this.

Cloud computing is the most emergent form of technology and with it; digital forensics using cloud computing systems is also an emergent technology (Dykstra & Sherman, 2011; Ma, et al., 2015; Ruan, 2013). Cloud computing complicates digital forensics because currently there is no reliable platform (Garfinkel, 2010; Ma, et al, 2015). Many propose to pursue this avenue quickly as they feel it can greatly improve the way investigations run using digital forensic technology (Dykstra & Sherman, 2011; Ma, et al., 2015). Continued focus on developing technology is necessary to begin establishing reliability and validity of digital forensic technology and software. Compatibility with criminal investigation also needs to be a part of the development of these tools. Future research needs to focus on reliable and valid software and technology as well as standardization and oversight within the industry of digital forensics; best practices need to be established.

Conclusion

Many of the ethical issues discussed in this chapter surround copying data, searching data, interpreting data and the methods used to accomplish these different analytical tasks. The lack of a governing body to monitor digital forensic technology and laboratories creates differential application between individual cases and jurisdictions. Many consider this bias important and necessary for inclusion in ethical training for digital forensic analysts. The other ethical considerations rest on the interpretation of the 4th Amendment and other legislation governing search and seizure of private property. The characteristics between the physical world and digital world are vastly different; this calls for more specificity in legislation regarding digital property and information. In order to evolve with technology in investigation rather than continue fighting against it, there is necessity to develop standards and oversight within the field of digital forensics. Along with this, legislation needs to address in detail the concepts of reasonableness and particularity within the specific context of a digital search. Continued research into reliable software and technology can only aid in this process and improve digital forensic analysis for criminal investigations.

Key terms and definitions

Acquisition: Deals with acquiring data in some capacity through copying or imaging data onto a new drive or into a new format.

Carving: Data carving is part of data recovery; users can identify data that has no file allocation by searching for clusters belonging to the file extension.

Data interception technology: Is similar to surveillance this technology collects data transactions that are intercepted before they meet the end user. Essentially this is real-time interception of conversations and data transfers.

Data viewing: This is the process of viewing data that has been copied and ready for search. Viewing data becomes problematic if it is a "general" search and even at times if it is not.

Decompressing: A process to extract data that has been compressed for disk space; this process can lead to broken or unreadable data.

Decrypting: Clarifies unclear (encrypted) data. Similar to decompression, this process can create data that is unusable or inaccurate to first inception of data.

Digital time-stamping: Is a technology that detects the time that digital transactions occurred and can also detect if digital time stamps have been edited or tampered.

Extraction: Refers to searching, viewing, or analyzing the data that is already acquired.

Keyword searching: This combats the quantity issues that data acquisition creates. Programs allow analysts to search for key words rather than going through each piece of data individually. Keyword searches arguably lead to bias in the search and possibly less exculpatory evidence.

References

Adams, C. W. (2008). Legal issues pertaining to the development of digital forensic tools. *Systematic Approaches to Digital Forensic Engineering, SADFE'08. Third International Workshop on* (pp. 123–132). IEEE.

Baryamureeba, V., & Tushabe, F. (2004). The enhanced digital investigation process model. *Proceedings of the Fourth Digital Forensic Research Workshop* (pp. 1–9).

Bedi, M. S. (2014). Social networks, government surveillance, and the fourth amendment mosaic theory. *Boston University Law Review, 94*(1809).

Beebe, N. (2009). Digital forensic research: The good, the bad and the unaddressed. *Advances in Digital Forensics V* (pp. 17–36). Springer Berlin Heidelberg.

Beebe, N. L., & Clark, J. G. (2005). A hierarchical, objectives-based framework for the digital investigations process. *Digital Investigation, 2*(2), 147–167.

Biggs, S., & Vidalis, S. (2009). Cloud computing: The impact on digital forensic investigations. *Internet Technology and Secured Transactions, 2009. ICITST 2009. International Conference for* (pp. 1–6). IEEE.

Broucek, V., & Turner, P. (2013). Technical, legal and ethical dilemmas: Distinguishing risks arising from malware and cyber-attack tools in the 'cloud'—A forensic computing perspective. *Journal of Computer Virology and Hacking Techniques, 9*(1), 27–33.

Buchholz, F., & Spafford, E. (2004). On the role of file system metadata in digital forensics. *Digital Investigation, 1*(4), 298–309.

Burnside, R. S. (1987). Electronic communications privacy act of 1986: The challenge of applying ambiguous statutory language to intricate telecommunication technologies. *The Rutgers Computer & Tech. LJ, 13*, 451.

Carrier, B. (2003). Defining digital forensic examination and analysis tools using abstraction layers. *International Journal of Digital Evidence, 1*(4), 1–12.

Carrier, B. D., & Grand, J. (2004). A hardware-based memory acquisition procedure for digital investigations. *Digital Investigation*, *1*(1), 50–60.

Carrier, B., & Spafford, E. H. (2004). An event-based digital forensic investigation framework. *Digital forensic research workshop* (pp. 11–13).

Casey, E. (2011). *Digital evidence and computer crime: Forensic science, computers, and the internet*. Academic press.

Cole, K. A., Gupta, S., Gurugubelli, D., & Rogers, M. K. (2015). A review of recent case law related to digital forensics: The current issues. *Proceedings of the Conference on Digital Forensics, Security and Law* (pp. 95–104).

Dykstra, J., & Sherman, A. T. (2011). Understanding issues in cloud forensics: Two hypothetical case studies. *Proceedings of the Conference on Digital Forensics, Security and Law* (pp. 45–54).

Eichenlaub, F. J. (2001). Carnivore: Taking a bite out of the fourth amendment. *NCL Rev.*, *80*, 315.

Elyas, M., Maynard, S. B., Ahmad, A., & Lonie, A. (2014). Towards a systemic framework for digital forensic readiness. *The Journal of Computer Information Systems*, *54*(3), 97.

Garfinkel, S. L. (2010). Digital forensics research: The next 10 years. *Digital Investigation*, *7*, S64–S73.

Garfinkel, S., Farrell, P., Roussev, V., & Dinolt, G. (2009). Bringing science to digital forensics with standardized forensic corpora. *Digital Investigation*, *6*, S2–S11.

Gilman, J. (2001). Carnivore: The uneasy relationship between the fourth amendment and electronic surveillance of internet communications. *CommLaw Conspectus*, *9*, 111.

Goldfoot, J. (2011). Physical computer and the fourth amendment. *Berkeley J. Crim. L.*, *16*, 112.

Goodman, J., Murphy, A., Streetman, M., & Sweet, M. (2001). Carnivore: Will it devour your privacy? *Duke Law & Technology Review*, *1*(1), 28.

Jekot, W. (2007). Computer forensics, search strategies, and the particularity requirement. *PGH. J. Tech. L. & Pol'y*, *7*, 5–9.

Kenneally, E. E., & Brown, C. L. (2005). Risk sensitive digital evidence collection. *Digital Investigation*, *2*(2), 101–119.

Kerr, O. S. (2005). Searches and seizures in a digital world. *Harv. L. Rev.*, 531–585.

Kerr, O. S. (2015). Fourth amendment and the global internet. *Stan. L. Rev.*, *67*, 285.

Kessler, G. C. (2004). An overview of steganography for the computer forensics examiner. *Forensic Science Communications*, *6*(3), 1–27.

Ma, H., Shen, G., Chen, M., & Zhang, J. (2015). The research of digital forensics technologies under cloud computing environment. *International Journal of Grid and Distributed Computing*, *8*(3), 127–134.

Martin, J. P., & Cendrowski, H. (2014). Electronic Communications Privacy Act. *Cloud Computing and Electronic Discovery*, 55–74.

Mocas, S. (2004). Building theoretical underpinnings for digital forensics research. *Digital Investigation*, *1*(1), 61–68.

Mulligan, D. K. (2003). Reasonable expectations in electronic communications: A critical perspective on the Electronic Communications Privacy Act. *Geo. Wash. L. Rev.*, *72*, 1557.

Nance, K., & Ryan, D. J. (2011). Legal aspects of digital forensics: A research agenda. *System Sciences (HICSS), 2011 44th Hawaii International Conference on* (pp. 1–6). IEEE.

Nelson, B., Phillips, A., & Steuart, C. (2014). *Guide to computer forensics and investigations.* Cengage Learning.

Nolan, R., O'sullivan, C., Branson, J., & Waits, C. (2005). *First responders guide to computer forensics* (No. CMU/SEI-2005-HB-001). Carnegie-Mellon Univ Pittsburgh PA Software Engineering Inst.

Pollitt, M. (2010). A history of digital forensics. *Advances in Digital Forensics VI* (pp. 3–15). Springer Berlin Heidelberg.

Reith, M., Carr, C., & Gunsch, G. (2002). An examination of digital forensic models. *International Journal of Digital Evidence, 1*(3), 1–12.

Rogers, M. (2003). The role of criminal profiling in the computer forensics process. *Computers & Security, 22*(4), 292–298.

Roussev, V. (2009). Hashing and data fingerprinting in digital forensics. *IEEE Security & Privacy*, (2), 49–55.

Ruan, K. (2013). Cybercrime and cloud forensics: Applications for investigation.

Salgado, R. P. (2005). Fourth amendment search and the power of the hash. *Harv. L. Rev. F., 119*, 38.

Slobogin, C. (2008). *Privacy at risk: The new government surveillance and the Fourth Amendment.* Chicago, IL: University of Chicago Press.

Solove, D. J. (2002). Digital dossiers and the dissipation of fourth amendment privacy. *Southern California Law Review, 75.*

Steere, R. S. (1998). Keeping private e-mail private: A proposal to modify the Electronic Communications Privacy Act. *Val. UL Rev., 33*, 231.

Vaughn, R. (2014). The more things change: An analysis of recent fourth amendment jurisprudence. *Criminal Justice Praxis*, 1.

Walls, R. J., Levine, B. N., Liberatore, M., & Shields, C. (2011). Effective digital forensics research is investigator-centric. *HotSec.*

Wang, Y., Cannady, J., & Rosenbluth, J. (2005). Foundations of computer forensics: A technology for the fight against computer crime. *Computer Law & Security Review, 21*(2), 119–127.

Wegman, J. (2005). Computer forensics: Admissibility of evidence in criminal cases. *Journal of Legal, Ethical and Regulatory Issues, 8*(1), 1–13.

Wolf, M., Shafer, A., & Gendron, M. (2006). Toward understanding digital forensics as a profession: Defining curricular needs. *Proceedings of the Conference on Digital Forensics, Security and Law* (pp. 57–66).

7 Emerging technologies in forensic anthropology

The potential utility and current limitations of 3D technologies

Heather M. Garvin, Alexandra R. Klales, and Sarah Furnier

Introduction

Forensic anthropologists are tasked with uncovering the identity of unknown individuals and providing insight into the circumstances surrounding their death, all from a set of skeletal remains. In general, the forensic anthropologist's toolbox has changed very little over the decades, consisting primarily of traditional measurement tools such as calipers and an osteometric board. In the last decade, however, certain three-dimensional (3D) technologies have been introduced to the field, either for case analysis or for forensic anthropological research. Despite their potential utilization in the discipline, these new technologies are not yet universally employed. Limiting factors include the costs associated with 3D technologies, a lack of standardized protocols in data collection and archiving, unknown accuracy rates, a lag in statistical methods of analysis, and ethical concerns regarding data access and sharing. The objectives of this chapter are to (1) explain how emerging 3D technologies can improve upon current forensic anthropological methods, (2) discuss current limitations of and ethical concerns regarding 3D technologies in forensic anthropology, and (3) present recommendations for dealing with these limitations.

Background

Before discussing the potential uses of 3D technologies in forensic anthropology, an overview of the field and the tools typically employed by forensic anthropologists is required. Forensic anthropology is defined as "the specialized subdiscipline of physical anthropology that applies the techniques of osteology and skeletal identification to problems of legal and public concern" (Kerley, 1978, p. 160). Historically, the forensic anthropologist's role was primarily restricted to the laboratory, where they used skeletal features to estimate the biological profile (sex, age, ancestry, and stature) of an unknown skeleton for identification purposes. Although the estimation of a biological profile remains an important component in the initial stages of an investigation (i.e., before a potential DNA comparison can be identified), the forensic anthropologist's role has now expanded. Forensic anthropologists are now frequently called upon to conduct skeletal trauma analyses and taphonomic interpretations (e.g., time-since-death), as well as to perform searches and recoveries at outdoor scenes (see Dirkmaat & Cabo, 2012, for a detailed history of this paradigm shift). Their goal is not only to assist with the identification of the unknown victim but also to help reconstruct the circumstances surrounding the individual's death.

Currently, very few forensic anthropologists work full time on forensic anthropological cases. Some are employed by medicolegal agencies (e.g., a medical examiner's office) or law enforcement and federal agencies (e.g., FBI). Their roles in many of these careers extend beyond forensic anthropological cases and include autopsy procedures, death scene investigation, or assisting with other forensic analyses (e.g., hair and fiber analyses). A small handful of forensic anthropologists work in museum settings (e.g., the Smithsonian Institution). The majority of forensic anthropologists work in academic settings, employed as faculty members at universities (Agostini & Gomez, 2009). As such, their primary work responsibility is teaching, with faculty research requirements varying by university. Thus, forensic anthropological casework is performed in addition to all other job requirements. In the academic setting, the search, recovery, and analysis of human skeletal remains are typically conducted free of charge as a community service, or for very minimal fees. As we will see, this contributes to funding issues, restricting the anthropologist's ability to purchase expensive equipment and software.

Regardless of their field of employment, forensic anthropologists also typically engage in human skeletal research. Newly proposed methods must be validated and tested on various samples, and error rates must be established in order to be in accordance with forensic standards set forth by the 1993 Daubert vs. Merrell Dow Pharmaceuticals Supreme Court ruling and the 2009 National Academy of Sciences (NAS) report regarding scientific rigor in the discipline of forensic science (Christensen & Crowder, 2009). Other research is aimed at documenting, explaining, and understanding human skeletal variation in order to create new methods for estimating biological profile parameters and interpreting an individual's life history. In either research scenario, a large collection of skeletal remains is required. In addition, the skeletal collection should be representative of the population in question (i.e., from the same ancestry group and temporal period), as skeletal features have been shown to vary across time and among groups. Finally, the skeletal collections must be well documented with reliable information on the age, sex, ancestry, stature, and cause of death for each individual. Unfortunately, there are only a few large, documented skeletal collections, which is a major limitation to forensic anthropological research. In the United States, the most utilized collections are the Terry Collection at the Smithsonian Institution National Museum of Natural History, Washington, D.C., and the Hamann-Todd Collection at the Cleveland Museum of Natural History, Cleveland, OH. Both of these collections house over 1,000 documented skeletons; however, these skeletons were acquired during the late 19th and early 20th centuries and typically were unclaimed dead from lower socioeconomic groups (Cobb, 1959; Hunt & Albanese, 2005). Thus, there are sample biases associated with these collections, and they may not be the most appropriate samples from which to derive modern forensic techniques (Jantz & Meadows Jantz, 2000; Meadows Jantz & Jantz, 1999; Ousley & Jantz, 1993). Large, more modern documented skeletal collections are scarce. The largest and most utilized modern skeletal collection in the United States is the William Bass Donated Skeleton Collection housed at the University of Tennessee, Knoxville. It consists of nearly 1,000 individuals, although they are primarily of European-American ancestry, making it difficult to study the full range of human variation (Komar & Grivas, 2008; "WM Bass Donated Skeletal Collection," n.d.).

In the field, forensic anthropologists have begun to use more advanced technologies in order to document outdoor crime scenes. For example, the Mercyhurst University Forensic Scene Recovery Team uses a high-grade Trimble R8 GNSS (global navigation satellite system), GPS, and electronic total station to obtain geographic spatial information, which can ultimately be mapped and analyzed using GIS (Geographic Information Systems) software. Additionally, geophysical detection instruments, such as a ground-penetrating radar or EM38, may be used to aid in the search for buried human remains (Conyers & Goodman, 1997). However, when moving from the field to the laboratory, the forensic anthropologist's toolbox for analyzing skeletal remains is less technologically advanced. The traditional tools of the trade consist of spreading and sliding calipers, which are used to collect skeletal measurements up to 15 cm in length, and an osteometric board, which is essentially a ruler embedded into a board with one fixed and one moveable vertical end that is used to collect longer measurements (Figure 7.1). Skeletal measurements can then be entered into linear regression or discriminant function equations to estimate the stature, sex, and ancestry of an unknown individual. The computer program FORDISC (Jantz & Ousley, 2005) assists with estimating these parameters. The program contains a database of skeletal measurements from individuals of known sex, stature, and ancestry, and uses discriminant function analyses to compare the measurements from the unknown individual to the database to predict group membership. Similarly, it uses linear regressions derived from the database of known individuals in order to predict stature. However, besides these measurement tools, the forensic anthropologist typically relies on visual (i.e., macroscopic) observations of nonmetric skeletal traits and previous experience to estimate sex and ancestry and to interpret any suspected trauma on the remains. Photo-documentation of the condition of the skeletal remains and of any unique or identifiable markers or regions of suspected trauma is also a common practice.

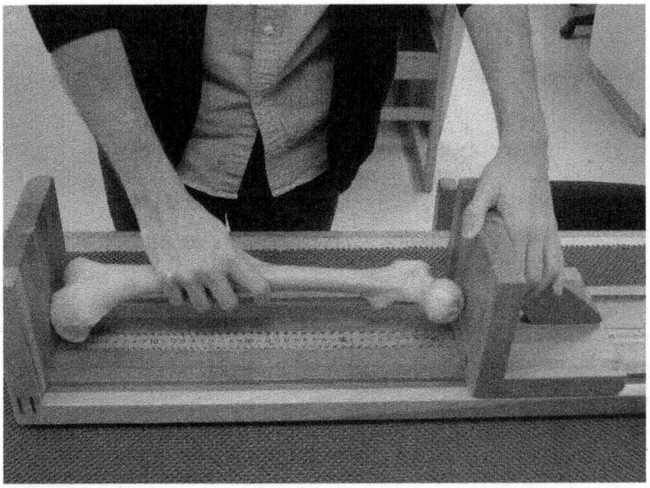

Figure 7.1 Osteometric board being used to collect femoral length.

These aforementioned tools accompany every practicing forensic anthropologist. In addition, some may have access to more advanced technology such as radiographic equipment, which can be utilized to evaluate the fusion of bones for age estimation, to document the dentition and dental restorations for identification purposes, and to evaluate any signs of trauma, such as indications of healing trauma, fracture patterns, or lead transfer from ballistic trauma. Microscopes are also used by some forensic anthropologists, most commonly to evaluate skeletal trauma, such as cut marks and fracture surfaces. A smaller minority of forensic anthropology laboratories may also have equipment to section and analyze skeletal remains for histological evaluation, or equipment to perform isotope analysis (e.g., mass spectrometer). High equipment expenses, the training and expertise required to run such equipment, and the destructive nature of these methods, however, mean that these tools remain relatively underutilized within the field.

Thus, the typical forensic anthropologist's toolbox has remained relatively low tech and has not changed much since the inception of the field. As other disciplines, such as engineering and medicine, have continued to advance; however, forensic anthropologists have begun to borrow technology from these fields. In particular, we have seen an increase in the use of 3D technologies in forensic anthropological casework and research, including the use of MicroScribe digitizers, 3D surface scanners, and computed tomography (CT) scanners. Photogrammetry methods of creating 3D models have also been recently introduced to the field. These emerging technologies have the potential to improve forensic anthropological case analyses, case- and data-archiving methods, access to skeletal material for research, and research methods and analyses.

Emerging 3D technologies in forensic anthropology

Digitizer

One of the first 3D technologies to be utilized in forensic anthropology was the digitizer. It is commonly referred to in the field as a MicroScribe. (MicroScribe is the most popular company selling digitizers to anthropology departments.) It has been used extensively in the field for over a decade. The MicroScribe digitizer is a portable piece of equipment that is used to collect 3D coordinates (*x*, *y*, and *z*) from landmarks on an object using a stylus that is attached to the digitizer by a long, mobile, mechanical arm (Figure 7.2). The landmark coordinates can be recorded directly into a spreadsheet, such as in Microsoft Excel, for further analyses. Because the 3D spatial data are preserved for each landmark, interlandmark distances can easily be calculated between the various combinations of landmarks. In this manner, the digitizer is most commonly used to record cranial measurements, taking the place of traditional spreading and sliding calipers. The software program 3Skull (Ousley, 2014) can be used in conjunction with a MicroScribe digitizer and will automatically transform digitizer data into traditional craniometric measurements for input into FORDISC (the software program previously described which is used for sex and ancestry estimation).

The advantages of the digitizer include easy, rapid data collection and automatic calculation of all possible interlandmark distances. Shape analyses can also be conducted directly from the landmark data using geometric morphometrics (GM). GM is a field of study focused on statistically analyzing shape data from Cartesian coordinate data; this type of analysis allows researchers to investigate and visualize shape variation within and between groups. Although these methods are not usually carried out on individual

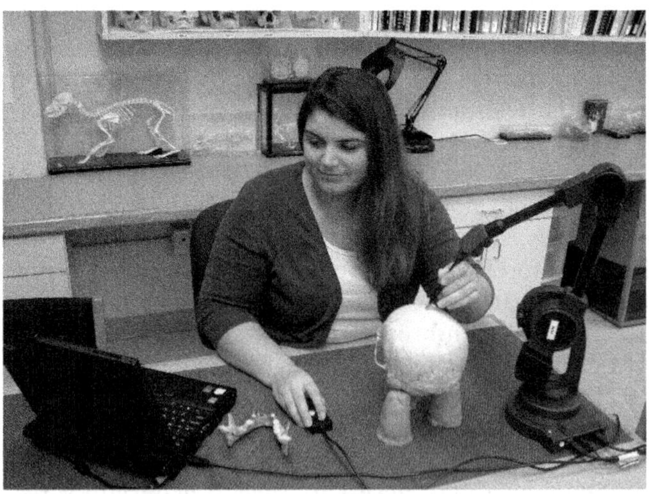

Figure 7.2 MicroScribe G2 digitizer being used to collect craniometric data.

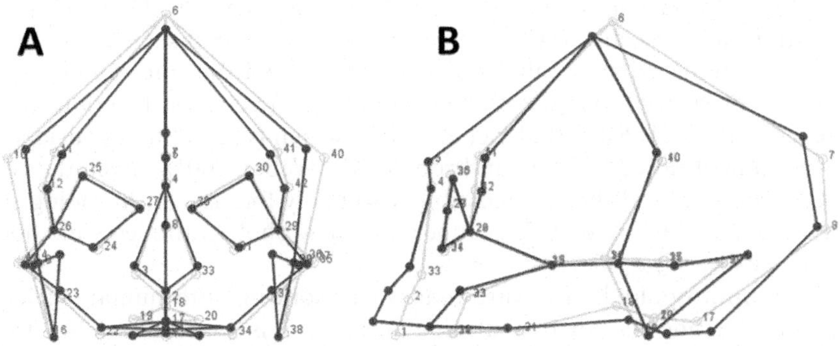

Figure 7.3 Example of a wireframe that can be created from landmarks captured with a MicroScribe digitizer. (a) Anterior view of skull; (b) left lateral view of skull. Blue wireframe represents a specific individual, which is being compared to a group average (teal wireframe). Note that it does not capture information about the surface of an object, only about the individual landmarks (dots) collected.

forensic anthropological cases, they are commonly employed in forensic anthropological research to evaluate sex and ancestry differences in cranial and pelvic shape (see, e.g., Bytheway & Ross, 2010; Kimmerle et al., 2008).

MicroScribe digitizers range in price from $8,000 to $11,000 depending on the accuracy required and the preferred degrees of freedom of the mechanical arm (i.e., how easily the arm rotates and can be moved around the object) (MicroScribe Pricing, 2015). Because the MicroScribe is portable, anthropologists can travel with it to skeletal collections to collect data. The repetitive movement and jostling that inevitably occurs during travelling, however, can occasionally create calibration issues and having the digitizer recalibrated can run upward of $4,000. At present, most institutions conducting forensic anthropological research own a MicroScribe digitizer for data-collection purposes. However, digitizers only preserve the shape of the object (i.e., skeletal element) in as much detail as the landmarks chosen (Figure 7.3). Surface and shape information between landmarks are not preserved. Thus, while digitizing is a good method for efficient and detailed data collection, it cannot be used to create a virtual reconstruction of the object of interest.

3D models

The remaining 3D technologies to be discussed in this chapter (surface scanners, photogrammetry, and CT), on the other hand, can be used to create 3D virtual models of skeletal elements that replicate entire bone surfaces, instead of individual landmarks. Over the last decade, the use of 3D virtual models in forensic research and case analyses has exponentially increased. A recent literature search reveals that since 2001 over 100 studies have been published in the *Journal of Forensic Sciences*, *Forensic Science International*, and the *American Journal of Physical Anthropology* in which 3D models (created from surface scans or CT) were used in some form of skeletal analysis. CT has been the primary catalyst for this transition from traditional osteological approaches to virtual methods and geometric reconstructions (e.g., Decker et al., 2011; Ramsthaler et al., 2010), which is sometimes considered a new subfield called "virtual anthropology" (Weber, 2001). Given the cost of CT scanners, and thus inaccessibility to many forensic anthropologists, the use of more affordable 3D scanners and new photogrammetry methods are increasing in popularity.

Every anthropologist is aware that when working with human skeletal remains, the time allotted to analyzing the remains is limited. Families want their remains of their loved one returned as soon as possible so that they can have closure and complete any funerary practices; they do not want their remains sitting in a laboratory for extended periods of time. Anthropologists are thus required to complete their analyses, write their reports, and then return the remains as soon as possible. If that case then goes to trial, the anthropologist must rely on their report and written and photographic documentation to recall important details and testify on the case. If the forensic anthropologist can create a 3D model of the skeletal elements (especially the more diagnostic elements, such as the skull, pelvis, or elements exhibiting trauma), they will continue to have access to a 3D replica of the remains after returning them. In this manner, 3D surface scans can be used to document and archive skeletal elements. The 3D models can even be utilized during testimony in order to illustrate points to the judge and jury. Instead of describing complex fracture patterns, a forensic anthropologist can actually show them in 3D. More and more institutions are investing in 3D printers as they continue to decrease in price and increase in quality. With a 3D printer, the forensic anthropologist can print out their 3D models to be used as a visual aid in court testimonies and for data archiving purposes. They can also be used if a method is thought to be destructive or damaging to the original skeletal element. For example, when a forensic anthropologist is tasked with matching a suspected tool to traumatic defects, they typically do so visually and with metric measurements for fear that any contact between the tool and bone could alter the original state of the skeletal defect. The 3D prints eliminate this issue and can be used to illustrate congruency between suspected tools and skeletal trauma.

Surface data and landmarks can also be collected from the virtual 3D models and utilized for case analyses and research purposes. Freeware (e.g., MeshLab) and for-cost (e.g., GeoMagic) software programs allow researchers to collect measurement data from 3D virtual models (GeoMagic, 2015; MeshLab, n.d.). A simple click of the mouse on two points of interest on the surface of an element returns a measurement value. Similarly, landmark coordinate data (*x*, *y*, and *z*) can also be easily collected. Thus, traditional osteometric analyses can be performed, as well as GM shape analyses.

The 3D models also present an opportunity to analyze new variables and develop novel methods. For example, measures of surface areas and volumes can be collected at the click of a button and used in skeletal GM analyses can be used to evaluate shape differences in curves and surfaces (e.g., Bookstein, 1997; Kieser et al., 2007). Some have even explored using entire 3D models in analyses and automating data collection and analysis procedures (Fatah et al., 2014). Furthermore, because the 3D models can be archived, as new measurements are defined or novel methods of analysis develop they can be applied to older cases for reanalysis.

3D models also present an opportunity to make virtual models of skeletal material available to a broader audience. At present, researchers must travel to the few large, documented skeletal collections in order to collect data for research or comparison. Skeletal collections limit the number of researchers at site at any one time, so data collection depends on space availability. In addition, there are travel and accommodation costs, bench fees, and logistics to consider. Depending on the data collected, the researcher may have to schedule days or even weeks away from their primary job responsibilities. If these collections created 3D models of their skeletal material, then it could be accessed worldwide by researchers without any of the aforementioned hindrances. Before this becomes a reality, there are a number of factors that must first be considered, such as scanning protocols and data access and sharing policies (to be discussed further later). However, if the true purpose of these skeletal collections is to promote research and scientific advancement, then a digital database of virtual skeletal remains is the next step. Besides saving researchers time and resources, a virtual collection would also prevent repeated handling of and damage to actual specimens, thus helping to preserve the skeletal collections. All of the 3D technologies discussed in this chapter are nondestructive.

Virtual skull models also present a new resource for facial reconstructions. Typically, facial reconstructions (i.e., reconstructing what the person looked like during life) are performed by forensic artists, not forensic anthropologists, but this is an area where the two fields may intertwine and collaborate. The skull serves as the foundation for forensic facial reconstructions. For two-dimensional (2D) manual reconstruction (i.e., drawings), the forensic artist uses photographs of the skull to approximate facial appearance. For 3D manual reconstructions, the skull itself is used for the reconstruction or a cast, usually of plaster, is made of the skull and used in its place. First, any damaged aspects of the skull or cast are filled and approximated. Then tissue depth markers are placed on various aspects of the skull or cast. The depth of the markers will vary depending on the sex and ancestry of the individual; therefore, the forensic anthropologist must provide the forensic artist with estimated demographic information prior to reconstruction. Next, the artist adds layers of clay, filling in the markers and following muscle orientations, thereby building up the soft tissue facial features of the individual. Features like

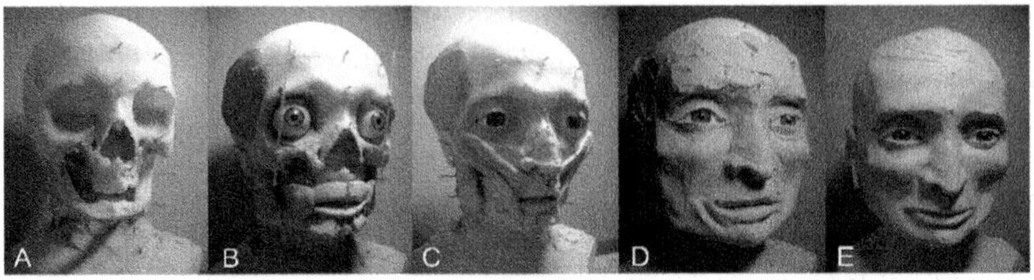

Figure 7.4 Forensic facial reconstruction using a 3D printed skull model. (a) 3D skull model printed by Dr. Robert Hoppa, University of Manitoba; (b–e) forensic facial reconstruction done by Dr. Jonathan Elias, Akhmim Mummy Studies Consortium. Photo credit: Jonathan Elias.

eye color, nose shape, or hair color and texture must be estimated since they cannot be predicted from the skeletal material. If the actual skull is used for the reconstruction, images of the artist's reconstruction are then taken from multiple angles before it is then deconstructed to remove the remains. As can be imagined, the methods of reconstruction and even the process of casting a skull can alter and even damage the skeletal remains. Thus, all forensic anthropological analyses and documentation (both written and photographic) must be completed prior to any reconstructions. Three-dimensional skull models alleviate these issues. Facial reconstructions can be completed virtually, allowing the researcher or artist to continuously update and modify the model as more information becomes available (e.g., hair color). Multiple virtual reconstructions can also be easily created, displaying variations in unknown variables such as eye and hair color. If a tangible model is preferred, the virtual skull model can be printed using a 3D printer. As with the casts, the printed model can then be used for the reconstruction, but in contrast, the 3D print has no risk of altering or damaging the remains (Figure 7.4). With either method (virtual or 3D print), the forensic anthropologist and artist are able to work simultaneously on the case, decreasing the amount of time before the information can be distributed to law enforcement.

3D surface scanner

There are a number of 3D surface scanners on the market, but the NextEngine Desktop scanner is by far the most widely used and is owned by numerous academic, research, and medicolegal institutions. The popularity of the NextEngine is, in part, due to its affordability, user-friendly interface, and overall accuracy. The scanner itself costs $2,995 and comes with a standard form of the ScanStudio software. A ScanStudio HD PRO upgrade, an additional cost of $995, is recommended and is necessary for collection of quality scans. Thus, for approximately $4,000 a forensic anthropologist can be equipped with all of the technology needed to create accurate 3D replicas of skeletal material. Higher-end surface scanners are available, such as the Breuckmann SmartScan, which uses structured patterns of white light to collect surface data; these scanners can cost upward of $200,000, hence the popularity of the NextEngine. The NextEngine is also portable and very easy to use. The scanner, itself, is a small box (dimensions: *L* = 8.8 in., *W* = 3.6 in., *H* = 10.9 in.) ("NextEngine 3D Scanner TechSpecs," n.d.). It attaches to a rotating base via a cable, and the item to be scanned is placed on the base (Figure 7.5). During the scanning process, the NextEngine uses a laser to collect surface spatial data and an embedded camera takes photographs so that the object's color (referred to as "texture") can be overlaid onto the surface model. The user can define how many scans they would like the NextEngine to take as the base rotates 360°. After all of the scans are complete, they can be automatically aligned and then merged together to create a final 3D model that can then be exported in various file formats (e.g., .ply, .obj, .stl, or .vrml) (Figure 7.6). The NextEngine reports an accuracy of up to 0.005 inches, collecting up to 268,000 points per square inch in the Macro mode with the upgraded PRO software. The user can control the quality and size of the scans via a number of settings. In order to collect data from all surfaces of an object, usually more than one 360° scan is required, and depending on the size of the object, number of scans, and points per inch being collected, it can be a fairly time consuming process. For example, to obtain a moderate quality 3D model of a human cranium, it takes a *minimum* of 30 minutes of scanning time, plus an additional 10 minutes of postprocessing time. However, the end product—a 3D virtual replica of the element, complete with the original surface coloration—is a valuable and permanent resource to forensic anthropologists.

Figure 7.5 The setup of the NextEngine scanner; the most popular 3D surface scanner utilized in forensic anthropology.

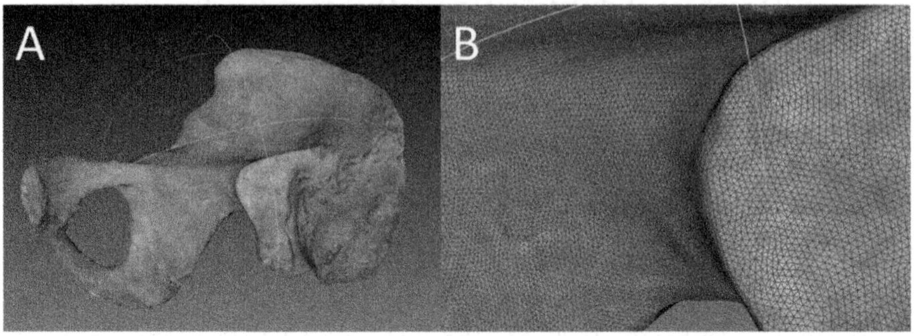

Figure 7.6 Screen capture of 3D models created with the NextEngine Desktop Scanner. (a) 3D model of os coxae scanned in Figure 7.5 with coloration/texture mapped on; (b) close-up illustrating wire mesh illustrating all of the surface points captured by the scanner.

Photogrammetry

Another affordable option for the production of 3D models is the use of photogrammetry. Although the concept of photogrammetry was developed well before 3D surface scanning was available, photogrammetry methods have recently become increasingly popular as cameras have become more advanced and more affordable. Recent publications demonstrate its versatility in several fields, including applications to biological anthropology (Katz & Friess, 2014; Kraus, 2007), though it is not yet common practice in forensic anthropology. Essentially, photogrammetry allows for the reconstruction of the position, orientation, shape, and size of an object from a series of digital photographs and thus requires no expensive hardware, other than a decent digital camera (Kraus, 2007). A number of fee-based and free software programs are available online (e.g., Acute3D, Agisoft PhotoScan, 123DCatch, 3DFlow) that use algorithms to more-or-less automatically align multiple photographs of an object into a 3D model. Further work still needs to be completed to test the accuracy of photogrammetry methods, especially in forensic anthropological contexts, but the method has the potential to achieve the benefits and utility of 3D surface scanning at a fraction of the cost. Like the surface scans, the photogrammetry models can retain the surface form and coloration (Figure 7.7), and potentially can be used to collect skeletal measurements, 3D landmarks, be subjected to geometric morphometric analyses, used to create archives and digital databases, or used for facial reconstructions. We may also see photogrammetry methods extended to the outdoor crime-scene, where digital photographs can be used to create virtual reconstructions of the skeletal remains *in situ* and the surrounding crime scene.

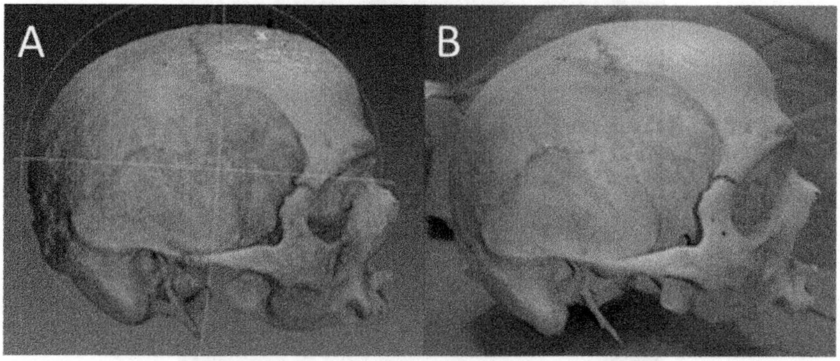

Figure 7.7 Example of photogrammetry 3D model (a) created using 15 photographs taken with a regular point-and-shoot digital camera and the Agisoft PhotoScan trial software. Note that this was a minimal number of photographs and that cranial base was not photographed in this example (hence, some of the distortion at the base). (b) Photograph of the specimen. Results suggest that with a better camera, increased number of photographs, and further experience with the program, that photogrammetry can produce a 3D model comparable to the actual specimen.

Computed tomography

An increasing number of forensic anthropologists are also taking advantage of CT technology. CT was original developed for medical fields. It is performed using an x-ray source and a detector that rotate around the object or person being imaged, allowing the acquisition of projection x-ray data from multiple angles. These projection data are reconstructed into a series of 2D image slices that can be "stacked" onto one another using multiplanar reconstruction. Three-dimensional models can then be generated using volumetric rendering (Figure 7.8). Because CT scanners use x-ray technology, it captures information on all materials present (i.e., soft tissue and bone) and is not restricted to only collecting surface information like the aforementioned 3D surface scanners and photogrammetry methods. Thresholds can be set to particular densities representative of the materials of interest (e.g., bone), and those elements can be isolated and segmented from the scan to create 3D virtual models. Thus, complete skeletonization of the remains is not required and 3D skeletal models can be created from either deceased or living individuals.

Overall, the continual technological advancements in CT imaging have promoted greater use of CT outside the medical realm and more in the analytical and educational arenas. Within forensic anthropology, CT was initially used to investigate bone biology (e.g., the micro-CT analyses of trabecular bone patterns); however, within the last decade research has been

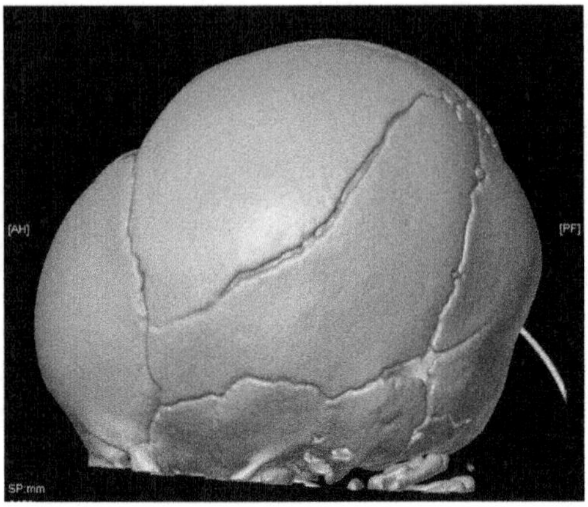

Figure 7.8 Example of a 3D model created from computed tomography scans. Scans were taken with all soft tissue *in situ* during postmortem examination. Three-dimensional model represents an infant skull with a healing fracture traversing the parietal bone.

more focused on 3D skeletal models created from the CT scans. As with the surface scanners and photogrammetry methods, a surface model can be created from the scans, although surface coloration (i.e., "texture") is not documented. The skeletal models do, however, retain all information regarding internal bone structure. Therefore, 3D bone models can include internal features such as medullary cavities, trabecular bone, endocranial surfaces, and sinus morphologies. These details may be important to document, particularly if there is trauma or pathology. Fracture patterns can vary in ectocranial and endocranial surfaces and with CT data both patterns can be documented without any alteration to the bone (e.g., removal of the calotte). Similarly, the ectocranial and endocranial surfaces of a bullet wound can be documented which can provide additional information regarding the direction of travel. Trabecular bone patterns, frontal sinus morphology, and dental morphology (including root shape and formation) captured from CT data can be used to make positive identifications just as they have traditionally been used to do so using 2D radiographs. Moreover, because CT scans represent 3D radiographs, body orientation at the time of scanning is less important as the scans can be manipulated digitally to match antemortem records for comparison and 2D radiographic issues with overlapping structures during image acquisition are minimized. In fact, specific elements of interest (such as sinuses, dentition, and surgical implants) can be isolated and segmented from the remaining model to be examined in detail.

Because CT data can be collected from living or fully fleshed deceased individuals, it presents an opportunity to create large virtual collections of modern documented skeletal material that can replace the commonly used

19th- and 20th-century historical collections. In addition, since most of the CT data are from clinical medical fields, demographic and life-history information is typically well documented for all individuals (this cannot be claimed for most historical collections). As will be discussed, however, CT scanners are very expensive and almost always outside the budget of forensic anthropologists, especially those working in academia. Thus, forensic anthropologists interested in using CT data, especially for research purposes where large sample sizes are required, rely on using scan data that have been collected previously for medical purposes.[1] At times, there may still be costs associated with the use of these data, for example, paying someone to anonymize the data, and a number of ethical concerns also arise in this situation (to be discussed in more detail in the following section). As a result, despite the fact that CT scans have become a common medical practice, access to this modality by forensic anthropologists is still very limited.

As with the surface scans and photogrammetry models, the 3D skeletal models created from CT scans can also be used for case archiving, visualization in court, facial reconstructions, and data analyses. Again, traditional osteometric measurements and landmark coordinate data can be collected digitally from the models for case analysis or research purposes. Unlike 3D scanner and photogrammetry data, CT data have been subjected to a number of accuracy studies. Estimations of biological profile parameters from CT data using techniques designed for dry bone specimens performed correctly with low associated error rates (Decker et al., 2011). Furthermore, tests of congruence between actual bone versus virtual or printed models for morphological and metric assessments indicate overall agreement between the two methodologies (e.g., Gamble & Hoppa, 2010; Robinson et al., 2008; Verhoff et al., 2008a). Unfortunately, many CT studies fail to report the specific machine utilized and scanning parameters (e.g., slice thickness or kVP and mAs settings). Slice thickness will directly affect the accuracy of the scan, as it controls the distance between slices and any information between the slices is just approximated based on the slice data. Furthermore, post-processing and manual segmentation of the image means, as Lynnerup (2009, p. 368) suggests, that CT scanned images and "3D renderings should not be viewed as a totally objective and 'true' representation of internal structures and tissues." Because of this inherent variation and subjectivity in postprocessing, more work is being done to establish or suggest standardized guidelines and methodologies for CT scanning.

Issues, controversies, problems: current limitations

With all of the potential benefits of 3D technology, why have these methods not yet been universally employed in forensic anthropology? The primary limiting factor to widespread acceptance and use of 3D technologies is the cost associated with acquiring and maintaining the equipment and software necessary to utilize the technology. Given that most forensic anthropologists work in academic settings and conduct forensic casework for free or for very minimal compensation, funds are extremely limited. They must rely on research grants for funding, which in today's pressing economic times are becoming more competitive and difficult to obtain. In addition, many grant solicitations commonly discourage funding for equipment and software. For example, in a recent review of a grant proposal submitted by the first author, it was stated that the author's institution may not be the best place to carry out the proposed research given the fact that the research required the purchase equipment and software not already available at that particular institution.

Of the aforementioned 3D technologies, photogrammetry is the cheapest, as it only requires a digital camera with decent resolution capacities, a computer, and postprocessing software (which range from free up to a couple hundred dollars); however, this is currently the least utilized and remains relatively untested in the field. Three-dimensional surface scanners are also fairly affordable (<$4,000) and thus are gaining popularity in forensic anthropology and related fields. However, despite the relatively low cost of many of the scanners, more advanced and expensive software is commonly required to process and analyze the scan data in great detail. This is because the scanners were originally developed for engineering fields and as a result, the processing software tends to be advanced reverse-engineering programs (e.g., GeoMagic and Rapidworks). These programs can cost thousands of dollars for a single license. Although forensic anthropologists do not need all of the advanced options within these software programs, more simplified and cheaper programs have yet to become available. Digitizers are around the same price as 3D surface scanners once postprocessing software is taken into consideration ($8,000–12,000), which combined with their simplicity, explains their frequency of use in forensic anthropology; however, digitizers cannot be used to create a virtual model of an object and thus have limited utility in comparison to the other 3D technologies.

CT is by far the most expensive of the 3D imaging technologies. CT equipment can range from $100,000 for used, older generation models, to over $1 million dollars for multiaxis, high-resolution micro-CT scanners. The high cost associated with owning and operating a CT machine prevents the majority of forensic anthropologists working in academic institution, museums, or medical examiner offices from owning a scanner themselves. Instead, access to a CT scanner owned by another institution must either be purchased by the day (e.g., $500/day at the Smithsonian Institution) or by the hour (e.g., $100–$200 per hour at the Texas State FACTS laboratory). Access to previously collected CT scans may still require an access fee or costs associated with pulling, anonymizing, and processing the data. For example, in a recent grant proposal involving the authors, $50/scan was negotiated with an institution for the de-identification and transfer of CT data. In some rare instances, manufacturing companies will allow scans to be completed *gratis* as a way to test their machines. Medical facilities have also occasionally donated their services and equipment depending upon project interest and collaboration. These services are easier to render if it is a single skull or skeleton that needs to be scanned, rather than a collection of elements or skeletons as might be required for research. Medical facilities may also grant access to CT scans previously collected for medical purposes to forensic anthropologists for research purposes, but as will be discussed shortly, this requires institutional approval from both parties and is not without associated ethical concerns.

The high expenses associated with 3D imaging equipment and software is likely because the technology was originally designed for STEM fields that have far greater budgetary resources than anthropology. Essentially, forensic anthropologists are "borrowing" technology from other fields, such as engineering and medical disciplines, and applying them to forensic anthropological research questions and cases. Besides cost, the adaptation of these technologies to forensic anthropology also means that the software and methods are not specifically designed for anthropological applications. There is a steep learning curve for many of the post-processing software programs (Grabherr et al., 2009); thus, only researchers that can dedicate a significant portion of their time to 3D technology and research use them as common practice. In addition, the predefined machines and software settings often must be manually altered for anthropological analyses. Frequently, to collect and analyze the full range of data, forensic anthropologists must jump between different platforms. Each technology and software has advantages, disadvantages and limitations (e.g., see the detailed description and comparison of surface scanners in Slizewski & Semal, 2009). This requires knowledge and expertise using an array of programs.

Because of the unique anthropological uses of 3D technology, novel statistical and analytical methods must also be developed. This takes time, and at present, the current 3D technology outperforms available statistical methods. The field of geometric morphometrics has made huge leaps in creating methods to statistically analyze curves and surfaces from biological forms, but creating these statistical methods to analyze 3D skeletal data requires knowledge of complex mathematical models and computational skills typically beyond the training of many forensic anthropologists. There is a new niche in the field for individuals with backgrounds in advanced statistics, computer science and programming, and engineering. Most educational tracks in forensic anthropology have not incorporated aspects of these disciplines into their recommended curricula. Furthermore, if individuals with expertise in these areas transfer to anthropology, they typically find that there is a significant decrease in pay and funding, thus discouraging such career moves.

An overall lack of consensus on the precision and accuracy rates for each of the 3D technology modalities, as it applies to human skeletal material, currently exists within forensic anthropology. Use of the MicroScribe digitizer is widely accepted, with numerous studies supporting the reliability of cranial landmark data (and derived osteometric measurements from the inter-landmark distances) collected using the MicroScribe (Sholts et al., 2011; Sládek et al., 2012; Stephen et al., 2015). There are also a number of studies looking at differences between CT scans and actual skeletal elements, and they generally report an overall congruency between the methods (e.g., Gamble & Hoppa, 2010; Robinson et al., 2008; Verhoff et al., 2008b). Accuracy results, however, will depend on the scanning parameters and at present, there are no standardized protocols for CT data collection in anthropology.

Friess (2012) provides a comprehensive summary of studies within the broader discipline of physical anthropology that have explored accuracy, precision, and observer error using 3D surface scanners. After reviewing the results presented in over 20 papers, Friess (2012: 16) concludes that "surface models replicate anthropological (hard tissue) objects with sufficient precision, but that the human operator remains the biggest source of error when it comes to extracting linear measurements or Cartesian coordinates." Precision and accuracy in data collected from 3D surface models can be comparable to that collected using traditional anthropological techniques. For example, Sholts et al. (2011) report a standard deviation of 1.05 mm when analyzing 42 craniometrics landmarks, as compared to 0.79 mm with the MicroScribe. Because of variations in the scanning parameters and research designs, however, universal accuracies cannot be provided. An accuracy rate obtained using one study's protocols may not be true when using a different set of parameters. Research into the application of photogrammetry methods to skeletal elements for forensic purposes are completely lacking. This is likely due to its more recent introduction to the field.

Studies evaluating the reliability of a specific piece of equipment to accurately represent a skeletal element can only attest to the reliability of that particular machine. Research analyzing variations in accuracy between equipment is desperately needed. In addition, each of the 3D modalities discussed above has a number of settings that can be altered by the user. For example, with the NextEngine scanner, the user can change the number of divisions, points per inch, whether texture is collected, whether alignment is conducted manually or automatically, whether model holes are automatically filled, and whether any noise reduction is complete—among many other options. Changes in settings will also affect the accuracy of the resultant model and data. Unfortunately, optimal settings for skeletal data collection using these 3D technologies have not been experimentally determined and standard operating procedures (SOPs) have yet to be developed (to be discussed further later). In addition, the reliability of the 3D data collected from virtual models may vary depending on their application. Even if osteometric landmarks can be reliably collected from virtual 3D models, it does not mean that the technology can accurately represent other skeletal characteristics, such as striation patterns left from a saw on a cut surface. The conclusions drawn from current studies, such as Friess (2012) and the studies within, are encouraging, but more research needs to be done examining precision, accuracy, and observer error for the specific tasks forensic anthropologists are conducting with 3D technology and how variables such as specific equipment, settings, and methodologies effect these accuracy rates.

There is also a need to make 3D virtual models more accessible to researchers. As previously mentioned, 3D models present the opportunity to create large, modern databases of virtual human skeletal material for research and comparative purposes. Without well-developed SOPs, however, the utility of such virtual collections would be limited. The first step in creating SOPs is to determine the most optimal settings for scanning and post-processing scans into 3D models. There is a definite compromise between 3D model quality and file size, which also needs to be considered. If the file sizes of the scans or models are too large, they can present issues with computing requirements, and storage requirements for databases. Experimental research should be conducted to determine the minimum required quality in 3D models for anthropological methods to be accurately applied. Creation of online databases of 3D models will require a significant amount of storage space to accommodate the large files sizes of 3D models. Continued advances in online data storage methods (e.g., cloud storage), create a real possibility for virtual skeletal collections. For example, Duke University curates an online data archive called MorphoSource comprised of 3D morphological data sets of various taxonomic species ("MorphoSource," n.d.). It is not limited to human remains, and at present only 45 specimens are attributed to *Homo sapiens,* many of which are a single element (e.g., a first metatarsal). However, MorphoSource illustrates the possibility of virtual human skeletal collections in the future. The project is funded by the National Science Foundation and the data are contributed by independent researchers. Because the 3D data that are uploaded were created by different individuals, using different equipment, software, and settings, there is a lack of uniformity in the quality of the scans available. Before a usable virtual collection of human skeletal material can be created, SOPs in 3D technology needs to be addressed and agreed upon.

There is also the significant time requirement for scanning and processing the scans that should be considered with the creation of a virtual database. Converting a tangible skeletal collection of over 2,000 complete skeletons (such as the Hamann-Todd Collection at the Cleveland Museum of Natural History) would take years of work. Institutions would also have to monitor access to the 3D data and limit access to qualified researchers in order to ensure that all data are used in appropriate and ethical manners. In addition, questions arise as to ownership and copyrights of the 3D material. If a researcher visits a skeletal collection and collects 3D data, who then owns the rights to the 3D material? Is it the owner(s) of the actual skeletal material, the facility acquiring the 3D data, or the researcher(s) coordinating the scan and subsequent research? Is the researcher permitted to pass that 3D data to others without soliciting permission from the original collection first? Many institutions now require researchers to provide the collection with a copy of all data collected (including 3D scans). At present, it seems that those institutions curate those data in a folder and it is not readily accessible to other researchers, which means the next researcher that comes to collect 3D data from those specimens, duplicates those efforts. Would it be appropriate for institutions to provide 3D data collected by one research to another researcher? And if so, is there a proper embargo time? In an attempt to rectify the new concerns of intellectual property brought up by "virtual collections," Elton and Cardini (2008) and Sumner and Riddle (2009) have set forth suggestions and guidelines for digital data storage, sharing, and access that can be applied to digital data in forensic anthropology. Further research and debate on these issues are likely to follow as 3D technologies continue to become more popular in forensic anthropology and related disciplines.

Solutions and recommendations

Because 3D technologies are already being incorporated into forensic anthropological work, the first steps that need to be taken are the development of standard practices and systematic evaluation of accuracy and precision rates. Digitizer data are probably the most widely accepted and validated of the 3D data discussed in this chapter. This is in part due to the simplicity of the digitizer (e.g., it is used to collect single, landmark coordinates and there are no complicated settings to choose from) and also because forensic anthropologists are using it to collect osteometric landmarks that have already been well defined in the field. When it comes to 3D models of bone surfaces, however, the number of options in equipment, settings, uses, type of data being collected, and methods of analysis leads to a lack of uniformity and unknown error rates. Current studies utilizing these emerging technologies need to report all settings and methodological details in their publications so that others can replicate the methods. Experimental research needs to be conducted to determine best practices and optimal settings when collecting and processing surface scanner, CT, and photogrammetry data. Once experimentally defined, SOPs can be distributed to members of the field. The development and dissemination of SOPs for these 3D instruments would mean that individual researchers would not need to spend extra time or effort figuring out ideal settings; instead, they could follow the proposed settings and preferences. Making the collection of 3D data easier and less time consuming would likely increase the number of facilities and users. It would also ensure the quality of the 3D data and models, allowing 3D models that were collected from multiple individuals or institutions to be combined into a single large-scale database for research and forensic analysis without any concern regarding varying degrees of quality and accuracy. The specific accuracy and error rates associated with models created using the SOP would be known (and published) in case questions arise in a court of law. This is an important step that needs to be made in order to comply with *Daubert* (1993) ruling and NAS report (NAS, 2009).

SOPs for the use and exchange of 3D virtual models should also be developed. This will require extensive discussions within the field between researchers as well as institutions. There should be no reason to duplicate efforts by making each new researcher recreate new 3D models; however, proper credit should be given to the technicians that put in the time to create the models. An agreed upon embargo time would allow researchers to publish their research prior to opening up access to the virtual models to other researchers. Considering the amount of time that it takes to create the 3D models, it might be appropriate to develop a method of noting the original technician in all subsequent studies using that data to provide accurate credit. Current skeletal collections should begin working toward virtualization. For collections such as the Hamann-Todd collection, which houses more than 2,000 skeletons, this might seem like an impossible feat ("Collection Areas & Databases," n.d.). Personnel would have to be dedicated to this process and equipment may need to be purchased, both of which would require additional funding. Contributions by researchers could help alleviate some of this work, but would require them to follow the same SOP in creating the 3D models. Protocols for 3D data access and sharing would need to be developed to ensure that only qualified researchers can gain access to the data. Collections and museums seem hesitant to make their skeletal remains so accessible, which is difficult to understand if their role is to further studies in the field. By converting skeletal collections to virtual collections, it would allow more researchers to utilize the collections and thus, support increased advancements in the field. Virtual collections would also provide a permanent archive of the remains, and as the use of already created 3D models does not require handling of the remains, can help preserve these collections longer. Ideally, the 3D models from all current skeletal collections could be combined with individual 3D models collected from forensic anthropological casework, along with the proper documentation regarding the 3D model settings and the individual's demographics and cause of death to create a single, massive database of skeletal resources. The first step in making this a reality, however, is creation of SOPs and discussions between researchers and institutions in order to work out the aforementioned issues.

There is also the need for increased interdisciplinary collaboration. As mentioned earlier, some of the limitations with using 3D technology in forensic anthropology are due to the fact that these resources were originally developed for other fields (e.g., engineering, medicine, graphic artistry, cinematography). Forensic anthropologists are adapting the technology for their needs, but as a field do not yet realize the full potential of these technologies. Individuals in these fields can provide forensic anthropologists with insight into 3D capabilities. By collaborating with researchers in these fields, anthropologists can build from their methods instead of reinventing methods of analysis. Anthropologists can work with software engineers, computer programmers and technology developers to create programs and analytical methods to answer questions specific to forensic anthropology. As forensic anthropology continues to incorporate more advanced technologies, it may become important to consider revising academic curricula in the field to include some interdisciplinary courses. Students with more advanced statistical backgrounds, knowledge in computer programming and experience with 3D technologies will fill a much-needed niche in the field. More collaborations with medical institutions can also lead to increased use of clinical 3D data, and possibly even the creation of virtual databases from such data.

Increased funding for technology and research is key to advancing these new technologies within forensic anthropology. Because forensic anthropology is not a very profitable field and most of the work is down in academic settings, forensic anthropologists rely mostly on federal grants (e.g., NIJ, NIH, and NSF) for equipment and research funding. Research grants that include new technology and equipment in their budgets should not be discredited due to their lack of current access to these resources. In fact, there should be more grant solicitations specifically for obtaining advanced technologies in forensic anthropology laboratories.

Along with increased funding is the need for increased awareness of these technologies and their potential utilization within the field. Some of these newer technologies, discussed above, were presented in a 2011 special workshop offered by the National Institute of Justice, through the National Forensic Science Technology Center, entitled *Advances in Forensic Anthropology Technology Transition.* The goal of the workshop was "to bring practicing forensic scientists up to speed on recent technologies, so they can be used in the field to assist in body identification, murder and missing persons investigations, and other legal and medical cases" (Shipman, 2011). Although this effort should be applauded, at the rate that technology is advancing, it is important to hold such workshops more frequently and to make them more publicly available. Perhaps an annual session on technological advancements should be held at national and international forensic meetings in order to get the information out to a wider audience. Forensic anthropologists should be encouraged to take continuing education courses that cover new technologies (and their institutions should offer them more time and resources to do such).

Public awareness of the utility of these technologies as well as their limitations, is also important. Popular media shows like Bones and CSI give the general public false expectations of what is possible with 3D technology. Anything less than what is presented in these fictional shows is perceived by family members of victims and jurors as a lack of effort by the forensic anthropologist. Similarly, jurors and the public need to understand that just because 3D technologies are higher-tech, it does not necessarily equate to better accuracies and results. These new resources have a lot of potential, but until methods utilizing 3D technologies can be thoroughly validated, traditional methods should form the basis of case analyses. In fact, it is possible that 3D technologies may never completely replace traditional methods, but instead may complement them by presenting novel methods and ways to obtain new information.

Ethical issues associated with this forensic area

Three-dimensional CT scan data collected from clinical settings is especially useful for examining modern human variation in forensic anthropological studies. In most cases, the data have already been collected and can be accessed retrospectively; thus, 3D models of modern individuals can be collected without incurring any additional physical risks to individuals or additional costs related to procuring the scans (although there may still be costs associated with accessing or de-identifying the data). Well-documented information (e.g., sex, age, ancestry, stature) accompanies clinical scan data. In clinical settings, the Health Insurance Portability and Accountability Act of 1996 (HIPAA Privacy Rule) and U.S. Departments of Health and Human Services and Food and Drug Administration Protection of Human Subject Regulations (Clinical Research and the HIPAA Privacy Rule, 2007) regulate the use of data from living human subjects. Thus, before gaining access to clinical data, most institutions require approval through an internal review board (IRB). The IRB reviews the proposed study and data that will be collected and determines whether the proposed study violates either of these regulations and whether there are any additional ethical concerns to consider. For example, to be in compliance with the HIPAA Privacy Rule, the IRB will want to see that all personal identifying information will be removed prior to the research, so that the data cannot be associated back to any single individual. This is generally not an issue in forensic anthropological research. Demographic information that is of interest to the anthropologist, such as age, sex, ancestry, and stature, are not considered identifiable information, and thus are not protected by the HIPAA Privacy Rule. This, combined with the fact that forensic anthropologists are generally using retrospective medical data which have previously been collected and thus are not using any methods which could harm individuals, usually equates to a simple IRB approval process. If there are any identifiable information (e.g., name, address, social security number, etc.) stored with the requested data, it can first be de-identified, so that the forensic anthropologist does not have access to this information. This process ensures the anonymity of the patient data; however, it should be noted that the patient does not actually have to provide consent to be included in the research if IRB approval is obtained. In most cases, the patient will never be aware that their data were used in research. It should also be noted that the above regulations apply to living human subjects; regulations are much more relaxed when dealing with deceased individuals (e.g., CT data collected from ME offices or forensic cases).

As the use of 3D models increases in fields of research, new regulations may need to be put in place to ensure that all data are utilized in an ethical manner. Three-dimensional models of pelves may not be very identifiable, but what about 3D models of skulls? Photographs of individuals are considered identifiable information, so one would assume that if the 3D skull models include the soft tissue of the face (such as in a CT scan), it would likewise fall under HIPAA regulations. This does not mean that the data cannot be used, but there may be additional stipulations to the data usage to ensure the confidentiality of the patient. But would a 3D model of a skull, without associated soft tissue, be considered identifiable information? If not, would there be anything keeping the researcher from 3D printing someone's skull? That 3D print could potentially be around forever. At present, IRBs handle each research proposal on a case-by-case manner to address such issues. The required data are collected from the 3D models and all subsequent analyses are conducted on that data (not the models *per se*), and the if the 3D models are to be used in any additional analyses beyond the original proposed project additional approval must first be obtained. Some institutions may not even allow the 3D models to leave their premises, and the researchers may be required to collect all necessary data from the models on-site.

But what about the potential use of clinical 3D models to create a virtual skeletal collection? If the skeletal elements are segmented from the models so that all soft-tissue is absent (and of course all other traditional identifiable information is removed from the records), would the use of that data to create a virtual skeletal collection be ethical? The authors would argue that in order to use 3D models in a permanent collection where researchers (with the appropriate approval) will continuously have access, consent should first be obtained from the individual or in the case of a deceased individual, from their living relatives. If given the option, we may be surprised to see how many individuals are willing to contribute their data to advance research in the field. An increasing number of people are donating their entire bodies to science, including their skeletons to forensic anthropological programs. If they can do this, why not set up a program with a clinical institute where you can give a patient the option to donate a virtual 3D model of their skeletons to similar purposes? Of course, finding ways to control access to such a database and ensure that 3D models do not fall into the wrong hands presents new dilemmas. This is a potential research avenue that deserves some discussion amongst researchers, medical institutions, and the public.

Three-dimensional printing of models brings about a different set of ethical concerns. For some reason, there seems to be a difference in our minds between a 3D model on a screen and a printed tangible 3D replica of an object, such as someone's skull. For most research purposes, a 3D print is not required. Besides educational purposes, the only scenario the authors can imagine necessitating a 3D print would be in forensic cases for facial reconstruction, evaluation of potential tools used to create traumatic defects, or for illustrative purposes during court testimonies. But how can you stop someone from making a 3D print if the 3D model is available online? Besides having a system to review research applicants prior to granting access to the 3D data, it may require the signage of detailed usage agreements with researchers that would limit the use and distribution of the data. Depending on the type of 3D model, permitted use may vary (e.g., since a pelvis is considered nonidentifiable perhaps it would be okay to create a 3D print of it, while allowance of a skull to be 3D printed may depend on whether it was donated for educational purposes). If a 3D print is made of elements of a forensic case for analysis or court testimony, should it be returned to the families once the case is complete or should it be retained for evidentiary purposes in the event that the case is revisited again in the future? The first step to answering these questions is to have these discussions on case and research ethics in light of new technology, which could then result in standard guidelines for institutions to follow.

Applications in diverse forensic settings

A few forensic institutions have begun to use CT scanning in conjunction with magnetic resonance imaging (MRI) to conduct virtual autopsies. As mentioned above, the CT scanning is good at differentiating between tissues of disparate densities (e.g., bone vs. soft tissue). MRI is better at documenting differences in soft tissue structures. By combining the two methods, forensic pathologists are able to investigate most aspects of the body in an effort to determine the cause and manner of death. This is a relatively new procedure and debates on the accuracy of such virtual autopsies compared to traditional autopsies in which the examiner cuts into the body continue (Bolliger et al., 2008; Jackowski et al., 2005; Thali et al., 2003, 2005; Verhoff et al., 2008b). Issues include the fact that virtual autopsies do not record coloration, which can be important when examining features such as bruising. In addition, it requires expensive equipment and experienced technicians. The benefit, however, is that it is noninvasive, which makes it more acceptable to family members, especially in religions or cultures that do not accept traditional autopsy procedures.

Crime scene technicians and crash scene reconstructionists have also adopted various forms of 3D surface scanning and photogrammetry technology. While at the scene of a recent plane crash, a state police officer that traditionally works with reconstructing crash scenes was able to 3D scan the entire crash scene in about a half hour (not including postprocessing time). In this manner, he was able to document the scene *in situ*, so that at any time investigators could go back and see a replica of what the original scene looked like prior to any recovery efforts. Metric data, such as distances, angles, and topography, can be collected from the 3D scan. Such technology has also been used for indoor crime scenes, which can allow for the reconstruction of such items as bullet trajectories. Whether at an indoor or outdoor crime scene, forensic technicians understand that once the scene is processed all of the spatial and contextual information is lost. By creating 3D models of the scene, this is no longer true. Forensic anthropologists could also use this technology to document outdoor recovery scenes, although at present the 3D scanners capable of capturing data on this larger, scene scale are more expensive (>$50,000), and thus generally outside of the budget of most forensic anthropologists. More experimental usage of affordable photogrammetry methods or collaborations with organizations that own this more expensive equipment could make this a real prospect for forensic archaeological recoveries.

As previously mentioned, 3D models of skulls can also assist with forensic facial reconstructions. In the future, it may be possible to include those virtual reconstructions as well as the actual 3D models of the skull or other identifiable elements in databases such as NAMUS, so that forensic departments across the country will readily have that data available to search their databases, thereby increasing the chances of identification. Three-dimensional CT scans of an individual's dentition can be electronically sent to a forensic odontologist, or printed as a 3D model and the model sent, for comparison to antemortem records. This would eliminate the need to physically section (with a Stryker saw) the dentition from the rest of the skull (as is done in some cases so that forensic odontologists have a clear view of the dentition and can work simultaneously with forensic anthropologists). Finally, the 3D scans and prints can be used in forensic educational contexts. For example, not many forensic anthropological programs have access to an abundance of skeletal remains, or examples of various skeletal traumas. Through collaboration between institutes, 3D models and prints can be used in place of the true skeletal elements so that students have an opportunity to learn not just from their own institution's resources but from others as well. The sharing of such knowledge will ensure that forensic anthropologists are getting the best training available and ultimately contributes to the continued advancement in our field.

Future research directions

Future research directions should focus on the current limitations of 3D technology in forensic anthropology, as discussed in the previous chapter sections. There is a need for experimental research to determine the accuracy of specific equipment when used for forensic anthropological methods as well as to determine the optimal settings for scanning and postprocessing procedures. Based on this research, SOPs and best practice guidelines can be developed, creating a possibility for virtual databases in the future. Through interdisciplinary collaborations, novel methods of analyses can be developed and applied to the 3D models. For example, a study by Fatah et al. (2014) explores the possibility of automating the collection of osteometric data from 3D models of skulls for sex estimation. By combining such automated methods with a virtual database, software programs similar to FORDISC can be developed to conduct forensic analyses on 3D models. Others have explored using 3D models of the pubic symphysis (an area of the pelvis forensic anthropologists evaluate to estimate the age of an individual) in order to create more objective, quantitative, and accurate methods of assessment than current methods of qualitative evaluation (Decker et al., 2011; Grabherr et al., 2009; Lottering et al., 2013; Telmon et al., 2005). There appears to be a general trend in using 3D data to convert more subjective traditional methods into more quantitative analyses, and thus minimizing human error associated with subjectivity (Fatah et al., 2014; Garvin & Ruff, 2012; Shearer et al., 2012). This progression is in line with the scientific responsibilities outlined by the Daubert ruling and NAS reports. Beyond the forensic anthropology laboratory, there is also a need to explore the potential use of 3D technologies at the scene. By capturing the 3D spatial location of remains, research can be conducted on taphonomic agents, such as the movement of remains due to topographic relief or by carnivores. Overall, the research potentials are endless if put in the hands of the innovative researcher.

Conclusion

Although digitizers are used widespread in forensic anthropology, the use of CT imaging, surface scanning, and photogrammetry are not as common in the discipline. These 3D technologies, which are capable of producing 3D models, have the potential to revolutionize the forensic sciences. Within forensic anthropology, virtual models of skeletal elements or recovery scenes can be used to permanently document cases, illustrate examples in court, and create novel methods of analysis. The ability to access 3D models from any computer and electronically distribute the data can mean increased collaboration between agencies with the potential to develop online virtual databases of 3D models for research and education purposes. Given the high costs associated with these technologies and the limited budgets of forensic anthropologists, however, their use is not likely to become common practice until the technology becomes more affordable or we see an increase in collaboration and technology sharing between laboratories, disciplines, and agencies. As use of these technologies increase in the field, more research needs to be conducted to determine accuracy rates and SOPs. Furthermore, open discussions regarding the ethical use of 3D data and policies regarding data curation and sharing should be conducted between disciplines, institutions, and the public.

Key terms and definitions

Biological profile: The estimation of an individual's age, sex, stature, and ancestry from their skeletal remains using both qualitative and quantitative methods.

Computed tomography: A form of radiology that combines x-ray images from multiple angles to create three-dimensional images of the structure being scanned.

Digitizer: a portable coordinate measurement machine that allows the operator to collect physical properties of a 3D object as *x*, *y*, and *z* coordinate data.

Facial reconstruction: Recreation of an individual's face, usually of someone who is unidentified, using their skeletal remains and soft tissue approximations. Reconstructions can be done in two or three dimensions.

Forensic anthropology: A subfield of biological anthropology that integrates the methods used in biological anthropology and archaeology to medicolegal death investigations and criminal law.

Geometric morphometrics: The quantitative analysis of morphological shape using Cartesian coordinates generated from landmarks, semilandmarks, outlines, and 3D surface renderings.

Photogrammetry: The use of photographs to measure and model real-world objects and scenes.

Surface scanner: A device that digitally collects three-dimensional data of an object's surface properties (e.g., shape or color) using lasers or light that can then be used to create 3D models of that object.

Taphonomy: The study of what happens to once-living organisms after death until the time of discovery. Taphonomy takes into account a number of biotic (living) and abiotic (nonliving) agents that impact the organism after death.

Thresholding: A simple method of image segmentation. In computed tomography, thresholding refers to rendering objects or surfaces that meet a specific density level.

Note

1 Researchers at the aforementioned Bass Skeletal Collection have begun to accumulate CT scans of the skeletal material accumulated in their collection. This is a great first step to making the collection available virtually to researchers. However, this is a very time-consuming process and the availability of these CT data are not currently advertised to outside researchers.

References

Agostini, G., & Gomez, E. (2009). Forensic anthropology academic and employment trends. *Proceedings of the American Academy of Forensic Sciences, 15*, 346.

Bolliger, S. A., Thali, M. J., Ross, S., Buck, U., Naether, S., & Vock, P. (2008). Virtual autopsy using imaging: Bridging radiologic and forensic sciences. A review of the Virtopsy and similar projects. *European Radiology, 18*(2), 273–282.

Bookstein, F. L. (1997). Landmark methods for forms without landmarks: Morphometrics of group differences in outline shape. *Medical Image Analysis, 1*(3), 225–243.

Bytheway, J. A., & Ross, A. H. (2010). A geometric morphometric approach to sex determination of the human adult os coxa. *Journal of Forensic Sciences, 55*(4), 859–864.

Christensen, A. M., & Crowder, C. M. (2009). Evidentiary standards for forensic anthropology. *Journal of Forensic Sciences*, *54*(6), 1211–1216.

Clinical Research and the HIPAA Privacy Rule. (2007). Retrieved November 24, 2015, from https://privacyruleandresearch.nih.gov/clin_research.asp.

Cobb, W. M. (1959). Thomas Wingate Todd, MB, Ch. B., FRCS (Eng.), 1885–1938. *Journal of the National Medical Association*, *51*(3), 233.

Collection Areas & Database. (n.d). Retrieved November 24, 2015, from www.cmnh.org/phys-anthro/collection-database.

Conyers, L. B., & Goodman, D. (1997). *Ground-penetrating radar: An Introduction for Archaeologists*. AltaMira Press, London.

Daubert v. Merrell Dow Pharmaceuticals, Inc. (1993). US Supreme Court 509.U.S.579,113S.Ct.2786, 125L. Ed.2d 469.

Decker, S. J., Davy-Jow, S. L., Ford, J. M., & Hilbelink, D. R. (2011). Virtual determination of sex: Metric and nonmetric traits of the adult pelvis from 3D computed tomography models. *Journal of Forensic Sciences, 56*(5), 1107–1114.

Dirkmaat, D. C., & Cabo, L. L. (2012). Embracing the new paradigm. In D. C. Dirkmaat (Ed.), *A Companion to Forensic Anthropology* (pp. 3–40): New York: Wiley-Blackwell.

Elton, S., & Cardini, A. (2008). Anthropology from the desk? The challenges of the emerging era of data sharing. *Journal of Anthropological Sciences, 86*, 209–212.

Fatah, E. E., Shirley, N. R., Jantz, R. L., & Mahfouz, M. R. (2014). Improving sex estimation from crania using a novel three-dimensional quantitative method. *Journal of Forensic Sciences*, *59*(3), 590–600.

Friess, M. (2012). Scratching the surface? The use of surface scanning in physical and paleoanthropology. *Journal of Anthropological Sciences*, *90*, 7–31.

Gamble, J., & Hoppa, R. D. (2010). *Congruence of metric and morphological approaches to determination of sex in the real, virtual and 3D pelvis*. Proceedings of the 38th Annual Meetings of the Canadian Association for Physical Anthropology, Saskatoon, SK.

Garvin, H. M., & Ruff, C. B. (2012). Sexual dimorphism in skeletal browridge and chin morphologies determined using a new quantitative method. *American Journal of Physical Anthropology*, *147*(4), 661–670.

GeoMagic. (2015). Retrieved November 24, 2015, from www.geomagic.com/en/.

Grabherr, S., Cooper, C., Ulrich-Bochsler, S., Uldin, T., Ross, S., Oesterhelweg, L., Bollinger, S., Christe, A., Schnyder, P., Mangin, P., & Thali, M. J. (2009). Estimation of sex and age of "virtual skeletons"–A feasibility study. *European Radiology*, *19*(2), 419–429.

Hunt, D. R., & Albanese, J. (2005). History and demographic composition of the Robert J. Terry anatomical collection. *American Journal of Physical Anthropology*, *127*(4), 406–417.

Jackowski, C., Sonnenschein, M., Thali, M. J., Aghayev, E., von Allmen, G., Yen, K., Dirnhofer, R., & Vock, P. (2005). Virtopsy: Postmortem minimally invasive angiography using cross section techniques—Implementation and preliminary results. *Journal of Forensic Sciences*, *50*(5), 1175–1186.

Jantz, R. L., & Meadows Jantz, L. (2000). Secular change in craniofacial morphology. *American Journal of Human Biology*, *12*(3), 327–338.

Jantz, R. L., & Ousley, S. D. (2005) *FORDISC 3: Computerized Forensic Discriminant Functions* (Version 3.1) [computer software]. The University of Tennessee, Knoxville.

Katz, D., & Friess, M. (2014). Technical note: 3D from standard digital photography of human crania—A preliminary assessment. *American Journal of Physical Anthropology, 154*(1), 152–158.

Kerley, E. R. (1978). Recent developments in forensic anthropology. *Yearbook of Physical Anthropology, 21*, 160–173.

Kieser, J. A., Bernal, V., Neil Waddell, J., & Raju, S. (2007). The uniqueness of the human anterior dentition: A geometric morphometric analysis. *Journal of Forensic Sciences, 52*(3), 671–677.

Kimmerle, E. H., Ross, A., & Slice, D. (2008). Sexual dimorphism in America: Geometric morphometric analysis of the craniofacial region. *Journal of Forensic Sciences, 53*(1), 54–57.

Komar, D. A., & Grivas, C. (2008). Manufactured populations: What do contemporary reference skeletal collections represent? A comparative study using the Maxwell Museum documented collection. *American Journal of Physical Anthropology, 137*(2), 224–233.

Kraus, K. (2007). *Photogrammetry: Geometry from Images and Laser Scans*. Berlin: Walter de Gruyter GmbH & Co.

Lottering, N., MacGregor, D. M., Meredith, M., Alston, C. L., & Gregory, L. S. (2013). Evaluation of the Suchey–Brooks method of age estimation in an Australian subpopulation using computed tomography of the pubic symphyseal surface. *American Journal of Physical Anthropology, 150*(3), 386–399.

Lynnerup, N. (2009) Methods in mummy research. *Anthropologischer Anzeiger, 67*(4), 357–84.

Meadows Jantz, L., & Jantz, R. L. (1999). Secular change in long bone length and proportion in the United States, 1800–1970. *American Journal of Physical Anthropology, 110*(1), 57–67.

MeshLab. (n.d.). Retrieved November 24, 2015, from http://meshlab.sourceforge.net/.

MicroScribe Pricing. (2015). Retrieved November 6, 2015, from www.3d-microscribe.com/MicroScan Pricing.htm.

MorphoSource. (n.d.). Retrieved June 27, 2018, from http://morphosource.org/.

National Academy of Sciences. (2009). *Strengthening Forensic Sciences in the United States: A Path Forward*. Washington, DC: The National Academies Press.

NextEngine 3D Scanner TechSpecs. (n.d.). Retrieved November 6, 2015, from www.nextengine.com/assets/pdf/scanner-techspecs-uhd.pdf.

Ousley, S. D. (2014). *3Skull* (Version 1.76) [computer software]. Available from http://math.mercyhurst.edu/~sousley/Software/.

Ousley, S. D., & Jantz, R. L. (1993). *Postcranial racial discriminant functions from the Forensic Data Bank* [Abstract]. Presented at the American Academy of Forensic Sciences Annual Meeting, Boston, MA.

Ramsthaler, F., Kettner, M., Gehl, A., & Verhoff, M. A. (2010). Digital forensic osteology: Morphological sexing of skeletal remains using volume-rendered cranial CT scans. *Forensic Science International, 195*, 148–152.

Robinson, C., Eisma, R., Morgan, B., Jeffery, A., Graham, E., Black, S., & Rutty, G. (2008). Anthropological measurement of lower limb and foot bones using multidetector computed tomography. *Journal of Forensic Sciences*, 53, 1289–1295.

Shearer, B. M., Sholts, S. B., Garvin, H. M., & Wärmländer, S. K. (2012). Sexual dimorphism in human browridge volume measured from 3D models of dry crania: A new digital morphometrics approach. *Forensic Science International, 222*(1), 400-e1.

Shipman, M. (2011). Advances in forensic anthropology: 3D-ID. *NC State News*, Retrieved November 5, 2015, from https://news.ncsu.edu/2011/08/wms-forensic-3d-id/.

Sholts, S. B., Flores, L., Walker, P. L., & Wärmländer, S. K. T. S. (2011). Comparison of coordinate measurement precision of different landmark types on human crania using a 3D laser scanner and a 3D digitiser: Implications for applications of digital morphometrics. *International Journal of Osteoarchaeology, 21*(5), 535–543.

Sládek, V., Galeta, P., & Sosna, D. (2012). Measuring human remains in the field: Grid technique, total station, or MicroScribe? *Forensic Science International, 221*(1), 16–22.

Slizewski, A., & Semal, P. (2009). Experiences with low and high cost 3D surface scanner. *Quartär, 56*, 131–138.

Stephen, A. J., Wegscheider, P. K., Nelson, A. J., & Dickey, J. P. (2015). Quantifying the precision and accuracy of the MicroScribe G2X three-dimensional digitizer. *Digital Applications in Archaeology and Cultural Heritage, 2*(1), 28–33.

Sumner, T. A., & Riddle, A. T. (2009). Remote anthropology: Reconciling research priorities with digital data sharing. *Journal of Anthropological Sciences, 87*, 219–221.

Telmon, N., Gaston, A., Chemla, P., Blanc, A., Joffre, F., & Rouge, D. (2005). Application of the Suchey-Brooks method to three-dimensional imaging of the pubic symphysis. *Journal of Forensic Sciences, 50*(3), 507–12.

Thali, M. J., Yen, K., Schweitzer, W., Vock, P., Boesch, C., Ozdoba, C., Schroth, G., Ith, M., Sonnenschein, M., Doernhoefer, T., & Scheuer, E. (2003). Virtopsy, a new imaging horizon in forensic pathology: Virtual autopsy by postmortem multislice computed tomography (MSCT) and magnetic resonance imaging (MRI)-A feasibility study. *Journal of Forensic Sciences, 48*(2), 386–403.

Thali, M. J., Braun, M., Buck, U., Aghayev, E., Jackowski, C., Vock, P., Sonnenschein, M., & Dirnhofer, R. (2005). VIRTOPSY—Scientific documentation, reconstruction and animation in forensic: Individual and real 3D data based geo-metric approach including optical body/object surface and radiological CT/MRI scanning. *Journal of Forensic Sciences, 50*(2), 428–42.

Verhoff, P. D. M., Obert, M., Harth, S., Reuß, C., Karger, B., Lazarova, B., & Traupe, H. (2008a). "Flat-Panel"-Computertomographie in der Rechtsmedizin. *Rechtsmedizin, 18*(4), 242–246.

Verhoff, M. A., Ramsthaler, F., Krähahn, J., Deml, U., Gille, R. J., Grabherr, S., Thali, M. J., & Kreutz, K. (2008b). Digital forensic osteology—Possibilities in cooperation with the Virtopsy® project. *Forensic Science International, 174*(2), 152–156.

Weber, G. W. (2001). Virtual anthropology (VA): A call for glasnost in paleoanthropology. *The Anatomical Record, 265*(4), 193–201.

WM Bass Donated Skeletal Collection. (n.d.). Retrieved November 17, 2015, from http://fac.utk.edu/collection.html.

8 3D laser scanning

John D. DeHaan

Background

Accurate and comprehensive documentation of scenes of fire, explosions, or traffic accidents is essential for effective investigations, accurate reconstructions, and expert conclusions in court testimony. Photographic documentation is much improved (and nearly mistake-proof) with today's high-resolution digital cameras and videos. Scene documentation, however, remains a record of physical dimensions, not just images. For decades, scene dimensions were taken manually with tape measures, often requiring two persons to take and record each measurement. Accuracies needed to be only in the range of ±1% (30 cm at 30 m). People often resorted to "paces" for long distances. Wheel-type measuring devices were more accurate and easier to use but were accurate only on smooth surfaces such as roads. Some agencies came to call on the services of land surveyors using a transit and a range pole/level rod and 100′ manual chain. The next step forward was to use a laser on a heavier, more stable base called a theodolite and a prism or reflector on the range pole to reduce the time needed to take each sighting. The total station is an electronic/optical system used in modern surveying and building construction. The concept was a major advance since the measurements of inclination/declination, angular displacement, and distance were entered by the operator onto a laptop or computer. Later, the data points could be plotted out to a graphic display in either plan-view (map) format or vertical (elevation) view. These systems have been widely used by vehicle accident and fire/explosion scene investigators for decades and are still in use today in some agencies. Because each data point has to be entered individually, recording a large scene or one involving many separate small items (such as an explosion scene) can be very time consuming. Some Total Station systems permit remote control of the data collection and entry functions (allowing one-person operations) but each data point type has to be marked by the operator. Some also include an integrated GPS (global positioning system) for accurate documentation of larger scenes. GPS data are sometimes important in determining the correct jurisdiction (city, county, state, or federal) responsible for the scene. The accurate use of the system requires considerable hours of operator training.

In 1993, the first handheld laser distance meter was introduced by Leica Geosystems. This permitted one person to take and record straight line distances and dimensions as long as there was a suitably reflective surface at the "far" end. Beginning in the 1990s, optical manufacturers devised ways of "automating" the sighting and recording (total station) process by using a scanning pulsed laser and a very fast data system to record the distance (by time of flight measurement), direction, and elevation/declination of every "point" in the scanned area. The sophisticated software then can create a virtual 3D image and overlay a scanned digital photographic image over for a realistic 3D image with the positional data embedded in it. It is these systems with which this chapter will deal.

The need for such comprehensive documentation has long been clear to forensic investigators in many disciplines. Accident reconstruction can today recreate the condition, position, dynamics, and reactions of vehicles in crashes very accurately, but only if the measurements of critical features are taken such as road surfaces; obstacles on the road or nearby; skid-or impact marks; inclines, declines, or camber changes to the road; lane and road markings; and any locations of buildings, trees, or other objects that could affect sight lines. Another category of criminal investigations for which dimensions and locations are critical are firearms incidents where view lines and projectile trajectories are essential to establish who was where and what were they shooting and in what direction. In these cases, data from 3D scans allow the accurate reconstruction of dimensions, distances, and directions throughout the space involved, and even in large exterior scenes. When linked to firearms evidence (projectiles and casings), witness descriptions, video and audio recordings, and firearm operation and ballistics, scanned image data can provide reliable information to test various hypotheses and confirm the best reconstruction (Grissim and Haag, 2008; Haag and Sturdivant, 2013). Because 3D laser scanning is rapid and noninvasive, it has been successfully applied to capture ephemeral evidence such as tool, tooth, and shoe impressions in soft, unstable substrates such as mud, sand, snow, or food (Komar, Davy-Jow, and Decker, 2012).

Explosion scenes usually include a large area (interior or exterior) littered with debris created by the explosion. This debris can include fragments of the target surface, the device (or accidental mechanism) responsible, container, or human victim (see Figure 8.1a, b for an example). The distribution of these fragments and their nature are critical elements for determining the "seat" of a condensed-phase (solid) explosive or the possible point of ignition for a diffuse-phase (gas or vapor) explosion. Even a small explosion can create hundreds of fragments scattered over a large area. In an ideal world, the investigator would like to document the location of each fragment at a scene and its possible identity (whether the incident was accidental or intentional). Practically speaking, the investigator usually has to choose the most potentially "valuable" fragments and document those before collecting them for analyses. Having the ability to scan the entire debris field and document the location of every fragment and then "tag" digitally the location of each item collected would be ideal if it could be accomplished in a reasonable time frame. Aerial or overhead views of an explosion scene can be the most useful in evaluating the origin of the blast pressure. At many scenes, this is not possible physically but a modern 3D laser scanner can

Figure 8.1 (a) Explosion scene (test): Large ambulance destroyed by 9lbs (4kg) of high explosive. Fragments scattered over a radius of nearly 200ft (62m). Leica three-dimensional laser scanner in foreground. (b) Windshields blown over 100ft (31m) from the vehicle. Other components were against fence in background nearly 200ft (62m) away. Photos courtesy of John D. DeHaan.

create a virtual overhead (bird's eye) view from the data it collects from various viewpoints at "eye" level (Gersbeck and Grissim, 2008).

While fire scenes can often be suitably documented by thorough photography, the fire investigator today is expected to do some type of analysis of the dynamics of the fire—its ignition location, first fuel package, development, growth rate, and ventilation conditions during the requisite

ROOM FIRE DATA

Room ____________________ **Room #** _______

Length __________

Width __________

Height __________

Note ceiling height changes __________

Floor Plan

Walls: Structure/material __________ Thickness _____ Covering ________ Sample? Y/N

Structure/material __________ Thickness _____ Covering ________ Sample? Y/N

Ceiling: Structure/material __________ Thickness _____ Covering ________ Sample? Y/N

Floor: Structure/material __________ Thickness _____ Covering ________ Sample? Y/N

Openings (door, window, other vents) into room (number on plan above):

Height (bottom to top of opening)	Sill Height	Soffit Depth (above opening)	Width	Open or Closed? Changes During Fire?
1.				
2.				
3.				
4.				
5.				
6.				
7.				

HVAC System:

Description ____________________

On/off prior to/during fire? __________

Furnishings (descriptions of major fuel items, including floor and wall coverings, draperies):

____

____

____

____

Time Line: Alarm time __________ FD arr. Time __________ Control time __________

Detection: ____________________

Pre-Fire Events: ____________________

Figure 8.2 Room fire data form. The dimensions of fire-damaged rooms (floors, walls, ceiling heights) and ventilation openings (doors, windows, HVAC) are critical to accurate fire reconstructions and are rapidly captured with a 3D laser scan. Courtesy of John D. DeHaan.

hypothesis testing of various scenarios.(DeHaan & Icove 2012) This analysis requires the dimensions of the room involved, the presence of architectural features that could affect the growth or detection of the fire (or its suppression), and the dimensions of doors, windows, or other openings that could supply air to a fire growing in the room. Capturing all of this data (as shown in Figure 8.2) requires considerable time if done by tape measure and note pad. A 3D laser scan can capture all of that information in a few minutes.

Basic principles

A laser can be used to measure distance by either measuring the time delay between an outgoing pulse and its return reflection or by measuring the shift in phase between the outgoing light and its return. Since light is traveling at 300,000 m/s, the time delay in a "pulse" system are typically on the order of 1 μs (for a 150m distance). A small handheld laser distance meter (available in a hardware store for about $125) relies on such technology. When the beam is scanned vertically by a rotating mirror, each pulse is tracked by its return angle. The pattern of data is like a single "bar" of the raster pattern of a television's cathode ray tube. When the head is rotated around a fixed vertical axis, the data are collected on the rotational position of the beam. As a rotation is completed, every point from which a return (reflected) beam is received is labeled as to its distance, position, and its reflectance strength. This scanning technology is sometimes called LIDAR (light/radar). See Figure 8.3 for examples of modern 3D laser scanners. Figure 8.4 is an example of a 3D laser scan of the same scene as Figure 8.1a using the reflectance of the laser beam as the visual image.

Early scanners (circa 1999) would capture about 800 points/second. As the data systems improved with faster speeds, so did the measurement rates. Today's pulse scanning systems can capture on the order of 1,000,000 points/second. Such rates allow for rapid rotational scans (about 1 minute for a full 360°) at moderate resolution. Higher resolutions require slower scans. Needless to say, such scans generate huge data files (typically several GB). These raw data are known as "point cloud" and become the basis for all analysis.

A high-resolution digital camera (either built-in to the scanner or as an external unit mounted on the scanning head) captures a true color image corresponding to the scan. The processing program can then produce either a "reflectance" image from the laser light return or a true-color image overlaid on the measurement data or any mixture in between. The program can also convert the data to a "wire frame" image (outline) or a 3D virtual reality image. Specialized processing programs are necessary to convert the data files to useful images.

The final images can be linked to GPS coordinates for accurate placement. Since each scan captures image data from surfaces facing the scanner, a single scan produces "shadows" where the laser light has not struck (or returned). The scanner is then repositioned one or more times and the area scanned again from another angle. The resulting individual scans can then

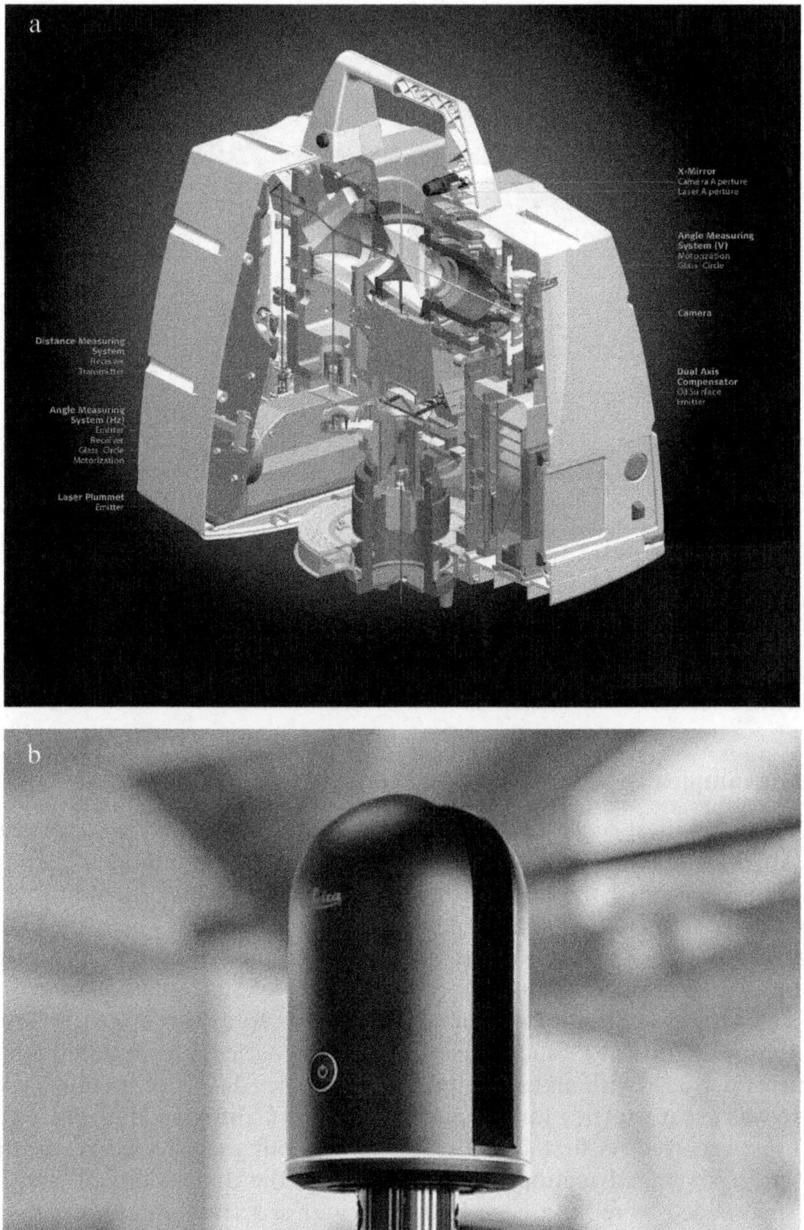

Figure 8.3 (a) Cut-away of state-of-the-art Leica 3D laser scanner with rotating mirror, built-in camera, dual axis compensator, laser plummet, and central panel with screen. Courtesy of Leica Geosystems USA. (b) The new Leica Geosystems BLK360 3D imaging scanner can record laser scan point cloud data, HDR (high dynamic resolution) photos, and thermal imagery all in about three minutes in a small, self-contained unit. Courtesy of Leica Geosystems USA.

Figure 8.4 Rendering of reflectance data from point cloud of same scene as Figure 8.1a. In True View©, every distance can be measured by clicking on the points of interest and a true-color image can be blended in. Courtesy of Precision Simulations, Inc., Grass Valley.

be integrated together to produce a fully 3D, geospatially correct, record of the entire scene.

Case examples

One of the earliest successes of 3D laser scanning in the public services was on April 29, 2007 when a tanker truck carrying 8,600 gallons of gasoline crashed and burned on a connector ramp between Oakland—San Francisco Bay Bridge and its connecting freeways (in an area called the MacArthur Maze). The ferocious fire was on an elevated connector under another one and it caused the collapse of some 168 ft. (50m) of steel and concrete overpass and the distortion of another 120+ ft. for a total gap of nearly 300 ft. (95m) (see Figure 8.5). Since these connectors carried an estimated 80,000 vehicles per day (with an estimated $6 million per day in economic losses), there was great urgency for a replacement. The California Highway Patrol had just received its first Leica scanners and put them to work scanning the entire damaged section. By mid-morning, the dimensions of the steel girders needed to rebuild were known to within a fraction of an inch and the replacement steel was ordered before the debris had been cleared away. The connector reopened on May 24, 2007, only 26 days after the accident (MTC & Caltrans, 2007; MTC, 2007).

While first used as tools to expedite the documentation of vehicle accident scenes, the scanners found rapid acceptance by public safety agencies in documenting building collapses, large fire and explosion scenes, and large crime scenes. Architects and historians discovered the

Figure 8.5 Three-dimensional laser scan of collapsed Bay Bridge overpass (2007). Data from these scans allowed immediate ordering of replacement beams. Courtesy of Leica Geosystems USA.

high-resolution images were ideal for preserving 3D architectural details (carvings, statues, columns, and even gargoyles) for replacement after earthquakes and fires (or even guiding the restoration of weather-ravaged features). Security agencies found 3D scans of public buildings to be useful in planning security arrangements and threat assessments. Forensic archeologists and historians could document historic or heritage sites as they were excavated with pinpoint accuracy. Historic crime scenes such as Dealey Plaza in Dallas could be documented for full recreations of the John F. Kennedy assassination, as laser scanning has provided critical data in a 2012 reexamination of the 1963crime. It supported accurate reconstruction of sight-lines and bullet trajectories in Dealey Plaza. Combined with extensive new analysis of the ballistics of the ammunition and weapon used and enhanced film recordings, analysts were able to solve many of the questions raised over the years and prove it was a single shooter (Haag and Sturdivant, 2013).

In an extensive evaluation of hazard mitigation in major wildland fires, NIST used 3D laser scanning, not only to aid in preserving the topography of the land and location of the houses in a community affected by a 2007 fire storm, but also the locations of trees and other natural fuels that contributed to the hazards of wind-blown embers (Maranghides et al., 2009).

An accidental fire in a large feed and fertilizer store in West, Texas on April 17, 2013 led to the detonation of an estimated 30 tons of ammonium nitrate fertilizer contaminated with charred wood ash. This explosion leveled numerous buildings and killed 15 people, including 11 firefighters. The debris field covered some 20 acres and the crater was some 93′ across and 30′ deep. The first incident investigators used a Total Station and after 2½ days had a nearly comprehensive mapping of the scene. An assisting agency brought in a Leica ScanStation™ and captured the same data in about less than 1day with a reproducibility of about ±1% (Kistner, 2015).

The Chandra Levy murder case in Washington, D.C. came to a critical juncture when her body was discovered in an overgrown portion of Rock Creek Park a year after her disappearance. The entire scene was laser scanned and then Leica TrueView™ software was used to display and integrate all digital media collected by various agencies, including GPS, Total Station, video, audio, and aerial photos for court display in the murder trial of Ingmar Guandique (Leica Geosystems, 2015).

The investigation and successful prosecution of Marco Topete for the murder of Yolo County Sheriff's Deputy Jose "Tony" Diaz depended upon extensive analysis of multiple bullet trajectories, physical evidence, and dashboard video and audio from the patrol car. The expert's reconstruction was anchored by Leica laser scanning data and was the key to proving that Diaz was ambushed by Topete who was lying in wait and fired 17 shots from an AR-15 assault rifle. Topete was found guilty of the June 2008 murder and sentenced to death (Precision Simulations, 2015).

Laser scanning was also used to document the interior of the large Galleria Shopping Mall in Rocklin California when an arson fire in the store room of a second-story shop involved the concealed utility spaces behind the stores on October 21, 2010. The fire, started by a mentally ill man in an alleged suicide attempt, did $55 million in damages, closing 44 of the 240 stores in the complex. He ultimately pled guilty, but the documentation was useful in damage and repair assessment (Kirk McKinzie, 2015).

One of the highest profile cases to first use laser scanning was the London Underground bombings of July 7, 2005. Scans were done, both within and outside the damaged coaches, while deceased victims were still present to aid in reconstruction of both victim location and device location (Murphy, 2006).

Wildland fires have been successfully documented with ground-based scanners as well as aerial systems. Wildland fires can involve hundreds of thousands of acres and laser scans have been used to map fire perimeters and surviving areas. These systems have been used as part of investigations on such major wildland fires as the Poe fire (northern California), Moonlight fire (Plumas County California), and the Witch and Guerito fires in southern California. Laser scanned images have been used for hypothesis testing in a variety of wildland fire cases. (Curtis, 2015) One example was the 1999 Pendola fire (11,000 acres) with $5.8 million in suppression costs where it was suggested that a 180′ tall tree fell and struck a power line some distance away. The tree in question had been cut into several sections but left at the scene. The individual sections were scanned and the images "reassembled" by a qualified forestry expert to recreate the whole tree. When the image was placed in the scanned scene (on a steep hillside), it was demonstrated that a tree strike was the ignition cause. The utility company settled with the government for $14.75 million without going to trial (PRnews, 2015).

A landmark case in which Leica Scan Station data and the expert witness analysis built upon it was admitted as scientific evidence was a civil proceeding—Bertoli v. City of Sebastopol in 2015 (Precision Simulations, 2015). The case focused on a pedestrian struck by a car in a crosswalk suffering major injuries. The incident appeared to have been caused by shadows of trees next to the otherwise bright roadway making it impossible for a driver to see people in the crosswalk. An independent consulting firm, Precision Simulations, scanned the roadway, buildings, and trees in high resolution. The scene could then be reconstructed for computer animations that included data on vehicle speeds, distances, and rates of deceleration and human factors from outside experts. The documentation of the scene by the police (photos and Total Station data) allowed accurate placement of physical evidence such as skid marks, blood stains, and points of rest for the victim and the vehicle. All of the experts involved collaborated using the 3D model. The case focused on the shadows cast by the trees along the road. The team documented the scene on the anniversary date and time of the incident and then modeled the effects of the shade cast if the trees had been trimmed according to city, county, and state codes. The plaintiff's attorney then argued that the data and its analysis were scientific exhibits (rather than demonstrative exhibits). The accuracy of laser scanning previously was the focus of a 2007 report by the University of California—Davis commissioned by Caltrans (California Dept. of Transportation). (Hiremagalur et al., (2007) This report verified the scientific reliability of Leica laser scan data. After an extensive Frye hearing, the judge ruled that the Leica Scan Station data and its analysis were scientific evidence accepted by the scientific community as reliable and accurate and therefore admissible. (Lyons, 2015)

A recent innovation is to use the point cloud data to produce a scale model with a 3D printer. This is particularly useful in preserving perishable evidence or evidence changed during its recovery or examination. This has been demonstrated in producing both 3D "printed" models and holographic images of a human body in a fire scene before it is recovered.

Legal and ethical issues

In Stephen Cordova v. City of Albuquerque, the U.S. District Court ruled that Leica ScanStation–based evidence met the standard of the Daubert decision for scientific evidence and was admitted as scientific evidence on behalf of the City of Albuquerque (M. Cunningham, 2013). Cordova had sued Albuquerque Police for medical costs when he was shot multiple times. The laser scan-based reconstruction supported the police officers' description of the events. (Cordova v. City of Albuquerque, 2013)

The Superior Court of District of Columbia ruled in a homicide trial that Leica scans of the scene were admissible in demonstrating that a defense witness could not have seen the act of violence due to an obstruction. The Leica scans were used to reconstruct and test the perspective of the purported witness (U.S. v. Leon Robinson and Shanika Robinson, 2014).

Table 8.1 Examples of cases of judicial proceedings where 3D laser scanner data admitted

Citation	*Court*	*Notes*
Stephen Cordova v. City of Albuquerque	U.S. District Court of New Mexico	Daubert Motion—Scan Station data admissible
U.S. v. Seneca Benjamin	Superior Court of District of Columbia	Aid to witness testimony
U.S. v. Leon Robinson and Shanika Robinson	Superior Court of District of Columbia	Demonstrated witness testimony unreliable
U.S. v. Ingmar Guandique	Superior Court of District of Columbia	Chandra Levy case—scene reconstruction
Fyfe v. State of Hawaii	Hawaiian Supreme Court	1999—first judicial admittance of scan data
Pinasco v. State of California	9th Circuit—Court of Appeals	
People v. Marco Topete	Yolo County Superior Court	Homicide of police officer
People v. David Levi Chavez	4th Judicial District Court—New Mexico	
Mitchell v. City and County of San Francisco	San Francisco Superior Court	
State of Tennessee v. Kaylan Bailey	Hamilton County (TN) Criminal Court	
People v. Anthony Wagner	San Bernardino County (CA) Superior Court	
State of Georgia v. Antonio Jerome Greenlee	Decatur County (GA) Superior Court	

Brandon Henslee was found guilty of murdering his half-brother in San Luis Obispo County California in September 2013, based in part on 3D laser scan analysis of the multiple crime scenes.

Joseph Pinasco sued the State of California and two CHP officers for the wrongful death of his son while in police custody in 2009. The reconstruction of the event aided the jury's determination in 2012 where the plaintiff was awarded over $2 million in compensation (Pinasco *et al.* v. State of CA). See Table 8.1 for some additional court cases in which 3D laser scanning was used.

Selection of equipment

Before using a 3D laser scanning system for the collection of reliable scientific data to be used as evidence, the operator must consider the suitability of the system for the applications. Current laser scanning systems vary in their capabilities and limitations and the user has to be aware of these, whether the application is preserving an image of a building or a feature, capturing images and dimensions of an explosion or fire investigation, or capturing data to be used in criminal trials involving lines of sight, environmental conditions, or bullet trajectories. Some designs are limited in their useful range. Capturing fire pattern images inside a small bedroom requires that the system will make accurate measurements on targets as close as 1 m. Wildland fire scenes, explosions, or plane crash sites may require useful ranges out to 200 m or greater. (It should be noted that the maximum range depends on the reflectivity of the target surface—the more laser light that is reflected, the greater the maximum range.) The wavelength of the laser's emission, or lack of optics protection, may make it unsuitable for use in daylight or rainy/foggy conditions. Most manufacturers use an "eye safe" wavelength so a scan can be carried out even when the room is occupied by investigators. Portability, tolerance of temperature extremes, and protection against moisture and dust are important considerations if the system is to be used in exterior situations.

Analysis of fire scenes

Fire investigators today are expected to conduct some sort of analysis of fire dynamics as part of their testing of hypotheses about nearly every fire. These hypotheses may be about the growth of the fire, its flame height, the depth of smoke level, what ventilation limits might be, or whether flashover (full room involvement) could have occurred. This analysis may involve simple mathematical calculations or computer fire models, but they all require measurements of the rooms involved, the ventilation openings, and contents. Such measurements can be made by hand using a tape measure and note pad, but that process is often time consuming and exposes the investigator to physical risks (falling debris, collapsing floors) or possibly hazardous environments. Laser scanning permits the collection of all necessary measurements—dimensions of the room, door, and window openings—and a record of large fuel packages such as chairs, sofas, and beds in just a few minutes. Chris Lautenberger of REAX Engineering has developed a simple subroutine that can input dimensional measurements conveniently into the most commonly used computer fire model—FDS (Fire Dynamic Simulator) developed by NIST. FDS requires extensive dimensional data and information on fuel packages within (Lautenberger, 2015). The REAX program will take the measurements of a fuel package and convert it to a "Lego block" type 3D character that the operator can characterize as to its actual fuel content. As fuel packages and wall-, floor-, and ceiling coverings are characterized (by physical appearance, investigator's notes, or analysis or reference samples), the thermal/or fire performance data on those products can then be added manually by the analyst. Whether doors or windows were open or closed at the time of the fire is also added manually by the analyst. The capture of accurate and complete data via LIDAR means all relevant hypotheses can be tested accurately.

Geospatial Measurement Solutions (GMS, LLC) has developed a similar program for taking point cloud data and putting all into computer fire models for wildland fires (Curtis, 2015). Since topography and fuel distribution play enormously complex roles in wildland fires, their postfire analysis (and prefire threat assessment) require a great deal of data that can really only be collected by 3D scanning. The work of GMS was essential in the previously cited study of the 2007 Witch and Guejito wildland fires in Southern California (Maranghides et al., 2013).

Laser scanning has the unique property of allowing rapid capture of critical data while minimizing exposure of the operator to hazardous or hostile environments. Leica ScanStations have been tested in Afghanistan at China Lake Naval Surface Warfare Center to demonstrate their ability to document the scenes of IED attacks on American forces (Gersbeck, 2008; Grissim and Haag, 2008). A Scan Station on a portable base can be left in seconds and the data transmitted by cable or wirelessly to the data system while the operator remains protected from enemy fire. The unit can be retrieved in a rapid drive-by if needed. The dust- and moisture-resistant properties of the scanner allow its operation in difficult weather. The rotational speed of the scanning head (and therefore the time required for a scan) is determined by the resolution required for the image. The higher the resolution, the slower the scan speed.

Scientific analysis and measurements

While reconstruction and analysis of fire scenes may not require the highest accuracy in distance and direction data, the analysis of shooting scenes does, as a minor variation in angle may mean a shooter could not have been in a position to accomplish a particular shot. Systematic error such as "Range Noise" on measurements of distance from the scanner to a target may mean that targets may be slightly closer or farther away than they actually are. The distances between other points in the scan depend on the angular accuracy of the scanner. Slight discrepancies between true angular displacement and the "measured" value are magnified by the distance. The greater the distance, the greater perceived error will be. Neither the range noise nor angular displacement error are self-canceling so the user must be very careful to allow for such uncertainties while using the data.

Recently the National Commission on Forensic Science has issued a recommendation that all forensic sciences service providers become accredited (NCFS, 2015). Forensic science service providers include crime-scene technicians, fire investigators, and blood pattern analysts, firearms/ballistics examiners, and digital and multimedia examiners. Such accreditation would be expected to include either ISO 17025 (testing) or ISO 17020 (inspection) certification for any instruments and technologies used in these disciplines. It will therefore be important to select equipment that will meet the necessary standards of accuracy. The specifications for 3D laser scanners published by different manufacturers vary in their content. The critical specifications for forensic uses include: accuracy, precision, angle resolution, and range noise (Walsh, 2015). Precision means the reproducibility of a measurement. Accuracy means the difference between a measurement and the true value (and is therefore critical). Accuracy is often reported as "mm/ppm" over the range (typically: 0.4mm rms at 10m, 10.5mm rms @ 50m). Range noise is the uncertainty of distance measurements from scanner to a target. Angular resolution is critical to establishing correct establishment of position. Angular accuracy is often reported in arc-sec (1/3,600 of a degree). Providing a test standard is routine to demonstrate the reliability of any technology. These may be reference samples of known polymers (for infrared spectrometry) or volatile hydrocarbons (for gas chromatographic analysis). In LIDAR data, a measurement standard should be used. Leica has developed a primary (NIST traceable) standard (a metal pole with targets at 1.0m separation ±120 μm) that can be used to certify horizontal and vertical accuracy (at 70m for instance) (NIST, 2013) (see Figure 8.6 for an example).

The inclusion of such a precise standard in every 3D laser scan will be the standard practice under the ISO 17020 guidelines to assure all users of the accuracy of critical data. (Shilling, 2015) The selection of the wrong LIDAR system may mean lack of accreditation under ISO 17025 or ISO 17020 or, more critically, having critical expert testimony excluded on the basis of being based on scientifically uncertain data. This is the difference between demonstrative evidence (which can be excluded relatively easily) and scientific testimony based on provable data. One of the *Daubert* issues that can

Figure 8.6 Leica standard distance reference pole in use at the Bay Bridge collapse. (Remains of the gasoline truck are under the collapse at the left side.) Courtesy of Leica Geosystems USA.

be challenged is the "known error rate" of any data or analysis. If the actual error rate cannot be documented, the court may be inclined to exclude it as scientific evidence. The scanning instrument used therefore must have a published error rate—for scientific data, this would specify a survey-grade laser scanner. Scanning systems that do not offer true accuracy data should be used only for capturing demonstrative evidence. The admissibility of data can then be argued on the basis of whether the data are collected by a scientific instrument or a device without provable accuracy.

Conclusions

Modern 3D laser scanning offers a number of significant benefits to investigations of fires, explosions, structural collapses, and shooting crime scenes. It allows for rapid and accurate documentation of even large scenes. The comprehensive capture of multiple dimensions can be accomplished with accuracy much more quickly than manual methods for both indoor and outdoor scenes. The scanning can be accomplished at the same time as other scene activities. The speed with which scenes can be documented reduces interruptions to traffic or business and can minimize exposure of the operator to hazardous environments. The data gathered enable post scene analysis of dynamics and indicators of many kinds and can be linked to other digital records. Three-dimensional virtual reality recreations allow experts and triers of fact to evaluate both demonstrative and scientific evidence. Data can be easily transferred to computer models for both structures and wildland fires and explosions.

Laser scan data and reconstructions based on it have been accepted in a wide variety of legal cases, both criminal and civil, across the United States. It has been found acceptable by both Frye and Daubert admissibility standards, if appropriate standards of accuracy are met. The proposed federal standards for forensic science service providers (including fire and crime scene investigators and firearms/ballistics analysts) will include requirements that equipment and techniques used will meet ISO 17020 or ISO 17025 standards. The superior accuracy of laser scanning technologies may be interpreted in future courts as required methods over the traditional methods of tape measures and pen. It is clear that 3D laser scanning techniques will become the "gold standard" of scene documentation for fires, explosions, and shooting incident scenes.

Key terms definitions

Virtual reality: A life-like or realistic image created by a computer that has the appearance of three dimensional.

LIDAR: light radar. The use of a light beam to measure distance (sometimes: light detection and ranging).

Total station: An electronic/optical device for measuring distance and position by reflectance of a laser light beam.

Fire: The self-sustaining, rapid oxidation of a fuel producing detectable heat and light; uncontrollable combustion.

Explosion: Sudden release or production of gases at high pressure capable of physical effects.

Accident: An unplanned or unintentional event.

Shooting: The discharge of a firearm.

Firearm: A mechanism used to propel a small projectile by rapid combustion of a fuel.

Documentation: Comprehensive recording of a physical scene including images, dimensions, location, and other features of interest.

Daubert: U.S. Supreme Court Decision on admissibility of scientific expert testing, *Daubert v. Merrell Dow Pharmaceuticals*, 1993.

References

Cordova v. Albuquerque U.S. District Court of the District of New Mexico, 11-CV-806 GBW/ACT, filed 09/30/13.

Cunningham, M., (2013), U.S. Federal Court Issues Daubert Ruling Affirming Scientific Validity of Leica ScanStation Evidence. *Ready Room.* Retrieved from Leica Geosystems website: http://psg.leica-geosystems.us/page/u-s-federal-court-issues-daubert-ruling-affirming-scientific-validity-of-leica-scanstation-evidence/

Curtis, C., (2015) CBC Geospatial Consulting, Inc., [RE: Computer Fire Models for Wildland Fires].

DeHaan, J.D., & Icove, D.J., (2012), *Kirk's Fire Investigation*, 7th Ed. Upper Saddle River, NJ, Brady Publishing.

Gersbeck, T.G., & Grissim, T. (2008), *Advancing the Process of Post Blast Investigation.* Paper presented at the American Academy of Forensic Sciences (AAFS) 60th Anniversary Scientific Meeting, Washington, D.C.

Grissim, T., & Haag, M., (2008,), *Technical Overview and Application of 3D Laser Scanning for Shooting Reconstruction and Crime Scene Investigation.* Paper presented at the American Academy of Forensic Sciences (AAFS) 60th Anniversary Scientific Meeting, Washington, D.C.

Haag, M., & Sturdivant, L., (2013, July 20), *The Ballistic Evidence in the Assassination of John Fitzgerald Kennedy.* Paper presented at the American Academy of Forensic Sciences, Washington, D.C.

Hiremagalur, J., Yen, K.S., Akin, K., Bui, T., Lasky, T.A., &Ravani, B., (2007), *Creating Standards and Specifications for the Use of Laser Scanning in Caltrans Projects.* University of California – Davis, Department of Mechanical and Aeronautical Engineering, Report Number F/CA/RI/2006/46.

Kistner, K., (2015), [Report on the West Texas Explosion].

Komar, D.A., Davy-Jow, S., &Decker, S.J., (2012),The Use of a 3-D Laser Scanner to Document Ephemeral Evidence at Crime Scenes and Postmortem Examinations. *Journal of Forensic Sciences, 57*(1), 188–191.

Lautenberger, C., (2015), [Leica 3D Laser Scan Data as Input to Fire Dynamic Simulator].

Leica Geosystems, (2015).

Leon Robinson and Shanika Robinson, Appellants, v. United States, Appellee., No. 11-CF-1443(District of Columbia Court of Appeals. 2014).

Lyons, W., (2015), California Case Sets Important Precedent for Leica ScanStation Data as Scientific Evidence. Retrieved from http://psg.leica-geosystems.us/page/california-case-sets-important-precedent-for-leica-scanstation-data-as-scientific-evidence/

Maranghides, A., McNamara, D., Mell, W., Trook, J., Toman, B., NIST Standards, S. E. D. Fire Research Division, Engineering Laboratory, (2013), A Case Study of a Community Affected by the Witch and Guejito Fires: Report #2: Evaluating the Effects of Hazard Mitigation Actions on Structure Ignitions. Retrieved from doi:10.6028/NIST.TN1796

McKinzie, K., (2012), New Forensic Fire Scene Documentation and Reconstruction Tools. *New Tools for Litigation, 7.* Retrieved from www.precisionsim.com/files/1314/0356/1728/NL_Vol7_FireSceneDocumentation_8.5X11_Email.pdf

MTC (Metropolitan Transportation Commission), (2007), Retrieved from www.mtc.ca.gov/news/info/2007/freeway_collapse.htm

MTC (Metropolitan Transportation Commission) and Caltrans, (2007), Bay Area Rapid Response to MacArthur Maze Meltdown. Retrieved from www.dot.ca.gov/dist4/mazedamage/docs/mazefactsheet_mtc.pdf

Murphy, P., (2006), *Report into the London Terrorist Attacks on 7 July 2005.* London: Stationary Office.

NCFS (National Commission on Forensic Science), (2015), Retrieved from www.justice.gov/ncfs/work-products

NIST - US Department of Commerce, (2013), Collaboration with Industry Leads to Improved Forensics Work and Industry Growth. Retrieved from www.nist.gov/pml/div683/artifact_calibration.cfm#.Vjq2HTEBd08.mendeley

Pinasco et al. v. State of CA et al. (United States District Court Eastern District of California 2012).

Precision Simulations (2015) 3-D Laser Scanning and Ballistic Trajectory Analysis, www.precisionsim.com/files/9914/0356/1727/NL_Vol.5_Topete

PRNews, (2009) Second Largest Settlement in a Forest Fire Case, Retrieved from www.prnewswire.com/news-releases/second-largest-settlement-in-a-forest-fire-case-62254167.html

Shilling, M., (2015), *Laser Scanner Error Resources*, Presented at NIST International Forensic Symposium, on Forensic Science Error Management, Washington D.C., July 23, 2015.

Walsh, G., (2015), *Measurement Errors with Point Clouds*, Presented at NIST International Symposium on Forensic Science Error Management, Washington D.C., July 22, 2015.

9 Computer fire models

John D. DeHaan

Introduction

Fire modeling of all kinds help us understand complex fire processes, such as the relationship of heat-release rate to other factors, such as ventilation or heat of vaporization. It can help relate postfire indicators to fire events and behaviors. It can help analyze unknown factors in fire—ventilation, ignition location, fuel type—and see what effects each has. This is much easier and cheaper than building life-size models and actually burning them. This process is at the heart of the scientific method—creating (or discerning) alternative hypotheses and testing them. This helps satisfy the court's demand for the scientific method in fire investigation. The "error" rates or, at least, the effect of unknown or unknowable variables on the estimation of fire processes such as size, rate of growth, and likelihood of flashover can be measured in some terms and relayed on to the court. (ASTM, 1980; Babrauskas, 1975)

Fire testing, whether small-scale (typically one-quarter) or full-scale, can reveal important data on temperatures, flame spread, fuel behavior, effects on target surfaces, smoke production, and ignition mechanisms. (ASTM, 2011a) Test results, if reliable, can be used to verify hypotheses about a particular fire or a general category of fires, or to create or support mathematical models.

Testing and modeling can also increase the reliability of fire codes by showing what works and what doesn't in limiting fire growth or smoke movement, detecting fires, or preventing deaths. (Buchanan, 2001, Hunt, 2000) That, in turn, allows fire codes to be more flexible (performance-based, rather than the sometimes arbitrary and erroneously limiting, prescriptive codes) to adapt to new architectural designs or materials. Such analyses can be applied to specific risk analysis issues. (Salley and Kassawar, 2010)

Mitler at NIST (Mitler, 1991) summarized the historical reasons for developing and applying fire modeling as:

- Avoiding repetitious full-scale testing,
- Helping designers and architects,

- Establishing flammability of materials,
- Increasing the flexibility and reliability of fire codes,
- Identifying needed fire research,
- Helping in fire investigations and litigation.

The impact of environmental limitations and financial restrictions on fire testing has made fire modeling a valuable supplement to most fire research programs. Factors such as the placement and combination of fuel items, changes in ventilation, and thickness of materials can be evaluated with fire models, reducing the need for repeated full-scale testing. Fire modeling can also supplement fire-testing programs when used as a screening tool for full-scale testing scenarios. (Hunt, 2000, McGrattan et al., 2014)

Fire models can be defined as a method using mathematical or computer calculations that describes a system or process related to fire development, including fire dynamics, fire spread, occupant exposure, and the effects of fire. The results produced by fire models can be compared with physical and eyewitness evidence to test working hypotheses (Icove, DeHaan and Haynes, 2013).

Mathematical models

There are several categories of mathematical and computer fire models, ranging from simple calculations, to spreadsheets, to complex computer code applications. The simple calculations were based on empirical tests where small fires were observed and data such as flame height, mass loss rate, temperatures, and smoke production rates were measured (Lawson and Quintiere, 1985). Various empirical mathematical relationships were drafted that most closely relate one factor to another.

For example:

Maximum size fire in a room with a single opening

$$\dot{Q}_{max} = 1370\ A_o\ (h_o)^{0.5}$$

where $\dot{Q}_{max}$ = maximum heat-release rate

A_o = area of openings (m^2)
h_o = height of opening (m)

Flame height (away from walls) $L_f = 0.23\dot{Q}^{2/5} - 1.02D$

where L_f = length of flame (m)

$\dot{Q}$ = heat-release rate (kW)
D = equivalent diameter of fuel package (m)

(Icove, DeHaan and Haynes, 2013, p. 85)

Flame height (with nearby walls)

$$L_f = 0.18(k\dot{Q})^{2/5}$$

where k = 1 if fire is away from all walls
k = 2 if fire is near a wall
k = 4 if wall is a corner

(Icove, DeHaan and Haynes, 2013)

Temperature within a flame plume can be calculated from:

$$T_f - T_a = 25\,(\dot{Q}c^{2/5} / (z - z_o))^{5/3}$$

where T_f = temperature of flame at height Z
$\dot{Q}_c$ = heat-release rate in convective portion
$z - z_o$ = height above virtual origin (m)

(Icove, DeHaan and Haynes, 2013, p. 113)

Temperatures within a flame plume or in the hot gas layer, smoke filling rates, rates of spread for a ceiling jet, radiant or convective heat flux (intensity) from a fire of given heat-release rate (heat output), and ventilation limits can all be predicted with relative accuracy from such equations (using only a pocket calculator with scientific notation). The computer calculation of compartment flashover was one of the first fire behaviors to be published. (Babrauskas, 1975)

When based on sound mathematical relationships and properly applied, such mathematical models can often assure that the scientific method has been satisfied and give the investigator reliable data on which to test hypotheses.

Spreadsheet models

The mathematical calculations described earlier are often incorporated into spreadsheet applications. One of the first such applications was created by Harold E. "Bud" Nelson (Nelson, 1990) as FPETOOL. It included a variety of useful fire dynamics calculations applicable to the analysis of simple fuel package—and compartment fires. It was updated and improved with a Technical Reference Guide by Deal. (Deal, 1995) FPETool consists of three main elements:

Fireform (Fire Formulas)—a collection of fire safety calculations (spreadsheets).

Makefire—a series of procedures to produce fire input data files for use with Fire Simulator.

Fire Simulator—an integrated set of equations (i.e., a model) designed to allow the user to create a fire case study in a Lotus format with specifications of: room and vent dimensions; fuel characteristics; ceiling, wall and floor materials; input fire to predict layer temperature, flashover, and tenability factors.

Because FPETOOL includes some simple fire dynamics routines, it is often considered to be a zone model.

Prof. Fred Mowrer, then of the University of Maryland, Department of Fire Protection Engineering, created a system of spreadsheet calculations that concentrated on the specifics of compartment fire analysis and included extensive property tables for typical materials first ignited (Mowrer, 1992; Milke and Mowrer, 2001, Mowrer, 2002) (Table 9.1).

Table 9.1 Mowrer's fire risk forum spreadsheets

Template	*Spreadsheet description*
ATRIATMP	Estimates the approximate average temperature rise in the hot gas layer that develops in a large open space such as an atrium
BUOYHEAD	Estimates the pressure differential, gas velocity, and unit mass flow rate caused by the buoyancy of the hot gases beneath a ceiling
BURNRATE	Estimates the burning rate history of a flammable liquid fire
CJTEMP	Estimates the temperature rise in an unconfined ceiling jet
DETACT	Estimates the response time of ceiling-mounted fire detectors
FLAMSPRED	Estimates the lateral flame spread rates on solid materials
FLASHOVR	Estimates the heat-release rate needed to cause flashover in a compartment with a single rectangular wall vent
FUELDATA	Contains thermophysical and burning rate data for common fuels
GASCONQS	Estimates gas species concentrations
IGNTIME	Estimates the time to ignite a thermally thick solid exposed to a constant heat flux
LAYDSCNT	Estimate the smoke layer interface position in a closed room due to entertainment
LAYERTMP	Estimates the average hot gas layer temperature in an enclosure with a single rectangular opening
MASSBAL	Estimates the mass flow rate through an enclosure with a single wall opening
MECHVENT	Estimates the fire conditions in a mechanically ventilated space without natural ventilation
PLUMEFIL	Estimates the volumetric rate of smoke flow in a fire plume
PLUMETMP	Estimates the temperature rise in an axisymmetric fire plume
RADIGN	Estimates the potential for radiant ignition of a combustible target
STACK	Estimates the mass flow rate through an enclosure
TEMPRISE	Estimates the average temperature rise in a closed room
THERMPRP	Contains thermal property data for 15 types of boundary materials

Sources: Mowrer, F.W., 2003, Spreadsheet Templates for Fire Dynamics Calculations, University of Maryland, Department of Fire Protection Engineering. Spreadsheet templates and documentation are posted on the ATF website: ATF_SS.xlsx.

Table 9.2 NRC fire dynamics spreadsheets

Chapter	*FDTs spreadsheet description*
2	Predicting hot gas layer temperature and smoke layer height in a room fire with natural ventilation
	Predicting hot gas layer temperature in a room fire with forced ventilation
	Predicting hot gas layer temperature in a room fire with door closed
3	Estimating burning characteristics of liquid pool fire, heat-release rate, burning duration, and flame height
4	Estimating wall fire flame height
5	Estimating radiant heat flux from fire to a target fuel at ground level under wind-free condition point source radiation model
	Estimating radiant heat flux from fire to a target fuel at ground level in presence of wind (tilted flame) solid flame radiation model
	Estimating thermal radiation from hydrocarbon fireballs
6	Estimating the ignition time of a target fuel exposed to a constant radiative heat flux
7	Estimating the full-scale heat-release rate of a cable tray fire
8	Estimating burning duration of solid combustibles
9	Estimating centerline temperature of buoyant fire plume
10	Estimating sprinkler response time
13	Calculating fire severity
14	Estimating pressure rise due to a fire in a closed compartment
15	Estimating pressure increase and explosive energy release associated with explosions
16	Calculating the rate of hydrogen gas generation in battery rooms
17	Estimating thickness of fire protection spray-applied coating for structural steel beams (substitution correlation)
	Estimating fire resistance time of steel beams protected by fire protection insulation (quasi-steady-state approach)
	Estimating fire resistance time of unprotected steel beams (quasi-steady-state approach)
18	Estimating visibility through smoke

Source: Iqbal and Salley (2004).

These templates, often referred to as FiREDSHEETS, incorporate a number of enclosure fire dynamics calculations used by FPETOOL (Nelson, 1990) and other suites.

A more industry-oriented set of spreadsheets was developed for the U.S. Nuclear Regulatory Commission fire protection inspection program. This was published as the *Fire Dynamics Tools* (Iqbal and Salley, 2004). Table 9.2 lists the spreadsheets included in it. These routines include pull-down menus of important values such as thermal conductivity, density, and ignition properties, and will convert SI units to English and vice versa.

Beyond the basic calculations, computer-based fire models, then, have come to include eight major overlapping categories, as shown in Table 9.3.

Table 9.3 Classes of computer fire models and commonly cited examples

Class of model	*Description*	*Example(s)*
Spreadsheet	Calculates mathematical solutions for interpretations of actual case data	FiREDSHEETS, NRC spreadsheets (Fire Dynamics Tools)
Zone	Calculates fire environment through two homogenous zones	FPETool[a], CFAST, ASET-B, BRANZFIRE, FireMD
Field	Calculates fire environment by solving conservation equations, usually with finite-elements mathematics	FDS, JASMINE, FLOW3D, SMARTFIRE, PHOENICS, SOFIE, fireFOAM
Postflashover	Calculates time-temperature history for energy, mass, and species and is useful in evaluating structural integrity in fire exposure	COMPF2, Ozone, SFIRE-4
Fire-protection performance	Calculates sprinkler and detector response times for specific fire exposures based on the response time index (RTI)	DETACT-QS, DETECT-T2, LAVENT
Thermal and structural response	Calculates structural fire endurance of a building using finite-element calculations	FIRES-T3, HEATING7, TASEF
Smoke movement	Calculates the dispersion of smoke and gaseous species	CONTAM96, Airnet, MFIRE
Egress	Calculates the evacuation times using stochastic modeling using smoke conditions, occupants, and egress variables	Allsafe, building EXODUS, EESCAPE, ELVAC, EVACNET, EXIT89, EXITT, EVACS, EXITT, Simplex, SIMULEX, WAYOUT

[a] FPETOOL is a collection of analytical tools about fire behavior and properties with simplifications as assumptions to make approximations rather than exact predictions using portable or desktop computers.

Sources: Updated from Bailey (2006), Friedman (1992), and Hunt (2000).

Zone models

Two-zone models are based on the concept that a fire in a room or enclosure can be described in terms of two unique zones where the conditions within each zone can be predicted. The two zones refer to two individual control volumes known as the upper and lower zone or layer. The upper zone of heated gases and by-products of combustion is penetrated only by the fire plume. These hot gases and smoke fill the ceiling layer, which then slowly

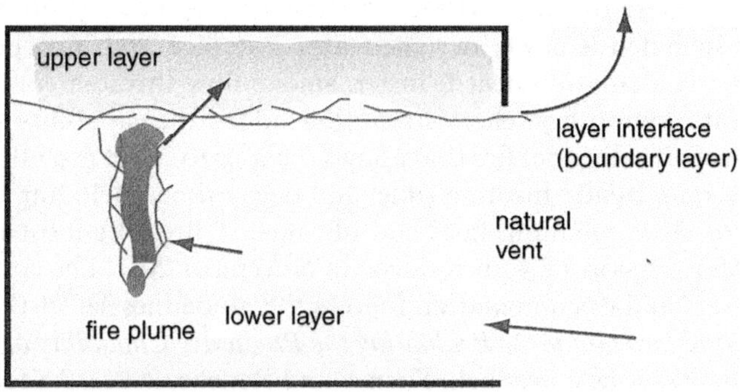

Figure 9.1 Schematic representation of a two-zone model of a room fire. Courtesy of NIST, from Forney and Moss, 1992.

descends. A boundary layer marks the intersection of the upper and lower areas. The only exchange represented in most models is between the upper and lower layers is due to the action of the fire plume acting as a pump.

Zone models use a set of differential equations for each zone solving for pressure, temperature, carbon dioxide, oxygen, water, and soot production. The accuracy of these calculations varies with not only the model but also the reliability of the input data describing the circumstances and room dimensions and, very importantly, the assumptions as to the heat-release rate of the materials burned in the fire. Complex two-zone models of up to 10 rooms typically take a few minutes to run using multi-core- class computer workstations or laptops. Figure 9.1 shows a schematic representation of the concepts behind a two-zone model.

A survey published in 1992 by the SFPE showed ASET-B and DETACT-QS to be the most widely used zone models at the time (Friedman, 1992). Today, CFAST, the Consolidated Model of Fire Growth and Smoke Transport, has become the most widely used zone model. CFAST combines the engineering calculations from the FIREFORM model and the user interface from FASTlite in a unique, comprehensive, zone modeling, and fire analysis tool (Peacock et al., 2015).

CFAST overview

The CFAST two-zone model is a heat-and mass-balance model based on not only the physics and chemistry but also the results of experimental and visual observations in actual fires. CFAST computes:

- Production of enthalpy (heat energy) and mass by burning objects;
- Buoyancy and forced transport of enthalpy and mass through horizontal and vertical vents;
- Temperatures, smoke optical densities, and species concentrations.

CFAST has been in use since its first public release in June 1990, with many enhancements over the years (Peacock, Reneke and Forney, 2015). CFAST version 7.0 included a newly designed user interface and incorporates improvements in calculating heat transfer, smoke flow through corridors, and more accurate combustion chemistry. NIST added the capability to define a general t^2 growth rate input fire that allows the user to select growth rate, peak heat-release rate, steady burning time, and decay time, including predefined constants for slow, medium, fast, and ultrafast t^2 fires. Its most recent version is CFAST Version 7.2.4, introduced in November 2017. The latest CFAST guide for users and documentation follows the guidelines set forth in *ASTM E1355-11: Standard Guide for Evaluating the Predictive Capability of Deterministic Fire Models* (Jones, Peacock, Forney and Reneke, 2005), ASTM, 2011b).

Because CFAST does not have a pyrolysis model to predict fire growth, a fuel source is entered or described by its fire signature (heat-release rate curve). The program converts this fuel information into two characteristics: enthalpy (heat) and mass. In an unconstrained (free-burning) fire, the burning of this fuel takes place with the fire plume. In constrained (oxygen-limited) fires, the pyrolyzed fuel may burn in the fire plume where there is sufficient oxygen, but it may also burn in the fire plume where there is sufficient oxygen, but it may also burn in the upper or lower layer of the room of fire origin, in the plume in the doorway leading to an adjoining room, or even in the layers or plumes in adjacent rooms. The numerical limits of data used in CFAST v. 6.2 are noted in Table 9.4.

Fire plumes and layers

As in previous mathematical models, the CFAST model instructs the fire plume to act as a "pump" to move heat energy and mass into the upper layer from the

Table 9.4 Summary of numerical limits in the software implementation of the CFAST model, version 6.2

Feature	*Maximum*
Simulation time (s)	86,400
Compartments	30
Object fires	31
Fire definitions in database file	30
Material thermal property definitions	125
Slabs in a single surface material	3
Fans in mechanical ventilation systems	5
Ducts in mechanical ventilation systems	60
Connections in compartments and mechanical ventilation systems	62
Independent mechanical ventilation systems	15
Targets	90
Data points in a history or spreadsheet file	900

Source: Peacock et al. (2005).

lower layer. Mixed with this transfer are inflows and outflows from horizontal or vertical vents (doors, windows, etc.), which are modeled as plumes.

Some assumptions of the model do not always hold in real fires. Some mixing of the upper and lower layers does take place at their interfaces. At cool wall surfaces, gases will flow downward as they lose heat and buoyancy. Also, heating and air-conditioning systems will cause mixing between the layers.

Horizontal flow of the fire from one room to the next occurs when the upper layer descends below the opening of an open vent. As the upper layer descends, pressure differentials between the connected rooms may cause air to flow into and out of a room or compartment in the opposite direction, resulting in the two flow conditions. Upward vertical flow may occur when the roof or ceiling of a room is opened.

Heat transfer

In the CFAST model, unique material properties of up to three layers can be defined for surfaces of the room (ceiling, walls, and flooring). This design consideration is useful, since in CFAST, heat transfers to the surfaces via convective heat transfer and through the surfaces via conductive heat transfer. Radiative heat transfer takes place among fire plumes, gas layers, and surfaces. In radiative heat transfer, emissivity is dominated primarily by species contributions (smoke, carbon dioxide, and water) within the gas layers. CFAST applies a combustion chemistry scheme that balances carbon, hydrogen, and oxygen in the room of fire origin among the lower-layer portion of the burning fire plume, the upper layer, and air entrained in the lower layer that is absorbed into the upper layer of the next connecting room.

Limitations

Zone models have their limitations. Six specific aspects of physics and chemistry either are not included or have very limited implementations in zone models: flame spread, heat-release rate, fire chemistry, smoke chemistry, realistic layer mixing, and suppression (Babrauskas, 1996). Some errors in species concentrations can result in errors in the distribution of enthalpy among the layers, a phenomenon that affects the accuracy temperatures, partly because heat losses due to window radiation are not presently incorporated in the model.

Even with these known limitations, zone models have been extremely successful in forensic fire reconstruction and litigation (Bukowski, 1991; DeWitt and Goff, 2000). The FIRE SIMULATOR program of FPETOOL was found to show good correlation with full-scale fire experiments. (Vettori and Madrzykowski, 2000) CFAST zone models have been successfully introduced in federal court litigation. These successes are a result of their correct applications by knowledgeable scientists and engineers, continued revalidation in actual fire testing, and applied research efforts, include verification and validation studies by the NRC (Salley and Kassawar, 2007a, 2007b, 2007c, 2007d, 2007e, 2007f, 2007g).

Field models

Field Models, the newest and most sophisticated of all deterministic models, rely on computational fluid dynamics (CFD) technology. These models are attractive, especially as a tool in litigation support and fire scene reconstruction, owing their ability to display the impact of information visually in three dimensions.

However, the downside of field models is that creating input data is typically very time- consuming and often requires powerful computer workstations to compute and display the results. The computation time to run field models may be days or weeks depending on the complexity of the model and the relative computational power of the computer system available.

Computational fluid dynamics models

CFD models estimate the fire environment by dividing the compartment into uniform small cells, or control volumes, instead of two zones. The program then simultaneously solves the conservation equations for combustion, radiation, and mass transport to and from each cell surface. The present shortcomings include the level of background and training needed to codify a model, particularly for complex layouts.

There are many advantages to using CFD models over zone models such as CFAST. The higher geometric resolution of CFD models makes the solutions more refined. Higher-speed laptops and workstations today allow CFD models to be run, whereas in the past, larger mainframe and minicomputers were required. Multiple workstations have been successfully linked together in parallel processing arrays to perform complex CFD calculations more quickly.

Because CFD models are used in companion areas such as fluid flow, combustion, and heat transfer, their underlying technology is generally more accepted over simpler and coarser models. CFD has a long history of use in forensic fire scene reconstructions and is gaining in popularity owing to commercially available and intuitive graphical interfaces (e.g., Smokeview (Forney, 2015) and PyroSim [Thunderhead, 2010, 2011]). CFD programs now outpace the capabilities of the zone models, which, for the most part, operate under maintenance modes only. Moreover, CFD programs such as the FDS make the technology economically feasible owing to its wide use, testing, and growing acceptance.

NIST CFD technology-based models

The recommended field model using CFD technology, owing to its availability at no cost, focus of research, and acceptance in the community is the Fire Dynamics Simulator (FDS). This model is available from NIST's Building and Fire Research Laboratory (McGrattan et al., 2010). The first version of the FDS was publicly released in February 2000 and has become a mainstay of the fire research and forensic communities. Its latest version is FDS 6.7.0, introduced in June 2018.

FDS is a fire-driven fluid flow model and numerically solves a form of the Navier-Stokes equations for thermally driven smoke and heat transport. (McGill, 2003; McGrattan et al., 2000, 2010) The companion visual program to FDS is SmokeView, which is a graphical interface that produces a variety of visual records of FDS calculations. These include concentration of species, temperature, heat fluxes, and temperatures of surfaces or on "slices" through the compartment in various animations. User guides for both FDS 6.3.0 (McGrattan et al., 2015), later versions, and SmokeView (Forney, 2010) are available at NIST's website, fire.nist.gov/.

Traditional use of FDS to date has been divided among the evaluation of smoke-handling systems, sprinkler and detector activation, and fire reconstructions. It is also being used to study fundamental fire dynamics problems encountered in both academic and industrial settings. It should be noted that heat and mass transfer in flash fires and gas/vapor deflagrations do not follow the buoyancy-driven flows used in FDS. Such events are not reliably modeled with FDS.

FDS, like other CFD models, breaks up the room (or fire area) into cells and calculates the mass heat energy and species transport (fluxes) into and out of each of the six faces of each cell (as seen in Figure 9.2). The smaller the cells are, the more sensitive the model is to movement and heat transfer, and therefore, the more realistic the model's predictions, but the smaller the cells the larger the number in any given space and the greater the computer power needed to track the mass and energy transfer from all of them. Unfortunately, FDS uses a rectilinear grid to describe the computational cells. This has made it slightly difficult to model sloping roofs, rounded tunnels, and curved walls without some form of approximation, however, the newest versions allow "sculpting" of both enclosures and fuel package shapes. The boundary conditions for material surfaces consists of assigned thermal constants that include information on their burning behavior.

Other features include several visualization modes such as tracer particle flow, animated contour slices of computed variables such as vector heat flux plots, animated surface data, and multiple isocontours. New visualization tools allow the user to compare fire burn patterns with computed uniform surface contour planes and temperatures and heat fluxes on target surfaces in the room. A typical Smokeview image is shown in Figure 9.3.

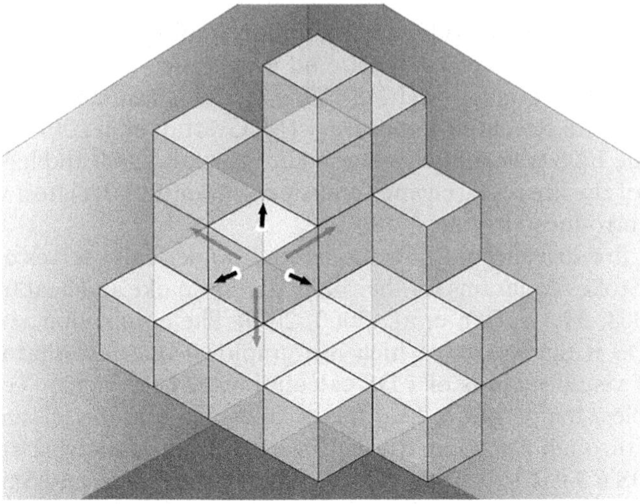

Figure 9.2 In CFD modeling, the interior volume of the fire compartment is broken up into rectilinear "control volumes" of a set dimension. The mass and heat transfer into and out of each volume (through all six of its surfaces) is then calculated repeatedly throughout the room. Newest versions of FDS allow for irregular, sloped volume surfaces more closely simulating real-world contours. Courtesy of Muffi Grinnell, MG Associates.

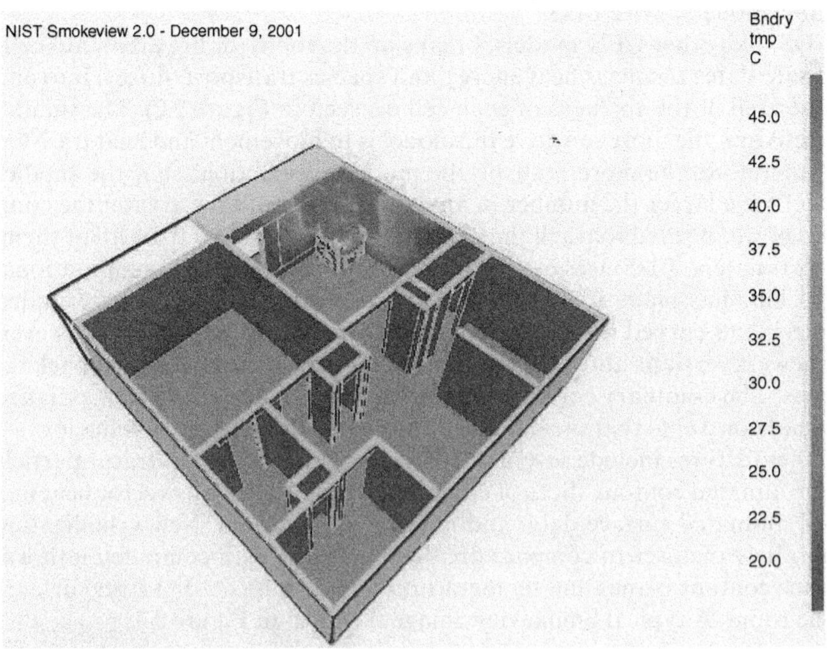

Figure 9.3 Typical SmokeView visualization of a fire in a multiroom occupancy showing heat flux impacts on walls adjacent to the purported area of origin (a recliner). Courtesy of NIST.

FM Global CFD fire modeling

FM Global has always served as a leader in the development of many of the fire protection engineering tools, fire dynamics relationships, fire growth and behavior models, and testing protocols. FM Global maintains a unique center for engaging in property loss-prevention strategies, scientific research, and product testing at their Research Campus in West Gloucester, Rhode Island (Bill and Dorofeev, 2011).

FM Global is presently developing a number of key physical models relating to fluid mechanics, heat transfer, and combustion into a new open-source software package called fireFOAM. This program is a CFD "toolbox" for fire and explosion modeling applications. To promote development and outside cooperation, FM Global has released its modeling technologies as an open source (openFOAM, 2012).

FireFOAM utilizes the finite volume method on arbitrarily unstructured meshes, and is highly scalable on massively parallel computers. The capabilities of fireFOAM include many complex simulations, including fire growth and suppression. For more information about fireFOAM, please refer to FM Global's website: www.fmglobal.com (openFOAM, 2012).

FLACS CFD model

The commercial FLACS CFD model, maintained by CMR GexCon, has historically been used for gas dispersion modeling and explosion calculations within many industries (GexCon, 2012). GexCon emphasizes the need for validation of CFD models such as FLACS. The software has many applications, including: dispersion of flammable gases and vapors (pre-blast), quantitative risk assessments, investigations into explosion venting of coupled and nonstandard vessels, predicting the effects of explosions, and modeling toxic gas dispersion.

FLACS has been successfully used in hypothesis testing during large industrial explosion investigations. One of the most impressive uses of FLACS documented by GexCon US, Bethesda, Maryland, was in their analysis of the November 22, 2006, explosion and fire at an ink and paint manufacturing facility explosion in Danvers, Massachusetts (Davis et al., 2010). Using FLACS made it possible to explore the chain of events leading to the explosion as well as to evaluate potential ignition sources within the facility. The investigators also compared the calculated overpressures of the exploding fuel/air cloud with observed internal and external blast damage, demonstrating how CFD tools can provide invaluable analyses for explosion investigations.

Wildland fire models

The most widely used model for wildland fire analysis and prediction is Behave Plus (Andrews, Devins and Seli, 2003; Andrews, 2013). Behave Plus is a Windows™-based computer program, supported today by the Fire Research and Management Exchange System at the Fire, Fuel and Smoke Science Program of the U.S. Forest Service at Missoula, Montana (FRAMES, 2015). It originated in 2002 as a revision of the original "Behave" program from 1984. Its version 5 was released in 2009, with annual updates since. It is described as a collection of mathematical models that describe fire behavior, fire effects, and the fire environment. The program simulates rate of fire spread, spotting distance, search height, tree mortality, fuel moisture, wind adjustment factor, and many other fire behaviors and effects, so it is commonly used to predict fire behavior in several situations (https://frames.gov/partner-sites, 11/4/15).

While primarily intended for use in wildland firefighting (predicting fire spread), it has also been used for planning abatement (treatment) of forest areas, assessing fuel hazards, and understanding fire behavior. While not intended for investigation of wildland fires, it has been used for hypothesis testing of possible ignition locations and spread factors. It has not been specifically validated for origin and cause determination applications. The Rocky Mountain Research Station also maintains other wildland fire computer models such as FARSITE (used to compute fire growth and behavior for long time periods under different conditions of weather, fuels and terrain), and a variety of hazard evaluation programs (FRAMES, 2015) (http://firelab.org/applications).

Validation

All fire models must be *validated* by comparing their predictions against the results of real fire tests if their results are to be relied on. This testing sometimes reveals flaws or blind spots—where some types of enclosures or conditions can produce predicted values that differ from observed fire behavior. If a model has not been shown to yield valid results for a particular type of fire, its predictions in an unknown scenario should not be accepted without question. For instance, FDS has been shown to be accurate in predicting growth of fires in large compartments, but data on how FDS predictions relate to real fires in very small compartments are just now being gathered and published.

It should be noted that FDS was derived from earlier work on modeling the growth and movement of smoke plumes from large stationary fires such as oil pools. This work, based on large eddy simulation, focused on atmospheric interaction with the smoke plumes and not on movement of the fire itself. More recently, FDS has been applied to studying wildland-urban interface (WUI) fires, considering the impact of vegetation fires on nearby structures. These developments depend on validation by comparison with actual incidents, since planned, real-world tests of such fires would be prohibitively expensive and dangerous. Some limited fire tests have been carried out, however.

Unfortunately, each year in the United States there are numerous opportunities to collect data from WUI fires *after* they occur (Maranghides et al., 2015). Because FDS predicts fire growth and spread based on the thermal characteristics of solid fuel surfaces, when fuel masses are porous (i.e., where fire can spread quickly *through* them as well as across them), the physics of heat transfer and ignition for that growth is too complex for FDS to accurately model such fires. The effects of wind-driven fire spread through such porous arrays complicate the process even further. Wildland fires in heavy brush or timber are almost always "crown fires" in their most destructive phase; that is, the fire spreads upward and through porous fuel arrays like leaves or needles, often driven by wind—either atmospheric (external) wind or fire-induced drafts. Although FDS has been validated against tests on flat grasslands (no vertical spread factors) with wind-aided spread only, it has not been shown to give accurate predictions for crown fires through heavier brush or timber.

The Dalmarnock tests

In 2006, the Dalmarnock fire tests were conducted in the United Kingdom, coordinated by the BRE Center for Fire Safety Engineering, University of Edinburgh. The tests consisted of a series of large-scale fire experiments conducted in a high-rise, 23-story building located in Dalmarnock, Glasgow.

Prior to the tests, seven teams were given mutual access to the available information on the compartment geometry, fuel packages, ignition source, and ventilation conditions of fire. The teams attempted to build models that would predict the fire scenarios. The primary intent of the tests was to see how accurate the teams' predictions might be. Unfortunately, their results showed a wide arrange of predictions, emphasizing the difficulties of modeling complex fire scenarios.

Rein et al. (2009) concluded that there was an inherent difficulty in modeling fire dynamics in complex fire scenarios like Dalmarnock, and the modelers' ability to accurately predict fire growth was poor. Several studies and one textbook describe these results in detail (Abecassis-Empis et al., 2008; BRE Center, 2012; Rein, Abecassis-Empis, and Carvel, 2007; Rein et al., 2009), Rein, John, and Torero, 2011).

However, it is interesting that a previous study by Rein et al. (2006) reported that when a combination of a first order, a zone, and a field model was used, the combined results of these three modeling approaches were in relatively good agreement, particularly in the early stages of fire. The researchers' work also indicated that this approach to modeling can be used as a first step toward confirming at least the order of magnitude of the results from more complex models.

Review of these tests, however, noted that the ignition sequence and actual maximum heat-release rate of major fuel package was not adequately considered in the room preparation and documentation supplied to the modelers. This led to some major differences between the temperature predictions and the actual fires. This points up the necessity of complete documentation. When those properties were well documented, modeling results much more closely fit the real fire.

Impact of verification and validation studies

Recent work sponsored in part by the NRC examined the **verification and validation (V&V)** of fire models, including CFAST and FDS (Salley and Kassawar, 2007a, 2007b, 2007c, 2007d, 2007e, 2007f, 2007g). Although the NRC's study centered on fire hazards specific to nuclear power plants, it addressed many of the concerns raised by those who think fire models are not accurate or appropriate for forensic fire scene reconstruction, such as in the previously mentioned Dalmarnock tests.

These concerns include the ability of the models to accurately predict common features of fires, such as upper-layer temperatures and heat fluxes. One of the features of this report was a comparison between actual fire test results and predictions of hand calculations, zone models, and field models. As shown in Figure 9.4 and in other studies, when the models are applied correctly, there is generally good agreement among them and the variability of real-world fires. The V&V reports are presented in seven publications. Volume 1 (Salley and Kassawar, 2007g), the main report, provides general background information, programmatic and technical overviews, and project insights and conclusions. Volumes 2 through 6 provide detailed discussions of the V&V of the Fire Dynamics Tools (FDTs) (Salley and Kassawar, 2007d); Fire-Induced Vulnerability Evaluation, Revision 1 (FIVE-Revl) (Salley and Kassawar, 2007e); Consolidated Model of Fire Growth and Smoke Transport (CFAST) (Salley and Kassawar, 2007a); MAGIC (Salley

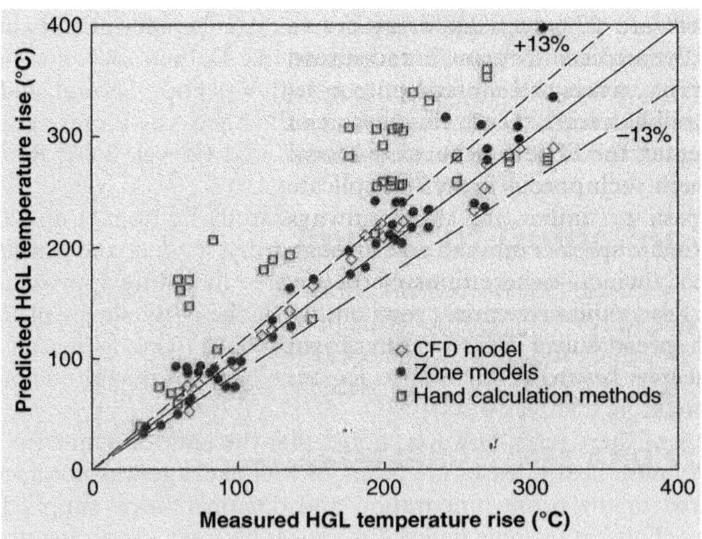

Figure 9.4 Comparison of hot gas layer (HGL) temperatures measured in full-scale tests against predictions from hand calculations, zone models, and FDS models. Courtesy of Nuclear Regulatory Commission.

and Kassawar, 2007f); and the Fire Dynamics Simulator (FDS) (Salley and Kassawar, 2007c). Volume 7 discusses in detail the uncertainty of the experiments used in the V&V study of these five fire models (Salley and Kassawar, 2007b).

Hypothesis testing

A number of studies were carried out by Carman that involved observations and data from real structure fires, full-scale fire tests, and the application of FDS to test hypotheses about the growth of fires, ventilation effects, and the production of postfire, physical fire patterns (Carman, 2009, 2010, 2011, 2013). These studies really exemplified the ideal application of advanced computer modeling to fire investigations. The unusual fire patterns observed at the scenes were replicated under controlled fire test conditions from which accurate data could be gathered (fuels, ventilation sources, and physical position). These data were used in hypothesis testing that resulted in supportable FDS simulations as conclusions as to the origin and development of these fires.

Using computer models

It is particularly important to point out to potential users of fire models that there are existing ASTM guides that deal with critical computer modeling issues. The most relevant to this discussion are:

ASTM E1355-11: Standard Guide for Evaluating the Predictive Capability of Deterministic Fire Models. Evaluates the predictive capacity of fire models by defining scenarios, validating assumptions, verifying the mathematical underpinnings of the model, and evaluating its accuracy (ASTM, 2011b). See also ISO 16730 (ISO, 2008).

ASTM E1591-07: Standard Guide for Obtaining Data for ASTM Deterministic Fire Models. Covers and documents available literature and data that are beneficial to modelers (ASTM, 2007b).

ASTM E1895-07: Standard Guide for Determining Uses and Limitations of Deterministic Fire Models. Examines the uses and limitations of fire models and addresses how to choose the most appropriate model for the situation (ASTM, 2007c). Note: Withdrawn in 2011.

ASTM E1472-07: Guide for Documenting Computer Software for Fire Models. Describes how a fire model is to be documented and includes a user's manual, a programmer's guide, mathematical routines, and installation and operation of the software (ASTM, 2007a). Note: Withdrawn in 2010.

All of these contain vital information and recommendations for someone contemplating the use of a fire model. As they are published by ASTM (and produced by Committee E-5 on Fire Standards), they represent peer-reviewed guides with which all fire model users should be familiar. They outline the documentation (of both the scene and the model) necessary for demonstrating the reliability of models used. They are intended for evaluation of zone models but are applicable to field models as well.

SFPE guidelines and standards

SFPE recently published *Guidelines for Substantiating a Fire Model for a Given Application*, Engineering Guide (SFPE, 2011). This guide supplements the ASTM standards and assists the user of a fire model in defining the problem, selecting a candidate model, interpreting the verification and validation studies for various models, and understanding user effects. An appendix covers fire-related phenomena with guidance as to application to specific models and interpreting the underlying key physics.

What Should Be Asked About Any Model Before Use?

- Is it applicable?
- Is it the right tool for the job?
- Does it give accurate results?
- How often does it predict events that do not occur in real fires?
- How sensitive is it to changes in input?
- What is its error rate?
- Has it been used to predict events in real fire tests?
- Has it been validated?
- Where did it come from?
- Where was it published?
- What supporting (or contradictory) data has been published?

If computer fire models of any kind are to be used, it is quickly realized that the input data they depend on are usually a lot more extensive than most fire investigators are used to gathering. This paucity of data is more often than not the result of careless or incomplete documentation of the scene. This has been recently addressed by the inclusion of a recommended data collection form (see Figure 9.5) (DeHaan and Icove, 2012; Icove, DeHaan and Haynes, 2013). It is a rare scene that is so completely destroyed that basic dimensions, structural and finish materials, and furnishings (type and placement) cannot be established by careful examination. Even in such instances, interviews, examination of nearby "exemplar" structures, or recovery of prefire photos or videos, can often fill in many of the missing pieces.

Performance-based fire codes

One concern for future legal proceedings is for the accuracy of the predictions of computer models for fire performance and fire prevention in new

ROOM FIRE DATA

Room ______________________________ **Room #** ________

Length ___________

Width ___________

Height ___________

Note ceiling height changes _____________

Floor Plan

Walls: Structure/material ____________ Thickness _____ Covering _________ Sample? Y/N

Structure/material ____________ Thickness _____ Covering _________ Sample? Y/N

Ceiling: Structure/material ____________ Thickness _____ Covering _________ Sample? Y/N

Floor: Structure/material ____________ Thickness _____ Covering _________ Sample? Y/N

Openings (door, window, other vents) into room (number on plan above):

Height (bottom to top of opening)	Sill Height	Soffit Depth (above opening)	Width	Open or Closed? Changes During Fire?
1.				
2.				
3.				
4.				
5.				
6.				
7.				

HVAC System:

Description ____

On/off prior to/during fire? ____________________

Furnishings (descriptions of major fuel items, including floor and wall coverings, draperies):

____

____

____

____

Time Line: Alarm time __________ FD arr. Time __________ Control time __________

Detection: ____

Pre-Fire Events: ____

Figure 9.5 Room fire data needed for accurate computer models. Courtesy of John D. DeHaan.

buildings with unusual features of structure or materials. The last decade or so has seen a worldwide revolution in building large structures with odd shapes and novel materials. These do not resemble the more traditional buildings whose fire prevention systems were designed by adherence to proscriptive building and fire codes (and some of those were tested by real fires and found wanting). How do people know if the new systems will work properly if they have never been installed in these odd new structures, let alone tested in even a scale test fire? That's putting a lot of faith in the computer models. What will happen if they are found wrong and people die as a result?

Assessment

As outlined in E1895, for instance, the user's first step should be to define the scope of the fire assessment and then determine if fire modeling is an appropriate tool (ASTM, 2007c). Then, the user should determine what models are available and are suitable to run on the available computer hardware considering the size and complexity of the problem. For the models being considered the available documentation should be acquired and evaluated in terms of guidance offered in E1472 (ASTM, 2007a), *Guide for Documenting Computer Software for Fire Models*. The limitations of the candidate models must be compared to the problem to be solved—one-room v. multi-room, preflashover v. postflashover.

While it is possible for existing models to be modified to deal with particular problems, any modifications must be made in cooperation with the original model developer and then subjected to suitable validation as outlined in E1355. Other tools, such as small-or large-scale fire tests or mathematical calculations, should be considered as well.

Once a model is selected, the following steps are recommended (Janssens and Birk, 2000):

1. Verify the known limitations of the model—room dimensions, fire size, or ventilation.
2. Determine the underlying assumptions (two-layer zone or CFD/field model) and assess their impact on the results.
3. Determine the characteristic variables.
4. Determine what input data are required and where it can be obtained.
5. Determine the rigor of the mathematics involved and check to see it will give an answer given the constraints of the problem.
6. Determine the extent of validation to establish its appropriateness for the problem. Validation processes are described in E1355.
7. If validation data are not available, sensitivity analysis must be conducted to establish the effect of changing critical variables.
8. Thoroughly document the model "run," including all input data, all assumptions made, and any and all modifications (including validation to support the accuracy of those modifications).

The documentation for a fire model should include a technical guide or user's manual (as described in E1472). The source code for the model should also be made available to any potential user. Some well-known programs, such as FPETOOL and FDS (Fire Dynamics Simulator) are available for downloading from NIST at no charge. The model fireFOAM is available at no cost from OPENFOAM. Other programs, such as BRANZFIRE, SIMULEX, and ASKFRS, must be purchased as a "user license" for a given period of time. The assumptions used, known numerical and physical limitations, and the physical and mathematical treatments used must also be made available to the user. An excellent source of information is the firemodelsurvey.com website where the information of about over 150 models is available. A summary of this information was published by Olenick and Carpenter, *An Updated International Survey* (Olenick and Carpenter, 2003).

A good example of a documentation guide for a fire-modeling package is *FPETOOL: Fire Protection Engineering Tools for Hazard Estimation,* by H.E. Nelson (Nelson, 1990). That guide describes the main elements of the package, its hardware and software requirements, and fundamental mathematics, underlying assumptions, and comparisons of FPETOOL (Fire Simulator routine) to fire test data. A separate technical reference guide was published as the Technical Reference Guide for FPETOOL, Version 3.2) (Deal, 1995).

The FIRM-QB zone model developed by Marc Janssens has a technical description, program description and user's manual all included in his excellent book, *An Introduction to Mathematical Fire Modeling* (Janssens and Birk, 2000). The entire program code and supporting documentation are included in a CD-ROM packaged with the book.

According to E1355, the evaluation process consists of four steps:

1 Define scenarios for which the evaluation is to be conducted.
2 Validate the theoretical basis and assumptions used in the model.
3 Verify the mathematical and numerical robustness of the model.
4 Evaluate/quantify the uncertainty and accuracy (ASTM, 2011b).

ASTM E1355 offers the following definitions regarding models:

Evaluation: The process of quantifying the accuracy of chosen results from a model when applied for a specific use.

Validation: The process of determining the correctness of the assumptions and governing equations implemented in a model when applied to the entire class of problems addressed by the model.

Verification: The process of determining the correctness of the solution of a system of governing equations in a model. Verification does not imply the solution of the correct set of governing equations, only that the given set of equations is solved correctly.

These steps are not isolated ones. As Janssens and Birk (2000) point out, "Step 4 is usually based on a comparison between model output and experimental method and provides an indirect method of validation (step 2) and verification (step 3) of a model for scenarios of interest (step 1). It is generally assumed that the model equations are solved correctly and the terms validation and evaluation are therefore often used interchangeably." It is very rare for anyone but the model's developer to spend the time necessary to carry out steps 2 and 3, although an independent reviewer or researcher may. Portions of the mathematics are sometimes compared to other analytical results. Step 4 can be carried out by comparing a model's predictive results to full-scale tests done specifically for that purpose, tests done by others and published in the literature, results of standard room fire tests done in accordance with ASTM E603, or even against observations or reconstructions of real fires (historical fire data).

A good example of comparing a compartment fire test against the predictions of mathematical calculations, zone models and a field model was published by Spearpoint, Mowrer, and McGrattan (1999). It showed how accurately calculations and zone models (Fire Simulator, FAST and FIRST) agreed with test data. The full model used was ES3D (a predecessor to FDS). Its predictions (including early fire development) were not as accurate as the calculations and indicated further development was needed.

Model uncertainty is based on repeated runs of similar data and sensitivity analyses to identify critical data. For complex model with many inputs, it is usually prohibitively costly to run repeated runs with different single inputs. Mathematical techniques have been used to streamline the process.

The accuracy with which FDS predicts temperatures and heat-release rates have been validated by large-scale fire tests. Testing has shown that FDS temperature predictions were within 15% measured temperatures, and heat-release rates were within 20% of measured values. Results, however, are often presented as ranges to account for some uncertainty. FDS is the primary fire-modeling tool used by NIST and has been used in major fire investigations involving large losses and deaths.

When model results are compared to a full-scale test fire, it is usually assumed that the real-fire data is the gold standard. The complexities of the room fire environment and the variables of turbulence in a large fire make it impossible, however, to get exactly the same measurements.

Fire modeling case studies

Since 1987, NIST has carried out investigations and reported on numerous major structure fires where fire modeling was used in the analysis (see Table 9.5 for a partial list). Some of these are discussed in Icove, DeHaan, and Haynes (2013).

In the UK, computer fire modeling provided the key to solving the puzzle of the King's Cross Underground Station Fire of November 18, 1987. In that incident, witnesses saw a relatively small localized fire on the wooden escalator some 21 m (68 ft.) from the top. Shortly after, a firestorm erupted from the escalator tunnel and incinerated the ticketing hall at the top, killing 31 people in all. The CFD model FLOW3D predicted a slope-driven, fire

Table 9.5 Representative NIST fire investigations using fire modeling

NIST report number or citation	*Case title and author*
NBSIR 87-3560 May 1987	*Engineering Analysis of the Early Stages of Fire Development: The Fire at the Dupont Plaza Hotel and Casino*, December 31, 1986 H.E. Nelson
NISTIR 90-4268; August 1990, vol. 1	*Full Scale Simulation of a Fatal Fire and Comparison of Results with Two Multiroom Models* R.S. Levine and H.E. Nelson
NISTIR 4665 September 1991	*Engineering Analysis of the Fire Development in the Hillhaven Nursing Home Fire, October 5, 1989* H.E. Nelson and K.M. Tu
Journal of Fire Protection Engineering 4, no. 4 (1992): 117–31	*Analysis of the Happyland Social Club Fire With HAZARD I* R.W. Bukowski and R.C. Spetzler
NISTIR 4489 June 1994	*Fire Growth Analysis of the Fire of March 20, 1990, Pulaski Building, 20 Massachusetts Avenue, NW, Washington, DC* H.E. Nelson
Fire Engineers Journal 56, no. 185 (November 1996): 14–17	*Modeling a Backdraft Incident: The 62 Watts Street (New York) Fire* R.W. Bukowksi
NISTIR 6030 June 1997	*Fire Investigation: An Analysis of the Waldbaum Fire, Brooklyn, New York, August 3, 1978* J.G. Quintiere
NISTIR 6510 April 2000	*Simulation of the Dynamics of the Fire at 3146 Cherry Road, NE, Washington D.C., May 30, 1999* D. Madrzykowski and R.L. Vettori
NISTIR 7137 May 2004	*Simulation of the Dynamics of the Fire in the Basement of a Hardware Store, New York, June 17, 2001* N.P. Bryner and S. Kerber
NISTIR 6923 October 2002	*Simulation of the Dynamics of a Fire in a One-Story Restaurant, Texas, February 14, 2000* R.L. Vettori, D. Madrzykowski, and W.D. Walton
NISTIR 6854 January 2002	*Simulation of the Dynamics of a Fire in a Two-Story Duplex, Iowa, December 22, 1999* D. Madrzykowski, G.P. Forney, and W.D. Walton
NIST SP 995 March 2003	*Flame Heights and Heat Release Rates of 1991 Kuwait Oil Field Fires* D. Evans, D. Madrzykowski, G.A. Haynes
NIST Special Pub. 1021 July 2004	*Cook County Administration Building Fire, 69 West Washington, Chicago, Illinois, October 17, 2003: Heat Release Rate Experiments and FDS Simulations* D. Madrzykowski and W.D. Walton

(*Continued*)

NIST report number or citation	*Case title and author*
NISTIR 7137 2004	*Simulation of the Dynamics of a Fire in the Basement of a Hardware Store, New York, June 17, 2001* N.P. Bryner and S. Kerber
NIST NCSTAR September 2005	*Reconstruction of the Fires in the World Trade Center Towers. Federal Building and Fire Safety Investigation of the World Trade Center Disaster* R.G. Gann, A. Hamins, K.B. McGrattan, G.W. Mullholland, H.E. Nelson, T.J. Ohlemiller, W.M. Pitts, and K.R. Prasad
Fire Technology 42, no. 4 (October 2006): 273–81	*Numerical Simulation of the Howard Street Tunnel Fire* K.B. McGrattan and A. Hamins
Fire Protection Engineering 31 (Summer 2006): 34–36	*NIST Station Nightclub Fire Investigation: Physical Simulation of the Fire* D. Madrzykowski, N.P. Bryner, and S.I. Kerber
NIST Special Pub. 1118, 2011	*Technical Study of the Sofa Super Store Fire, South Carolina, June 18, 2007* N. Bryner, S. Fuss, B. Klein, and A. Putorti
NIST TN 1729, 2012	*Simulation of the Dynamics of a Wind-Driven Fire in a Ranch-Style House, Texas* A. Barowy and D. Madrzykowski

spread up the wood-lined escalator. Reduced scale models demonstrated the mechanism. The original fire was apparently accidentally started by a dropped cigarette or match contacting grease-soaked debris on the escalator tracks. As a result if this tragic incident, escalators and their tunnels were made of noncombustible materials and all smoking was banned throughout the system (Moodie and Jagger, 1992).

The tragic loss of three firefighters in a residential fire in Pittsburgh PA prompted fire engineers to produce an FDS model of a multi-story building. The FDS model with Smokeview allowed an estimate to be made of the temperatures, and oxygen- and carbon monoxide concentrations in various rooms during the fire. The results were then used to estimate the likely survival times of the firefighters based on their toxicological measurements. (Christensen and Icove, 2004)

Fire modeling has been used to answer many important questions raised during fire investigations, such as:

What was the probable cause of the fire (i.e., can several possible causes be eliminated?)
How long did it take to activate sprinklers or smoke detectors?
Why didn't the occupants of the building survive the fire?
What were possible mechanisms for rapid fire spread?

How much time elapsed between ignition and flashover in the room of origin?
What were likely smoke and toxic gas conditions in various areas?
Was an accelerant used in the fire and how would it have contributed?
Did a faulty building design (or negligence) or a failed detection or suppression system contribute to the fire or loss of life?
What change to policy, building- or fire-codes would prevent incidents like this in the future?

Icove, DeHaan and Haynes (2013) provides an extensive selection of fire cases in which computer fire modeling was used to help answer just these sorts of questions.

Considerations in legal proceedings

When one is presented with computer fire model results in an adversarial (court) context, the following points may be useful.

Accuracy—The accuracy of input data—initial fire HRR, growth rate—is **critical** to the accuracy and reliability of the final result. "Garbage in—gospel out" is the risk with computer models. Are data arbitrary? Are they correct for the scenario in question? The accuracy required varies with the model and what is being tested. It can be assessed by sensitivity analysis—varying some inputs (such as room dimensions, ventilation openings or the heat-release rate or duration of the initial fire).

Assumptions—What assumptions were made by the user to fill the gaps? Incompleteness of data from the scene is the major reason for most failed computer model attempts. What default values does the model insert if data are not available? Will those default values make a difference (i.e., what is the model's sensitivity to those values)?

Impression—How are the data presented? Is it in the form of reviewable printed output or a single dramatic action cartoon? SMOKEVIEW will show "movement" of flames and smoke that is a stop-action representation of a "temperature" surface or smoke concentration. Other models (or users) refrain from showing smoke or flame movement because it is, to some extent, too complex, and too random to show accurately.

Correctness—Is it the right model for the job? What is the question the investigator wants to answer? What is the question the model was *intended* to answer (temperature, smoke filling, and species concentration)? What are the limitations of the model—number of rooms, fire growth, size of fire, ventilation, and time? Will this model address those issues correctly in the problem at hand? Is information about conditions in a specific location at a specific time needed? If so, a zone model may not be able to give an appropriate answer.

Evaluation/Validation—Was the model created and validated for a particular scenario (small fire in a big room) and was it being used here for a different scenario without proper (published) evaluation?

Fine-tuning—When a comparison to a test fire is offered, the number of model runs should be evaluated. Was the model run with changes in input data to get the model to "match" the real fire? (The FDS model was "adjusted" several times until its events coincided with the video recording taken during the tragic Station Nightclub fire in Rhode Island in 2003.)

User qualified?—Did the user have the correct documentation (user's guides, technical manuals)? How much experience did the user have with this model? Were other models considered or used? What steps did the user take that make sure the model was correct and correctly used (e.g., reviewing published evaluations)?

When fire test results are offered, many of the same issues arise. While there are some ASTM and NFPA guidelines for fire tests, there are many valid tests that cannot follow a specific guide due to the variables present of issues to be tested. The questions should be:

- What was the issue to be tested—what was the objective of the test?
- Was this a test in which particular variables were changed while others were held constant?
- Was this a demonstration rather than a controlled test?
- How was data collected, assembled, analyzed, and reported?
- How many times was the test repeated to establish its reproducibility?
- How do the "test" conditions vary from the actual (or purported) fire conditions?
- Were all important variables controlled and documented?
- If this were a reduced-scale model, what corrections were applied for factors that were not scalable by linear reduction (ventilation, velocity, radiant heat/distance, and material response, for instance)?

Summary

It is misleading to offer a computer fire model or a recreation based upon it as proof of how a particular fire actually occurred. The complexity of compartment fires involving multiple large fuel packages or an extended time of development introduces too many variables for even an advanced model to accommodate.

Appropriate uses for computer models include the following:

- Testing hypotheses, not proving causation
- Validating or explaining postfire indicators
- Estimating time lines
- Evaluating human factors and fire/smoke conditions

Precision of fire calculations or fire models:

- **Not** to second decimal point!
- Best results: predictions accurate (duplicative of real world results) to ±30%
- Predictions for complex scenes or prolonged fires—order of magnitude ranges

For fire tests:

- What was the intent of the test or demonstration?
- Were important variables identified and controlled?
- Was the collection and analysis of data done correctly?

Prefire conditions such as:

- The type and placement of fuel packages; floor, wall, and ceiling materials and coverings; dimensions of rooms; and sizes, sill and soffit heights of all vents are essential to any accurate model. The more estimates that have to be made, the less reliable the results will be. Without critical dimensions, any modeling is fancy guesswork, and such guesswork should play no role in scientific fire investigations.

In summary, it is misleading for a fire expert to claim that even the result of the best computer model proves that the fire in question happened in the manner shown. Fires large enough to be of legal interest are far too complex with too many unpredictable variables. Visual results like animations and colorful graphics can be very useful in understanding and explaining a fire event. They are persuasive but easily misinterpreted (intentionally or accidentally) and must be carefully evaluated before their use in court proceedings.

Judicial consideration:

- Does probative value outweigh potential bias or misunderstanding, especially when animations or graphical interpretations are offered (the "I saw it on TV, therefore it must be true" logic)?

Does any model used pass the major tests?

- Sensitivity
- Accuracy
- Published
- Tested (validated)
- Used by qualified expert
- Were sufficient and correct input data gathered and entered?

- What assumptions were made about the prefire conditions, ignition, and initial fire?
- What default decisions were made?
- Is this a fair and impartial analysis?

The most common flaw today in the use of fire models is the lack of data and incomplete documentation of the original scene (including the ignition and fire properties of fuels in the room).

Key term definitions

CFAST: Consolidated model of fire and smoke transport. An advanced computer zone model for the study of fires and their effects.

CFD: Computational fluid dynamics—mathematical representations of the flow and mixing of gases or liquids.

FDS: Fire dynamics simulator—an advanced computer (CFD) model created by NIST to study fires and their effects.

Fire test: Any evaluation carried out by setting fire (reduced scale or full scale).

fireFOAM: An advanced suite of fire models created by FMGlobal and offered in an open source.

FPETOOL: A suite of mathematical models, spreadsheets and zone models created by Harold E. "Bud" Nelson of NIST.

Mathematical models: Simple calculations empirically derived from observing and measuring properties of fires and material response to them.

Zone model: A computer model of compartment fires based on the simplification of the fire as a layer of hot gases overlying a layer of normal air with the fire acting as a pump for heat and combustion products.

References

Abecassis-Empis, C., Rezka, P., Steinhaus, T., Cowlard, A., Biteau, H., Welch, S., Rein, G., & Torero, J.L., (2008). Characterisation of Dalmarnock fire test one. *Experimental Thermal and Fluid Science* 32(7): 1334–43, doi:10.1016/j.exptherm flusci.2007.11.006.

Andrews, P.L., (2013), Current status and future needs of the behave plus fire modeling system. *International Journal of Wildland Fire* 23(1): 21–33.

Andrews, P.L., Devins, C.D.,& Seli, R.C., (2003), Behave Plus Fire Modeling System v. 2.0 Users Guide, Gen. Tech. Rep. RMRS-GTR-106WWW, Ogden, UT, Dept. of Agri. Forest Service, Rocky Mountain Research Station.

ASTM, (1980), Estimating room flashover potential. *Fire Technology* 16(2): 94–103. doi:10.1007/bf02481843.

ASTM, (2007a), *ASTM E1472–07: Standard guide for documenting computer software for fire models (withdrawn 2011)*. West Conshohocken, PA: ASTM International.

ASTM, (2007b), *ASTM E1591–07: Standard guide for obtaining data for deterministic fire models (withdrawn 2011).* West Conshohocken, PA: ASTM International.

ASTM, (2007c), *ASTM E1895–07: Standard guide for determining uses and limitations of deterministic fire models (withdrawn 2011).* West Conshohocken, PA: ASTM International.

ASTM, (2011a), *ASTM E119-11: Standard test methods for fire tests of building construction and materials.* West Conshohocken, PA: ASTM International.

ASTM, (2011b), *ASTM E1355-11: Standard guide for evaluating the predictive capability of deterministic fire models.* West Conshohocken, PA: ASTM International.

Babrauskas, V., (1975), *COMPF: A program for calculating post-flashover fire temperatures.* Berkeley: Fire Research Group, University of California.

Babrauskas, V., (1996), Fire modeling tools for FSE: Are they good enough? *Journal of Fire Protection Engineering* 8(2): 87–96.

Bill, R.G., & Dorofeev, S., (2011), *An overview of fire modeling at FM Global.* Proceedings: Fire and Materials 2011.Interscience Communications, London.

BRE-Center, (2012), *The Dalmarnock fire tests*, 2006, UK. Retrieved February 5, 2012, from www.see.ed.ac.uk/fire/dalmarnock.html/.

Buchanan, A.H., (2001), *Structural design for fire safety.* Chichester: Wiley.

Bukowski, R.W. (1991), *Fire models: the future is now*, Fire Journal 85(2): 60–69.

Carman, S.W., (2009), *Progressive burn pattern developments in post-flashover fires.* Proceedings: Fire and Materials, 11th International Conference, January 2009. Interscience Communications, London.

Carman, S.W., (2010), *Clean-burn fire patterns: A new prospective for interpretation.* Proceedings: Interflam, July 2010. Interscience Communications, London.

Carman, S.W., (2011), *Investigation of an elevated fire – perspectives on the Z-factor.* Proceedings: Fire and Materials, 12th International Conference, January 2011. Interscience Communications, London, 757–767.

Carman, S.W., (2013), *Investigating multi-compartment fire behavior of elevated origins.* Proceedings: Fire and Materials 13th International Conference. January 2013. Interscience Communications, London, 769–779.

Christensen, A., & Icove, D.J., (2004). To the investigation of carbon monoxide exposure in the deaths of three Pittsburgh fire fighters. *Journal of Forensic Sciences*49(1): 104–107.

Davis, S.G., Engel, D., Gavelli, F., Hinze, P., & Hansen, O.R., (2010), *Advanced methods for determining the origin of vapor cloud explosions case study: 2006 Danvers explosion investigation.* Paper presented at the International Symposium on Fire Investigation Science and Technology (ISFI), September 27–29, College Park, MD.

Deal, S., (1995) Technical Reference Guide for FPETool-Version 3.2, NIST, Gaithersburg, MD.

DeHaan, J.D., & Icove, D.J., (2012), *Kirk's fire investigation*, 7th ed. Upper Saddle River, NJ: Pearson-Prentice Hall.

DeWitt, W.E., & Goff, D.W., (2000), Forensic engineering assessment of FAST and FASTLite fire modeling software. *National Academy of Forensic Engineers*, December 2000: 9–19.

Forney, G.P., (2015), *Smokeview: A tool for visualizing fire dynamics simulation data. Volume I: User's guide.* NIST Publication 1017 – 1, Sixth Edition. Gaithersburg, MD: National Institute of Standards and Technology.

Forney, G.P. & Moss, W.F., (1992) *Analyzing and exploiting numerical characteristics of zone fire models*, NISTIR 4763, National Institute of Standards and Technology, Gaithersburg, MD.

FRAMES, (2015), U.S. Forest Service, Rocky Mountain Research Station, Missoula, Montana. Retrieved from www.firelab.org/applications.

Friedman, R., (1992), An international survey of computer models for fire and smoke. *SFPE Journal of Fire Protection Engineering* 4(3): 81–92.

GexCon, (2012), GexCon US. Retrieved February 5, 2012, from www.gexcon.com/.

Hunt, S., (2000), Computer fire models. *NFPA Section News* 1(2): 7–9.

Icove, D.J., DeHaan, J.D. & Haynes, G.A., (2013), *Forensic Fire Scene Reconstruction*, 3rd ed. Upper Saddle River, NJ: Pearson-Prentice Hall.

Iqbal, N., & Salley, M.H., (2004), Fire Dynamics Tools (FDTs): Quantitative fire hazard analysis methods for the U.S. Nuclear Regulatory Commission Fire Protection Inspection Program. Washington, DC.

ISO, (2008), ISO 16730: Fire safety engineering – Assessment, verification and validation of calculation methods. International Organization for Standards (ISO), Geneva, Switzerland.

Janssens, M.L., & Birk, D.M., (2000). *An introduction to mathematical fire modeling*, 2nd ed. Lancaster, PA: Technomic.

Jones, W.W., Peacock, R.D., Forney, G.P., & Reneke, P.A., (2005), CFAST consolidated model of fire growth and smoke transport (version 6): technical reference guide, NIST, Gaithersburg, MD, December 2005

Lawson, J.R., &Quintiere, J.G., (1985), Slide-rule estimates of fire growth. *Fire Technology* 21(4): 267–292.

Maranghides, A., McNamara, D., Vihnanek, R., Restaino, J., & Leland, C., (2015) A case study of a community affected by the Waldo fire – event timeline and defensive actions: NIST TN 1910, NIST, Gaithersburg MD, November 2015 (http://dx.doi.org/10.6028/NIST.TN.1910)

McGill, D., (2003), *Fire Dynamics Simulator, FDS 683, participant's handbook*. Toronto, Ontario: Seneca College, School of Fire Protection.

McGrattan, K., Baum, H., Rehm, R., Hamins, A., & Forney, G., (2000), *Fire Dynamics Simulator: Technical reference guide, NISTIR 6467*. Gaithersburg, MD: NIST.

McGrattan, K., Baum, H., Rehm, R., Mell, W., McDermott, R., Hostikka, S., & Floyd, J., (2010), *Fire Dynamics Simulator (version 5) technical reference guide*. NIST Special Publication 1019-6. Gaithersburg, MD: National Institute of Standards and Technology.

McGrattan, K., Hostikka, S., McDermott, R., Floyd, J., Weinschenk, C., & Overholt, K., (2015), *Fire Dynamics Simulator user's guide (version 6.3.0)*. NIST Special Publication 1019, Sixth Edition. Gaithersburg, MD: National Institute of Standards and Technology.

McGrattan, K., McDermott, R., Forney, G., Overholt, K., &Weinschenk, C., (2014), *Fire modeling for the fire research, fire protection and fire service communities*. Retrieved October 30, 2015, from SFPE, http://magazine.sfpe.org/content/fire-modeling-fire-research-fire-protection-and-fire-service-communities, 10/30/15.

Milke, J.A., & Mowrer, F.W., (2001), *Application of fire behavior and compartment fire models seminar*. Paper presented at the Tennessee Valley Society of Fire Protection Engineers (TVSFPE), September 27–28, Oak Ridge, TN.

Mitler, H.E., (1991), *Mathematical modeling of enclosure fires*. Gaithersburg, MD: National Institute of Standards and Technology.

Moodie, K., & Jagger, S.F., (1992), The King's Cross fire: Results and analysis from the scale model tests. *Fire Safety Journal* 18 (1): 83–103, doi:10.1016/0379-7112(92)90049-i.

Mowrer, F.W., (1992), *Methods of quantitative fire hazard analysis.* Boston, MA: Society of Fire Protection Engineers, prepared for Electric Power Research Institute (EPRI).

Mowrer, F.W., (2003), *Spreadsheet templates for fire dynamics calculations.* College Park, MD: University of Maryland.

Nelson, H.E., (1990), *FPETool: Fire protection engineering tools for hazard estimation.* Gaithersburg, MD: National Institute of Standards and Technology.

NFPA, (2014), *NFPA 921: Guide for fire and explosion investigations.* Quincy, MA: National Fire Protection Association.

Olenick, S.M., & Carpenter, D.J., (2003), An updated international survey of computer models for fire and smoke . *Fire Protection Engineering* 13: 87–110.

OpenFOAM, (2012), *FM research: Open source fire modeling.* Retrieved October 30, 2015, from www.fmglobal.com.

Peacock, R.D., Reneke, P.A., & Forney, G.P., (2015), *CFAST: Consolidated model of fire growth and smoke transport (version 7). Volume 2: User's guide.* NIST Technical Note 1889v2. Gaithersburg, MD: National Institute of Standards and Technology.

Rein, G., Abecassis-Empis, C., & Carvel, R., (2007), *The Dalmarnock fire tests: Experiments and modeling.* Edinburgh: University of Edinburgh.

Rein, G., Bar-Ilan, A., Fernandez-Pello, A.C., & Alvarez, N., (2006), A comparison of three models for the simulation of accidental fires. *Journal of Fire Protection Engineering* 16 (3): 183–209.

Rein, G., John, W., & Torero, J., (2011), *Modelling of the growth phase of Dalmarnock fire test one.* Proceedings: Fire and Materials 2011. Interscience Communication, London.

Rein, G., Torero, J.L., Jahn, W., Stern-Gottfried, J., Ryder, N.L., Desanghere, S., Lazaro, M., et al., (2009), Roundrobin study of a priori modeling predictions of the Dalmarnock fire test one. *Fire Safety Journal* 44 (4): 590–602. doi:10.1016/j.firesaf.2008.12.008.

Salley, M.H., & Kassawar, R.P., (2007a), *Verification and validation of selected fire models for nuclear power plant applications: Consolidated Fire Growth and Smoke Transport Model (CFAST).* Washington, DC: U.S. Nuclear Regulatory Commission.

Salley, M.H., & Kassawar, R.P., (2007b), *Verification and validation of selected fire models for nuclear power plant applications: Experimental uncertainty.* Washington, DC: U.S. Nuclear Regulatory Commission.

Salley, M.H., & Kassawar, R.P., (2007c), *Verification and validation of selected fire models for nuclear power plant applications: Fire Dynamics Simulator (FDS).* Washington, DC: U.S. Nuclear Regulatory Commission.

Salley, M.H., & Kassawar, R.P., (2007d), *Verification and validation of selected fire models for nuclear power plant applications: Fire Dynamics Tools (FDTs).* Washington DC: U.S. Nuclear Regulatory Commission.

Salley, M.H., & Kassawar, R.P., (2007e), *Verification and validation of selected fire models for nuclear power plant applications: Fire Induced Vulnerability Evaluation (FIVE-Rev1).* Washington, DC: U.S. Nuclear Regulatory Commission.

Salley, M.H., & Kassawar, R.P., (2007f), *Verification and validation of selected fire models for nuclear power plant applications: MAGIC.* Washington, DC: U.S. Nuclear Regulatory Commission.

Salley, M.H., & Kassawar, R.P., (2007g), *Verification and validation of selected fire models for nuclear power plant applications: Main report.* Washington, DC: U.S. Nuclear Regulatory Commission.

Salley, M.H., & Kassawar, R.P., (2010), *Methods for applying risk analysis to fire scenarios (MARIAFIRES)-2008.* NRC-RES/EPRI Fire PRA Workshop, vol. 1. Washington, DC: U.S. Nuclear Regulatory Commission.

SFPE, (2011), *Guidelines for substantiating a fire model for a given application.* G.06 Bethesda, MD: Society of Fire Protection Engineers.

Spearpoint, M., Mowrer, F.W., & McGrattan, K., (1990). *Simulation of a compartment flashover fire using hand calculations, zone models and a field model.* Proceedings of ICFRE. Chicago, IL, 3–14.

Sutula, J., (2002) Applications of the fire dynamics simulator in fire protection engineering consulting. *Fire Protection Engineering* 14: 33–43.

Thunderhead, (2010), *PyroSim example guide.* Manhattan, KS: Thunderhead Engineering.

Thunderhead, (2011), *PyroSim user manual.* Manhattan, KS: Thunderhead Engineering.

Vasudevan, R., (2004) Forensic engineering analysis of fires using Fire Dynamics Simulator (FDS) modeling program. *Journal of the National Academy of Forensic Engineers* 21(2): 79–86.

Vettori, R. & Madrzykowski, D.M., Comparison of FPETool: FIRE SIMULATOR with data from full scale experiments, NISTIR 6470. NIST, Gaithersburg MD, February 2000.

Part III

Corollary factors and prevention trends in forensic science arenas

10 The evolution of spatial forensics into forensic architecture

Applying CPTED and criminal target selection

Gregory Saville

Introduction

This chapter proposes spatial forensics as a form of forensic architecture based on crime prevention through environmental design (CPTED). It describes the crime prevention framework of CPTED and associated crime and place theories. It then illustrates how spatial forensic analysts apply CPTED principles to the analysis of crime events, with particular emphasis on how the social and physical environment influences target search and selection decisions prior to the crime event. Finally, the chapter outlines a spatial forensic methodology called the crime event site analysis that spatial analysts can deploy in order to collect relevant data.

CPTED opens a new investigative door for retroactively investigating crime sites to help solve, prosecute, or otherwise shed new light onto criminal events. A growing number of CPTED practitioners appear at criminal and civil trials each year to assess the crime opportunities created, or prevented, by urban design. As such, CPTED represents a new form of spatial analysis in forensic architecture and a novel tool for the forensic scientist.

Existing forms of forensic architecture sometimes includes methods at a microscale, such as forensic engineering on construction and architectural techniques and related to personal injury liability (Noon, 2000). The current chapter explores methods that operate at both the microscale of the crime scene itself and at a meso level that includes surrounding neighborhoods and streets. The priority at this level of analysis is to employ architecture, urban planning, and urban design to examine specific crime event locations and CPTED is ideally suited for this purpose. Aside from macro level crime mapping research (Harries, 1971, 1999) and geographical profiling (Rossmo, 1995, 2000) examining serial offenders (a small subset of the offender population), this is the first systematic attempt to describe the practice of spatial forensics at the crime scene.

CPTED emerges from 40 years of research into offender movement behavior, target selection, and criminal proxemic choice that, until now, helped urban designers, architects and crime prevention specialists minimize crime opportunities. Incorporating CPTED methods into spatial forensics allows the forensic scientist to use those same methods to assess factors such as the ease of target access, the extent of offender planning, and the vicarious liability of property owners when offenders choose one property or victim over another.

At the center of spatial forensics is the proposition of legal foreseeability which is the idea that certain crimes can reasonably be predicted, a proposition that suggests there is a refined criminological science of human behavior to predict crime patterns. That is not in fact the case. Human behavior is influenced by so many intricate factors that precise predictability is, at least at the present time, a distant goal. However, that is not to say we know nothing of both criminal and victim behavioral patterns, quite the opposite.

The past few decades have seen the growth of a growing body of research in criminal behavior as it relates to the physical and social place of crime events (Clarke & Cornish, 1985; Saville & Murdie, 1988; Sherman, Gartin, & Buerger, 1989). This research clusters under a number of titles including situational crime prevention (Clarke, 1980, 1983), environmental criminology (Brantingham & Brantingham, 1981), and the routine activity theory (Cohen & Felson, 1979). All these criminological theories emerged after the formulation of CPTED and today CPTED is, arguably, among the most dominant crime and place prevention practices (Saville, in press). CPTED has a 40-year legacy of practice and evaluation and a preponderance of evidence shows its effectiveness, to varying degrees, in reducing crime opportunities (Cozens, Saville, & Hillier, 2005). That is why CPTED practitioners are appearing more frequently as experts in court proceedings.

At the beginning of CPTED is the notion that the physical and social context of crime, most often determined by architecture and site planning, has significant impact on how a criminal chooses a location and target. Those criminal choice decisions are in turn constrained, or enhanced, by the social and spatial dynamics of the environments in which offenders and victims find themselves at the time of the crime. It is the analysis of those sociospatial dynamics that spatial forensics contributes to forensic science.

Thus, as this chapter describes, spatial forensics seeks an understanding of the motives and opportunities at crime scenes and in turn, involves two essential requirements:

1. Spatial forensic analysts must employ a comprehensive knowledge of traditional and modern CPTED, including the sociospatial patterns at crime sites;
2. Spatial forensic analysts must employ proper crime event site analysis, a process of systematic collection of spatial, social, and behavioral data at the crime scene.

This chapter describes those essential conditions and then lays the groundwork for future evolution of forensic spatial analysis as a practice in forensic science.

Conceptual background

Crime prevention through environmental design

The core of CPTED is the idea that a space can be defended from crime, or what architect Oscar Newman called defensible space (Jeffery, 1971; Newman, 1972). In the original formulation of CPTED, now termed first-generation CPTED, there are four principles that hinge around the central idea that small-scale places such as parks, schools, sidewalks, and building foyers, can be made safer if legitimate users of that space (such as residents) take a personal interest and ownership over that space. By claiming that territory, the users of that space make it more difficult for offenders to offend with impunity, thereby reducing the opportunity for crime.

The four basic principles (Atlas, 2013) include the following:

Territorial reinforcement—Architects and planners achieve this in many ways from signage and landscaping, to dividing space into public and private hierarchies. For example, a public sidewalk feeds into an apartment property through a semiprivate walkway demarcated with landscaping. The walkway is clearly visible through glass windows on the apartment front foyer. Signage and lights further demarcate the front entranceway as private property. Urban design with hierarchy of space helps reinforce territorial controls. Opportunities for crimes are lower in such places and offenders tend to avoid territorial controls that might lead to their apprehension.

Access controls—Property owners and facility managers often locate vehicle traffic controls onto their parking lots or gates at the entrance to their property. Access controls further define space and reduce opportunities for crime. Offenders who bypass access controls must use extra planning or take extra efforts in order to commit their crimes and these efforts leave additional clues at crime scenes, such as marks from forced entry.

Surveillance—CPTED designers use an array of surveillance methods to watch over areas. Natural surveillance is the most common, achieved by trimming hedges to improve sightlines or designing windows to overlook risky areas. Mechanical methods of surveillance include installing enhanced lighting or closed circuit television (CCTV) in public areas. Organized surveillance involves police/security patrols, volunteer patrols, or security at checkpoints. Offenders usually avoid surveillance areas; however, some develop sophisticated methods to avoid CCTV view areas or evade the casual surveillance of neighbors. When that happens, it suggests a particular kind of offender and motive.

Management and maintenance—Crime targets and locations that are unkempt and poorly cared for provide a particular message to others that no one seems to care about the area. Offenders seeking easy targets interpret this lack of care for an easier opportunity to steal or rob. By ensuring areas are clean, well maintained, and well managed, the opportunity for street crime diminishes thereby deterring many kinds of offenders.

Levels of rationality

The basic CPTED principles not only help defend spaces against crime, but they point to the kinds of decisions that offenders must make to offend in one place versus another. Criminal location decision-making is therefore a main focus of spatial forensics.

Spatial forensics adopts the view that there is a rationality to the psychology of criminal motivation, itself a vast and complex field of study. The psychology of human rationality has itself led to the forensic fields such as psychological profiling and geographical profiling (Rossmo, 2000). Geographical profiling is somewhat different from the spatial forensics proposed here because the former builds macroscale geographical probability maps of potential serial offender residence areas. Spatial forensics builds an environmental picture of a specific place from existing physical conditions of the crime scene and then describes how that space presented opportunities for expressive and instrumental offenders.

The idea of "planned" versus "unplanned" crime, in criminological and forensic circles known respectively as instrumental versus expressive crimes (Godwin, 2000), suggests dichotomous criminal types. In some cases, criminals simply wait for the opportunity or as some criminologists claim, opportunity makes the thief (Felson & Clarke, 1998). In other cases, criminals strike out in the heat of the moment when their rational mind is clouded by intoxication, passion, or insanity. This does not mean there are no patterns in different cases, only that the instrumental and expressive dichotomy is probably more elastic than first thought. It is more probable that the division between instrumental and expressive crime exists on a sliding continuum influenced by a myriad of social, psychological, and cultural influences. Consequently, from a purely scientific point of view, most offenders probably have a range of predispositions toward expressive or instrumental motives (Helfgott, 2008).

One new area of research provides some exciting possibilities for enhancing crime and place theories, particularly defensible space. It is known as guardianship (Reynald, 2010; Hollis-Peel et al., 2011) and it is defined as "the physical or symbolic presence of an individual or group of individuals that acts either intentionally or unintentionally to deter a potential criminal event" (Hollis-Peel et al., 2011, p. 53). This is an expansion of the original eyes-on-the-street concept from Jane Jacobs' book *The Death and Life of Great American Cities* (Jacobs, 1961). The extent to which guardianship impacts defensible space will ultimately influence the decisions made by both expressive and instrumental offenders.

For the analyst, this means that spatial forensics requires a clear understanding of the sociospatial patterns emerging at the crime site as well as a comprehensive process of data collection. While significant literature has existed for some time regarding sociospatial patterns of crime (Rengert, 1989; Sampson, Raudenbush, & Earls, 1997) and CPTED (Cozens, Saville, &

Hillier, 2005), spatial forensics is at embryonic stages of evolution and therefore this chapter suggests some directions for the future development and application of the field.

Expressive offenders

The expressive offender, one who acts out of intoxicated, emotional, or mental impulse, is obviously less likely to travel a rational decision-making chain prior to the offense. This is the case in domestic violence when the volatile dynamics of an intimate relationship may be the single most important element driving violent behavior. Location is a factor in domestic violence since it happens within the privacy of a household, but questions about where those houses are located and the physical configuration of that architecture is a minor factor in the offender location choice. However, offenders influenced by expressive motives are more likely less deterred by defensible space at a crime scene and that too helps the spatial analyst assess the crime event.

Instrumental offenders

By contrast, instrumental crimes committed on the street or in semipublic places such as parks, parking lots, shopping malls, corner stores, and exterior properties have very different profiles. The offender seeks a target with low chance of apprehension and high chance of success. Therefore, there will be at least four phases of a crime:

Phase 1—A preliminary phase when criminal motives are triggered,
Phase 2—A second phase when target search takes place,
Phase 3—A third stage when the target is selected,
Phase 4—A final stage when the crime is committed.

In Phase 1, social and psychological interventions target crime motives such as parenting training early childhood school programs. Job training and poverty reduction are other motive reducing strategies. In Phase 4, apprehended offenders receive prison rehabilitation programs or other interventions for substance abuse. Spatial forensics looks specifically to Phases 2 and 3 of the search process and that is also where traditional CPTED applies.

One reason Phases 2 and 3 are ideal for assessing opportunity at a crime scene is that target search and target selection are constrained by the physical and social environment where crime targets exist, for example in the isolated parking lot where cars are stolen. That is due to Tobler's First Law of Geography (Tobler, 1970), which states that no urban place is isolated from another due to the interconnections between social, cultural, and economic activities in urban life. Therefore, the relationships between those activities provide a template for analyzing spatial behavior. This template is the basis upon which crime and place theories operate in criminology, theories such as environmental criminology (Brantingham & Brantingham, 1981) and routine activity theory (Felson & Cohen, 1980). These theories, and the CPTED practice that informs them, are the primary means by which spatial analysts attempt to understand offender search and target selection.

Controversies with crime and place theories

There are two controversies affecting spatial forensics worth mention. The first involves the legal position of spatial analysis in the courtroom, particularly in relation to what is known as the Daubert challenge. In a Daubert challenge, forensic experts are challenged to prove their opinions by demonstrating a robust research methodology and by illustrating how those findings are scientifically valid.

The second controversy regards the conceptual limits of crime and place theories. While crime and place theories are now commonplace in the criminological literature, scholars are addressing shortfalls and are upgrading the basic tenets of those theories with social studies. Forensic analysts must therefore adapt their practices to incorporate those upgrades.

Daubert challenges

The legal principle of a Daubert challenge (Berger, 2005) arises from a 1993 U.S. Supreme Court case, Daubert v. Merrell Dow Pharmaceuticals, Inc. (509 U.S. 579). It arose regarding the admissibility of expert testimony and it basically requires the expert to provide his or her research methodology and proof that a particular opinion is scientifically valid.

CPTED expertise is no different in that it must withstand a Daubert challenge in court. Therefore, each expert must outline some comprehensive method by which he or she collect data and then apply that data to a relevant theory. CPTED practice over the past few decades is rife with methods and research approaches, some robust and others less so. Federal Rules of Evidence provide some guidelines suggesting that data from the crime event site analysis should appear along with expert opinions within a written report. That means all crime maps, photographs, interview findings, and so forth, will all appear in this report along with the environmental conditions and spatial patterns identified in the analysis.

Unfortunately, the Federal Rules of Evidence do not describe specifically how the analysis takes place nor what type of analysis should accompany those opinions. For example, the guidelines do not describe what type of evidence constitutes good scientific practice.

Fortunately the International CPTED Association (International CPTED Association, 2015), the only global organization working toward professionalization in the CPTED field, has adopted a formal certification program that includes a research competency. This chapter uses principles of that certification program as well as a well-developed risk assessment method described in the literature (Atlas & Saville, 2013) to create the crime event site analysis. Specifics regarding different data collection methods follow in the section below titled The Crime Event Site Analysis.

Conceptual limits of opportunity-based theories

While crime and place theories provide the preliminary basis for assessing opportunity, that turns out to be a limitation. Because crime and place theories are methodologically preoccupied with crime opportunity they end up unable to describe the relationship of crime motives and social conditions leading to those opportunities in the first place (Bouhana, 2013). Consequently, crime and place theories describe where crime locations occur but are as yet unable to predict where they will occur in future. There are some promising attempts to create computerized probability algorithms to predict where police calls for service might arise (Perry et al., 2013), but these are in early stages of development and have not yet been rigorously evaluated in the literature.

As yet, crime and place theories remain descriptive. It is somewhat like the theory of evolution that describes how species change due to the environment but cannot predict exactly how they will evolve in future. Yet evolution remains a powerful scientific theory, it provides a general framework for understanding systems in biology, and in a few subbranches of biology, it offers some predictive insight (Braude, 1997). Similarly, crime and place theories are descriptive theories offering insight of use to spatial analysts. While most crime and place theories may fail the predictive test of other scientific theories (Sutton, 2014), they do provide a general framework for understanding offender behavior.

Fortunately, crime and place theories are now being revised through a concept called "collective efficacy" (Sampson, Raudenbush, & Earls, 1997), or in CPTED terms, second-generation CPTED (Cleveland & Saville, 1998). Second-generation CPTED is a social version of crime and place theories with specific strategies (Cozens, 2014) such as the extent to which social cohesion in a place influences the willingness of residents to defend their space. Spatial analysts can now bolster opportunity measurements by surveying crimes scenes and interviewing residents for signs of social cohesion and collective efficacy. These sociospatial patterns provide a more realistic appreciation of crime site selection by offenders.

For example, well-lit areas may not in themselves signal a defended space if residents are afraid of street gang members and there is no social cohesion to help them work together. In those cases, well-lit public areas may increase opportunities for crime and not the reverse, a process known as offensible space (Atlas, 2008). By employing contemporary second-generation CPTED assessments, the spatial analyst avoids false positive or false negative errors in interpretation.

Sociospatial analysis

Spatial forensic analysis applies the conceptual tools outlined previously and collects data (described later) to decipher characteristics of the crime scene event. In some cases, the concepts are easy to translate onto crime scenes, but sociospatial patterns are not always obvious. For example, the idea that instrumental offenders engage in search patterns is not new in criminology or in the security field. Ethnographic studies confirm the theory (Rengert & Groff, 2011; Armitage, 2013). In some cases, they learn target search skills early and their target hunting patterns follow predictable patterns, for example, their choice of large and anonymous parking lots with little security.

However, in other cases an offender travels only short distances from their own homes in search of targets. They may not victimize others around their own home, particularly if they have a sense of social cohesion with their neighbors and family members. Therefore, they will travel short distances to areas where they can offend with impunity such as neighborhoods with low social cohesion where residents are too fearful or disconnected with their neighbors to challenge outsiders.

Crime and place theories, as well as social studies of high crime neighborhoods, provide examples of these sociospatial patterns that the spatial analyst can apply to decipher crime scenes. By comparing the data with some of those sociospatial patterns, such as the defensibility of crime locations, the analyst can determine which patterns apply at that crime scene. There are numerous sociospatial patterns spatial forensic analysts will examine, including the following six.

Territorial controls

Defensible spaces are those in which legitimate neighbors feel a sense of personal ownership, responsibility, or control over a particular location or property (Newman, 1972, 1980). They will watch those spaces or occupy them for work, recreational, or leisure activities. In traditional CPTED, architects may achieve this by facing residence or shop windows toward risky areas, such as children's playgrounds. Landscape designers may achieve this by trimming hedges and trees to enhance natural sightlines. Lighting engineers may achieve this by utilizing specific lighting types and intensities. Place programmers achieve this by activating spaces with place-making activities, such as art and music festivals or sporting events.

Spatial analysts analyze the physical and social features of a crime site, interview residents, and conduct daytime and nighttime visual surveys to determine the extent to which crime site locations were "defended" at the time of the crime. Territorial controls are also affected by urban design. For example, entrapment spots where offenders can easily hide from victims and walkways may bring potential victims to, or away from risky areas. These movement predictors have a major impact on crime opportunities.

Obviously well controlled and defensible spaces are far less attractive to instrumental offenders than those areas with weak territorial controls and understanding these territorial controls may provide insight into the target selection decisions of offenders.

Target search patterns

Very few expressive or instrumental offenders commit crimes in a totally random fashion without a reason. That is why crime is clustered in time and space across the urban landscape. There are crime hotspots and places where crime is rare and one of the reasons for those patterns is that offenders must locate places where defensible space is absent and where targets are plentiful. A visual site analysis using CPTED principles reveals the extent to which areas are defensible or vulnerable.

For expressive offenders the search may amount to nothing more than a laneway outside a pub during which a drunken brawl turns deadly and kills the victim. For instrumental offenders the search may amount to a sophisticated and lengthy pattern of hunting for target rich environments over days, months, and even years of experience. In the following prison interview, one CPTED expert spoke to a murderer on death row about his methods.

Case study #1—hunting patterns at the mall

John R. Roberts is among the most experienced security specialists in the country providing CPTED expertise in court cases. He has spent many years studying crime and place research and examining crime scenes In addition to CPTED reviews on site, he also collects other information about cases. On occasion, it is possible to speak to offenders to obtain information about their target choices. In the case below, he describes an interview with one habitual offender about his rationale for site selection.

> I had reviewed the file of course, spending hours going through the details concerning the abduction of the wife and mother from the Wal-Mart parking lot in broad daylight and of her subsequent rape and murder. I had read the investigative files, the court transcripts.
>
> Death row at a federal penitentiary is an appropriately grim place under any circumstances. It seemed particularly so on a cold February morning as I trudged through the snow on my way to keep an appointment with a killer.
>
> Over a short period of time he had become a prolific 'opportunistic offender' i.e. like most criminals he sought targets that

offered a quick and easy take with little risk to him. Over a period of a decade, he had frequented Wal-Mart parking lots, stealing purses and packages, developing cons and scams to get cash for "returns" on stolen items, negotiating bad checks and more.

Offender: "I'd been breaking into cars and stuff since I was about twelve. You just go there [Wal-Mart] and if you park, you can just watch people pull up. Like some people, they will put stuff in the trunk. And if you sit there and watch the people, you know which ones put stuff in their trunk or got stuff in their cars."

Not only did they find a safe haven and targets of opportunity in perpetrating crime at Wal-Mart stores across the country, Wal-Mart even served as a safe refuge for them to spend the night, sleeping undisturbed in their car in the store parking lot.

Offender: "Like if I was driving and I was falling asleep, we would pull over. It was sort of like I knew that was a place that we could make money breaking into cars."

Roberts: "In fact, you slept the night and the next morning broke into a truck parked right next to you?"

Offender: "Yeah."

Roberts: "So, Wal-Mart is a 24 hour opportunity?"

Offender: "Yeah. We didn't want to go nowhere where there was security."

(Roberts, 2007)

Journey to crime

Journey-to-crime research is well established in the crime and place literature (Kent, Leitner, & Curtis, 2006; Block, Galara, & Brice, 2007). One of the earliest crime and place concepts was that there are distinct patterns regarding the journey that the offender undertakes in order to commit crime, patterns such as distance, frequency, and direction of certain routes over others (Frank, Andresen, & Brantingham, 2012).

With expressive offenders, the journey represents immediate opportunities that present themselves at the time of the crime, for example, a drunken bar patron with an inclination to steal items might discover an unlocked and unoccupied vehicle parked as leaves the bar. The journey to crime in that case is the route from the bar and along adjacent sidewalks along which vehicles are parked on the street. Street lighting levels may or may not have any impact on this crime. With the instrumental offender, the journey to crime is more deliberate. For example, carjackers seeking a particular vehicle to steal might search large parking lots that have poor lighting, no CCTV, and weak security. While planning their crime, these offenders will carefully examine the opportunities for crime by searching target rich areas. In this instance, street lighting will have a significant impact on this crime. To examine these patterns, spatial analysts examine routes to and from bars to parking lots, including the lighting levels, along those routes.

Distance from offender residence

One well-studied factor in the crime and place literature is the offender residence and the distance he or she travels to their crime site (Van Daele & Beken, 2011). Police crime analysts have developed distance profiles for different kinds of crimes, most of which are very short distances for many types of crimes (Frisbie, 1977; Harries, 1999).

According to one researcher, "although journeys to crime vary among crime times and with the demographic characteristics of offenders, targets or victims tend to be chosen around the offender's home, place of work, or other often-visited locations. If your home is burglarized, the chances are that the burglar is a not-too-distant neighbor" (Harries, 1999, p. 28).

Risky facilities

Studies of crime locations and targets consistently identify crime hotspots around some facilities and targets over others (Taylor, 2002; Eck, Clarke, & Guerette, 2007). For example, some vehicles models of vehicles are more popular to thieves and instrumental offenders will search areas rich in those targets such as mall parking lots where those vehicles cluster. A high-risk vehicle stolen in a non-target rich area may suggest to the spatial analyst the presence of either an opportunistic offender or an instrumental offender who followed the victim along specific routes prior to committing the crime. Both possibilities offer investigative options for locating witnesses and identifying offender methods (thieves who drive to the crime scene versus walking or biking).

Risky facility research also examines buildings and land uses, for example, low-cost motels located in certain areas (LeBeau, 2011). The proximity of risky facilities to crime sites has a significant impact on whether defensible space strategies at the crime scene will impact decisions that offender make.

Offensible spaces

CPTED architect Randal A. Atlas (1990) describes the opposite of defensible spaces as those areas where organized criminals, such as gangs, assume territorial controls for their own purposes to claim their turf on a particular street. In such places the clear sightlines, trimmed trees, and access controls do not protect legitimate users from crime, they help gang members protect their turf from rival gangs. Atlas calls these places offensible space and they exist wherever social and legal controls have collapsed, such as in high crime vulnerable neighborhoods.

Police officers in the area and residents will help spatial analysts garner local knowledge, analyze offensible spaces, conduct site visits, interviews, create crime maps, and visual inspections to further add to the profile of the social conditions in a neighborhood. This area profile is necessary in order to determine whether territorial controls contributed, or detracted, from crime opportunities at a crime site.

Recommendations for practice

As the previous discussion illustrates, spatial forensic analysts need a comprehensive understanding of crime and place theories and the controversies surrounding those theories. They also need practical experience in the application of CPTED since the crime scene event will invariably provide examples of crime opportunities attractive to both instrumental and expressive offenders.

The most important element of any crime scene analysis is the means by which the spatial forensic analyst collects data. That is where Daubert legal challenges arise and it is based on the rigor of the data collection methodology and standard practices in the field. What is required is a crime event site analysis method?

The field of CPTED contains no formally recognized standards for collecting data, other than those outlined in the International CPTED Association certification program mentioned earlier. However, since formal ICA CPTED certification is not yet required of forensic spatial analysts (or for that matter of any CPTED practitioners), the following section proposes a number of data collection steps in a specific format. It is the start of a formal methodology for data collection in spatial forensic cases. These steps appear in the CPTED literature and they are drawn from a risk assessment matrix originally presented at conferences of the International CPTED Association and now appearing in the literature (Gamman & Pascoe, 2008; Atlas & Saville, 2013). They are called the crime event site analysis.

The crime event site analysis

Aerial photography and satellite GPS data

These photos and GPS maps illustrate the crime scene in relation to other urban features such as nearby roadways, land uses, and pathways that might lead to the crime scene thereby providing easily accessible movement predictors. Depending on the date of the photos, it may be possible to identify vegetation features not visible from the ground (such as tree clusters) that may provide hiding spots for offenders.

Two other advantages of aerial and GPS photos is that they are commonly available online through mapping services and they can provide historical land use data of value to the analyst. For example, because spatial forensic analysts may be brought into cases long after the event, a crime scene may have changed significantly since the time of the incident. Fences may have been installed, new buildings and street features like landscaping may have been redesigned, and security CCTV may have been installed. Older photos can reveal much about changes at the crime scene.

Crime data and crime mapping

Across the United States, most law-enforcement agencies submit reports of crime to either federal or state-based uniform crime reporting programs, administered nationally by the FBI. In fact, in recent years the International Association of Crime Analysts guides the training and professionalization of police crime analysis. This means that local organizations typically have crime data collated for analysis and in most jurisdictions that data is readily available for forensic analysis.

While many online services now publish neighborhood crime maps, for example through real estate websites to help home purchasers, these services use unreliable or old data unsuitable for analysis. It is far better to use police data directly and to that end most police agencies are now posting neighborhood crime data and maps on their own websites.

The crime data and maps are invaluable tools for the forensic architect. They give an indication, not only of reported crime patterns at the crime scene location, but also of surrounding crime scenes. Nearby crime hotspots can tell the analyst the relative crime risk of a location and, by extension, the role of security tactics like fences and CCTV.

Case study #2—attempted murder at public housing

In one recent Florida attempted murder case, a large public housing townhome project was encircled by a 10-foot security fence; however, the front entranceway was open and uncontrolled. Property managers and architects decided that an interior senior's center was better served by locating a city-bus stop in the center of the project and therefore kept the front entranceway uncontrolled and open as city busses would not enter a security checkpoint as part of their regular service.

It was an architectural decision with serious consequences as it did not take in to account the fact the adjacent neighborhood was a persistent crime hotspot with gang activity and shootings. In spite of the encircling security fence to protect residents, gang members had easy access to the property through the front entrance and regularly dealt drugs unimpeded on the site. This ultimately led to a teenage girl being shot through the window of her ground-level apartment by drug dealers.

By analyzing the crime statistical trends and crime maps of surrounding hotspots the CPTED analyst was able to determine that this was not a crime requiring much instrumental planning. The architectural design of this facility permitted easy opportunity to an on-site target. Although drug dealers claimed this turf as their own, in later interviews they said a security kiosk at the front entranceway would have deterred their entrance (a common feature in other housing townhome projects in that city). This was ultimately discovered to be an expressive, heat of the moment crime by a drug dealer.

Visual field analysis—daytime

There is no substitute for site surveys and visual analysis by trained CPTED observers. Although multiple methods of data collection must accompany a site visit, no forensic CPTED analysis is thorough without a visual analysis.

Often called the site survey, a visual analysis will typically comprise assessments of basic CPTED principles. Some analysts' employ a written survey with specific checklist questions (Atlas, 2014); however, there is as yet no formal survey accepted by CPTED practitioners. Some employ logbooks with written summaries. In all cases, the analyst will collect a photographic survey of the site in order to document the visual field analysis. This includes a series of high-resolution photographs with 360° coverage at the scene of the event. It will also include photographs of adjacent land uses and buildings to assess sightlines, lighting impacts from adjacent illumination, and other architectural features.

The analyst will examine CPTED principles including the environmental conditions of the building, at the property, or other spatial elements at the crime scene. These include territorial reinforcement, access controls, and surveillance modalities including natural surveillance from trimmed hedges to the field views of CCTV.

Another CPTED principle comprising part of the visual analysis is the movement predictor. This involves assessing the regular movement of people in hallways, trails, or pedestrian paths around a crime scene. Movement predictors may indicate the level of crime opportunity and whether offenders might expect potential witnesses. Seasoned or so-called "professional" offenders regularly describe seeking targets that avoid potential witnesses (Armitage, 2015). That is the mark of an instrumental criminal. Therefore, if the crime occurred when abundant witnesses were evident (as a result of well-worn pathways or well-lit sidewalks with nearby residential windows and occupied porches), it suggests the possibility of an unplanned, expressive crime, such as the intoxicated robber looking for an easy target, when there is little regard to apprehension.

Movement predictors tell the analyst about the distance to potential crime locations and crime generators, such as trouble bars and drug-dealing hotspots. The proximity of such locations provides one obvious source of potential offenders, which is why police canvas them during their investigation. Movement predictors also provide the CPTED analyst with pedestrian and vehicular travel routes to and from a crime scene, which in itself suggests potential motives for offenders.

For example, consider professional burglars who drive to a crime location, but park around the corner from the target area and then walk to the crime scene. They have a choice which route they take to the crime scene. The public sidewalk has plenty of eyes on that street, but due to such routine use by pedestrians raises little suspicion (at least with smaller items of stolen property). If the offender route varies from the public movement predictor, it may well indicate intention to steal larger items not easily concealed, items like VCR and computers demanded by professional fences to fit illegal shopping orders.

Visual field analysis—nighttime

The nighttime visual analysis is similar to the daytime assessment employing the basic CPTED principles. However, because many crimes occur in the evening, it is important to assess conditions in darkness. Nighttime considerations include the effect of available lighting on the opportunity for crime and the perceptions of people on lighting conditions, which will necessitate interviews with local residents. As with daytime analysis, survey checklists, logbook notes, and photographic surveys are part of the data collection.

Lighting occupies a large part of CPTED practice, especially lighting engineering. While urban designers and architects use light illuminance values prescribed from the Institute for Electronic and Electrical Engineers (IEEE), such standards rarely consider contemporary research on crime and lighting (Welsh & Farrington, 2002).

Further, while the standards prescribe illuminance in foot-candle levels, those are inadequate to describe light distribution patterns (technically called dispersion patterns and uniformity ratios) from different kinds of fixtures such as full cutoff versus fully shielded fixtures.

CPTED analysts can review lighting sketches, called photometric plans, to assess illuminance and dispersion at each square foot of a site but a photometric plan does not reveal what actually happens in practice. Lights are subject to elements like rain and snow; they begin to age and lose efficiency, or when the canopy of deciduous trees obstructs them in summer. Thus, the nighttime visual analysis is a critical part of the data collection for a crime scene.

Table 10.1 Typical color rendition index values for different light sources

Incandescent/halogen spotlight	100
Metal halide street lights	96
Warm-white fluorescent hallway lights	51
High-pressure sodium street lights	24
Low-pressure sodium alley lights	−44

Source: Eye Lighting International Website: www.eyelighting.com/resources/lighting-technology-education/general-lighting-basics/r9-color-rendering-value/ [Accessed Nov. 7, 2015].

A visual field analysis at night is also important for assessing the qualitative impact from different light sources. That is because color rendition (the color rendition index, or CRI) varies by light source. The CRI has an impact on the behavior of offenders and victims in evening.

Sodium lighting, for example, is among the most common exterior lighting types used in street lighting. But sodium lighting, particularly low-pressure sodium lights that sometimes appear in parking lots, has a notoriously poor CRI resulting in an inability to render any colors—an important factor for police to consider when interviewing witnesses at nighttime crime scenes.

Site interviews

It stands to reason that victims and witnesses represent among the most compelling data sources available to shed light on crime events. While they may not always be available, all site visits should incorporate interviews with people from the immediate, and surrounding, location. Interviews can be informal or formal depending on the status of the case (interviewing victims in the midst of a criminal trial should only be completed with the approval of attorneys involved in the case). But in all cases interviews should be recorded either electronically, or with detailed field notes at the time of the contact, and then incorporated into the analysis.

During interviews, CPTED analysts seek information from people involved at the scene and people around the scene. That information includes observations about the event, the activities of people around the site before the event such as how people move around the site, whether people respond to territorial controls and surveillance at the site, and the impact of environmental features such as lighting, CCTV, and fencing.

Ethical issues

There is one major ethical issue this chapter presents. Forensic science is based on the premise that sound scientific methods are applied to the study of crime scenes. They require detailed expertise in an area of study and a rigorous data collection methodology, including the documentation of research results for examine by other experts. Some CPTED practitioners currently provide expert testimony in courtrooms do not follow these basic tenets. They offer expert opinions after a cursory review of the crime site and with little more data than a CPTED survey. This places their evidence in jeopardy of a Daubert challenge and it discredits the importance of spatial data in crime cases.

This chapter recommends the evolution of spatial forensic analysis into forensic science under the rubric of forensic architecture. It recommends that forensic spatial analysts utilize the crime event site analysis steps proposed here and that spatial analysts must be intensely familiar with the crime and place theories that underpin the field. It also recommends that forensic spatial analysts have extensive experience with CPTED practice, possibly including formal certification with the International CPTED Association or an advanced degree specializing in CPTED-related topics.

Future research directions

There are a number of future research directions emerging in crime and place theories that may affect forensic spatial analysis. If anything, the popularity of crime and place theories is increasing in the criminological community. However, popularity of a theory should never outweigh the evidence that supports the basic constructs of that theory. Therefore, the following themes represent areas for fruitful future research.

1. The relationship between the social dynamics of a place (such as guardianship, social cohesion, and fear perceptions) and how they affect spatial decisions by:
   a. Expressive offenders
      i. At environments with high levels of defensible space
      ii. At environments with low levels of defensible space
   b. Instrumental offenders
      i. At environments with high levels of defensible space
      ii. At environments with low levels of defensible space
2. The specific extent to which measurable environmental factors at a location deter instrumental and expressive offenders. These environmental factors include:
   a. Lighting luminance levels, types of lights, and CRI
   b. Distance from windows or view areas where actual, or potential, people servile a location
   c. CCTV coverage, including number of cameras, distance of cameras to the location, and visibility of the camera to public view
   d. The density of people in, and around, a location

3 The relationship between the positive social and cultural activities in an area, what planners call placemaking, and the likelihood that expressive or instrumental offenders will be deterred from crime. This can be measured by examining

   a Areas where placemaking activities occur
      i At intense, daily levels with large and small numbers of people
      ii At intermediate, weekly levels with large and small numbers of people
      iii At modest, monthly levels with environments large and small numbers of people
      iv At environments with low levels of defensible space
   b Areas where placemaking takes the form of
      i Physical changes to the environment such as street benches, signs, landscaping, and lighting
      ii Recreational strategies such as music events, festivals, community gardens
      iii Social or economic strategies such as news vendors, farmers markets, and public auditoriums for presentations

Conclusion

This chapter provides the outline of the theory and practice of forensic spatial analysis. Forensic architecture already occupies a place in forensic science, and it is proposed here that a spatial form of analysis also provides a new way to assess crime scenes and crime motives. Unlike other forms of criminal or geographical profiling, the spatial analyst focuses on a microlevel at the crime scene and employs a very specific CPTED method to analyze crime opportunities.

The chapter also describes some of the sociospatial patterns emerging from the crime and place literature, as well as some of the limitations of that research. Daubert challenges require the spatial forensic analyst to adopt a rigorous data-collection methodology prior to offering any opinions or conclusions about a crime event. That data-collection methodology is termed here the crime event site analysis.

When a murder, rape, or drive-by shooting occurs in a neighborhood, the responding law-enforcement officers protect the scene and await the arrival of the forensic team. Samples are collected, photographs taken, and interviews are completed. However, until recently, the focus of that forensic analysis has been on collecting physical evidence for the offence itself. The advent of forensic spatial analysis introduces a very different perspective to crime scene analysis—offender movement, target selection, opportunities for crime, and the influence of physical defensibility on the crime scene.

Key terms and definitions

Spatial forensic analysis: A form of forensic architecture based on CPTED in which the spatial, architectural, and urban design elements of a crime scene are evaluated to uncover potential sociospatial patterns.

CPTED: Crime prevention through environmental design, which is a crime prevention practice that reduces physical opportunities for crime by modifying the physical environment.

Daubert challenge: A legal challenge to prove that the opinions offered by courtroom experts demonstrate a robust research methodology and illustrate how those findings are scientifically valid.

Sociospatial patterns: The social and physical characteristics of a place that influence how offenders search for, and select, targets. Characteristics include territorial controls, risky facilities, and offensible spaces.

Defensible space: A term coined by architect Oscar Newman in his book by the same name. The capacity of a place to be defended from crime through territorial controls such as urban design, access controls, and surveillance.

Crime event site analysis: A series of data-collection steps used by a spatial forensic analyst at a crime scene.

References

Armitage, R. (2013). *Crime Prevention through Housing Design: Policy and Practice.* New York: Palgrave MacMillan.

Armitage, R. (2015). Why my house? Exploring offender perspectives on risk and protective factors in residential housing design. Presentation to the 2015 Conference of the International CPTED Association. Calgary, Canada.

Atlas, R. I. (1990). "Offensible space" – Law and order obstruction through environmental design. *Proceedings of the Human Factors and Ergonomics Society Annual Meeting*, 34 (7), 570–574.

Atlas, R. I. (2013). *21st Century Security and CPTED: Designing for Critical Infrastructure Protection and Crime Prevention, Second Edition.* Boca Raton, FL: CRC Press.

Atlas, R. I., and Saville, G. (2013). Measuring success. In R. Atlas (Ed.), *21st Century Security and CPTED: Designing for Critical Infrastructure Protection and Crime Prevention, Second Edition.* Boca Raton, FL: CRC Press, pp. 837–850.

Berger, M. (2005). What has a decade of Daubert wrought? *American Journal of Public Health*, 95 (S1), S59–S65.

Block, R., Galary, A., and Brice, D. (2007). The journey to crime: Victims and offenders converge in violent index offences in Chicago. *Security Journal*, 20, 123–137.

Bouhana, N. (2013). The reasoning criminal vs. Homer Simpson: Conceptual challenges for crime science. *Frontiers of Human Neuroscience*, 7, 682.

Braude, S. (1997). The predictive power of evolutionary biology and the discovery of eusociality in the naked mole rat. *Reports of the National Center for Science Education*, 17 (4), 12–15.

Brantingham, P. J., and Brantingham, P. L. (1981). *Environmental Criminology*. Long Grove, IL: Waveland Press.

Clarke, R. V. G., and Cornish, D. (1985). Modeling offenders decisions: A framework for research and policy. In M. Tonry and N. Morris (Eds.), *Crime and Justice: An Annual Review of Research, Volume 6*. Chicago, IL: University of Chicago Press.

Cleveland, G., and Saville, G. (2003). An introduction to 2nd generation CPTED: Part 2. *CPTED Perspective*, 6, 7–10.

Cozens, P. (2014). *Think Crime: Using Evidence, Theory and Crime Prevention through Environmental Design (CPTED) for Planning Safer Cities*. Quinns Rock: Praxis Education Books.

Cozens, P. M., Saville, G., and Hillier, B. (2005). Crime prevention through environmental design (CPTED): A review and modern bibliography. *Property Management*, 23 (5), 328–356.Eck, J. E., Clarke, R. V., and Guerette, R. T. (2007). Risky facilities: Crime concentration in homogeneous sets of establishments and facilities. *Crime Prevention Studies*, 21, 225–264.

Felson, M., and Clarke, R. V. G. (1998). *Opportunity Makes the Thief: Police Research Series, Paper 98*. Policing and Reducing Crime Unit, Research, Development and Statistics Directorate, London, U.K., Home Office.

Felson, M., and Cohen, L. E. (1980). Human ecology and crime: A routine activity approach. *Human Ecology*, 8 (4), 389–405.

Frank, R., Andresen, M. A., and Brantingham, P. L. (2012). Criminal directionality and the structure of urban form. *Journal of Environmental Psychology*, 32 (1), 37–42.

Frisbie, D. W., Fishbine, G., Hintz, R., Joelson, M., and Nutter, J.B. (1977). *Crime in Minneapolis: Proposals for Prevention*. St. Paul, MN: Community Crime Prevention Project, Governor's Commission on Crime Prevention and Control.

Godwin, G. M. (2000). *Criminal Psychology and Forensic Technology: A Collaborative Approach to Effective Profiling*. New York: CRC Press.

Harries, K. (1971). The geography of American crime, 1968. *Journal of Geography*, 70, 204–213.

Harries, K. (1999). *Mapping Crime: Principle and Practice*. Washington, DC: National Institute of Justice, U.S. Department of Justice.

Helfgott, J. (2008). *Criminal Behavior: Theories, Typologies and Criminal Justice*. Thousand Oaks, CA: Sage.

Hollis-Peel, M. E., Reynald, D. M., van Bavel, V., Eiffers, H., and Welsh, B. C. (2011). Guardianship for crime prevention: A critical review of the literature. *Crime, Law, Social Change*, 56 (1), 53–70.

International CPTED Association. (2015). *The ICA Certification Program*. Retrieved from www.cpted.net/Certification

Jeffery, C. R. (1971). *Crime Prevention through Environmental Design*. Beverly Hills, CA: Sage Publications.

Kent, J., Leitner, M., and Curtis, A. (2006). Evaluating the usefulness of functional distance measures when calibrating journey-to-crime distance decay functions. *Computers, Environment and Urban Systems*, 20 (2), 181–200.

LeBeau, J. (2011). *Measuring, analyzing, and visualizing the criminality of place: The example of hotels and motels*. Paper presented at the 11th Annual Crime Mapping Research Conference. Miami, FL, October 19–21.

Lu, Y. (2003). Getting away with the stolen vehicle: An investigation of journey-after-crime. *The Professional Geographer*, 55 (4), 422–433.

Lynch, R. (2013). Forensic architecture: An introduction and case studies. *Proceedings of the American Academy of Forensic Sciences*, Volume XIX, Washington, DC.

Newman, O. (1972). *Defensible Space: Crime Prevention through Urban Design*. New York: MacMillan.

Newman, O. (1980). *Community of Interest*. New York: Anchor Press/Doubleday.

Noon, R. K. (2000). *Forensic Engineering*. Boca Raton, FL: CRC Press.

Perry, W. L., McInnis, B., Price, C. C., Smith, S., and Hollywood, J. S. (2013). *Predictive Policing: The Role of Crime Forecasting in Law Enforcement Operations*. Santa Monica, CA: Rand Corporation.

Rengert, G. (1989). Behavioral geography and criminal behavior. In D. Evans and D. T. Herbert (Eds.), *The Geography of Crime* (pp. 161–175). London: Routledge.

Rengert, G., and Groff, E. (2011). *Residential Burglary: How the Urban Environment and Our Lifestyle Play a Contributing Role*, 3rd Edition. Springfield, IL: Charles C. Thomas Publishers.

Reynald, D. (2010). Guardians on guardianship: Factors affecting the willingness to supervise, the ability to detect potential offenders, and the willingness to intervene. *Journal of Research in Crime and Delinquency*, 47 (3), 3358–3390.

Roberts, J. R. (2007). *Target Wal-Mart*. Retrieved from www.jrrobertssecurity.com/pdfs/target-wal-mart.pdf

Rossmo, D. K. (1995). Overview: Multivariate spatial profiles as a tool in crime investigation. In C. R. Block, M. Dabdoub, and S. Fregly (Eds.), *Crime Analysis through Computer Mapping*. Washington, DC: Police Executive Research Forum, pp. 65–97.

Rossmo, D. K. (2000). *Geographic Profiling*. Boca Raton, FL: CRC Press.

Sampson, R., Raudenbush, S. W., and Earls, F. (1997). Neighborhoods and violent crime: A multilevel study of collective efficacy. *Science*, 277 (5328), 918–924.

Saville, G. (In press). The missing link in CPTED theory. In B. Teasdale and M. Bradley (Eds.), *Preventing Crime and Violence*. New York: Springer Publications.

Saville, G., and Murdie, R. (1988). The spatial analysis of motor vehicle theft: A case study in Peel Region. *Journal of Police Science and Administration*, 16 (2), 125–134.

Sherman, L., Gartin, P., and Buerger, M. (1989). Hot spots of predatory crime: Routine activities and the criminology of place. *Criminology*, 27 (1), 27–55.

Sutton, M. (2014). Fencing/receiving stolen goods. In G. J. N. Bruinsma and D. L. Weisburd (Eds.), *Encyclopedia of Criminology and Criminal Justice*. New York: Springer Science and Business Media.

Taylor, N. (2002). *Robbery against Service Stations and Pharmacies: Recent Trends*. Canberra: Australian Institute for Criminology.

Tobler, W. (1970). A computer movie simulating urban growth in the Detroit region. *Economic Geography*, 46 (2), 234–240.

Van Daele, Stijn, and Vander Beken, T. (2011). Outbound offending: The journey to crime and crime sprees. *Journal of Environmental Psychology*, 31 (1), 70–78.

Welsh, B. C., and Farrington, D. (2002). Surveillance for crime prevention in public space: Results and policy choices in Britain and America. *Criminology and Public Policy*, 3 (2), 497–526.

11 Emerging trends in technology and forensic psychological roots of radicalization and lone wolf terrorists

Jessica Mueller and Ronn Johnson

Introduction

The technological advances in recent years have made it easier for people around the world to connect with one another. Since the first World Trade Center bombing in 1993 and the Oklahoma City bombing in 1995, the world has entered into a new age of terrorism that is more remarkably media oriented (Nacos, 2002). In the past, most, if not all, acts of terrorism result in vast amounts of publicity and news reporting. The growing use of the Internet and modern communication has made it easier to report on terrorist attacks, but it is also easier for terrorist groups to spread propaganda, recruit fighters for their cause, and to transmit their message more easily and freely (Weimann, 2015). Rather than the rallies and flyers of the past, we have message boards, social media, and YouTube videos. Terrorist groups are growing at an alarming rate due to the ability to amass media attention and to interact with individuals on a personal level. Much of the material on the Internet is aimed at recruiting new members to try to increase the size of the group or encourage individuals to engage in individual acts of terror (Silke, 2010).

Lone wolf terrorism has been considered the fastest-growing form of terrorism (Weimann, 2012, 2014a, 2014b, 2015) and is considered to be a global issue. Between 1990 and 2013, the United States was the most targeted country, representing 63% of all global lone wolf attacks followed by the United Kingdom, Germany, and other Western countries (Teich, 2013). According to Phillips (2011), a lone wolf terrorist may be more deadly and dangerous than a terrorist organization in part because lone wolves often plan and engage in attacks which reap the greatest payoff and all of their resources go into a singular attack. This chapter will focus on how lone wolves radicalize on the Internet and how emerging trends in technology can facilitate the radicalization process.

Lone wolf terrorism

Helfstein (2012) defines a lone wolf terrorist as an individual who plans or perpetrates an attack without a prior history of contact with social groups or organizations that operate with the aim of advancing shared radical political goals. Helfstein (2012) argues that lone wolves adopt a radical ideology on his or her own, without external influence, and then acts or attempts to act on those self-acquired beliefs. Spaaij (2010) defined lone wolf terrorism as terrorist attacks carried out by those who operate individually, do not belong to a terrorist group or network, and whose modus operandi are conceived by the individual without any direct outside command. The definition of lone wolf terrorism continues to be shaped as time goes on, in part due to the continuous development of technology. We now know that lone wolves may connect with others and engage in online conversations with other like-minded individuals to gather information or seek validation of their beliefs. These individuals may have some loose affiliations to terrorist groups or have been inspired by the beliefs, ideology, or political reasoning of a terrorist group.

Lone wolves are inspired by diverse ideologies, but Jihadist motivations appear to be rising. Researchers have explored motivations of lone wolf attacks; however, results appear to be conflicting. Spaaij (2010) evaluated 72 cases of lone wolf terrorism to determine the ideological motivation behind the attacks. According to Spaaij (2010), white supremacy ideals, Islam, and antiabortion views motivated lone wolves in the United States. In other countries, nationalism/separatism and white supremacy were the primary reasons uncovered. However, in 30 cases, the lone wolves motivation for the attacks were unknown. Gill, Horgan, and Deckert (2014) evaluated 119 lone wolf terrorists from the United States and Europe and found that among ideological motivations, religion was among the highest (43%). Lone wolves affiliated with right-wing ideologies (34%), antiabortion or environmental campaigning (18%), and others (5%) make up other lone wolf ideological motivations. Although these numbers do differ, it shows that both religion and right-wing extremism are the two largest motivation categories for lone wolf terrorism that have been identified.

Researchers argue group identification is an important reason for sacrificing oneself for the greater good of loved ones or group members, whether the actual act is carried out alone or with others. Lone wolf terrorists act on his or her own and are not directed by a terrorist group leader to commit a terrorist act (Moskalenko & McCauley, 2011; Phillips, 2011, 2012; Spaaij, 2010). A terrorist group may discourage individuals to act without direction from the group leaders, but the individual's self-interest overrides the group's interest (Moskalenko & McCauley, 2011). Moskalenko and McCauley (2011) stated, "identification with a group is cheap, action is expensive." In other words, it is easy to identify with a group and provide the group with funding or resources, but engaging in a terrorist act costs an individual their life or risks potential prison time. Often, the lone wolf may not identify with a group and may be motivated by a personal purpose (Phillips, 2011; Spaaij, 2010).

Lone wolves are continuing to challenge the police and intelligence organizations, as they are extremely difficult to detect and defend against (Weimann, 2012). Compared to members of a terrorist group, lone wolves are more difficult to identify since most of them do not share their plans and they come from different ideological and religious backgrounds. Lone wolves tend to connect and communicate with others involved in a terrorist group; however, they may keep their identity private and collect information to help them carry out their bigger plan.

According to Phillips (2011), lone wolf terrorists may be more deadly and dangerous than a terrorist organization. This could be because lone wolf terrorists assess a target or plan an attack, weighing the probability of success verse getting captured. Phillips (2011) states that lone wolf terrorists are likely to engage in assassinations, armed attacks, bombings, hostage takings, or unconventional methods depending on the individual's level of risk. Armed attacks or bombings typically provide the maximum payoff desired because both can reach many victims. Like most terrorist attacks, committed by lone wolves or a terrorist group, the motivation for each attack is open to interpretation.

Phillips (2012) describes two different types of lone wolf terrorists, serial and spree. The spree lone wolf is considered to be an individual who carries out a series of attacks over a period of time. Ted Kaczynski, also known as the Unabomber, is an example of a serial lone wolf. Kaczynski sent bombs to former professors and American Airlines over the course of almost 20 years. Kaczynski made sure he remained undetected so that he could carry out attacks as part of his larger plan. The spree lone wolf commits a sudden spree of violence in a very short period. Anders Breivik, a Norwegian lone wolf, is an example of spree lone wolf. On a single day in 2011, Breivik bombed the government buildings in Oslo and committed a mass shooting resulting in the death of 77 people total.

Researchers are trying to further classify different types of lone wolves. For example, Corner, Gill, and Mason (2016) examined different types of lone wolves including lone-mass murders (who kill four of more people in a 24-hour period, but did not have an identified ideological motivation), lone-actor terrorists (those who carry out an attack that is ideologically motivated, but who is not directed by a terrorist group to engage in the attack), solo-actor terrorists (who committed an act of terrorism by themselves, but were directed and controlled by a larger terrorist group), and lone dyads (a group of two terrorists). Weimann (2014b) simply states that lone wolves may be an individual or a small group of lone wolves. Recent examples of lone wolves include the Boston Marathon Bombers Dzhokar and Tamerlan Tsarnaev, the Pulse Nightclub Shooter Omar Mateen, the couple responsible for the San Bernardino attack, Syed Rizwan Farook and Tashfeen Malik, and arguably Ahmad Rahami, who was responsible for the September 2016 New York and New Jersey bombings.

Precursors to becoming a lone wolf terrorist

While there are dozens of precursors to radicalization, Wilner and Dubouloz (2015) describe three primary factors related to the radicalization of Westerners which includes sociopolitical alienation, religiosity and globalization, and anger over a state's foreign policy. The sociopolitical reasoning is the most commonly cited reason for radicalization and homegrown terrorism. When an individual is unable to assimilate to his or her country's identity or political system, he or she may seek out other like-minded individuals or those who share similar views. An individual who feels alienated will seek out a group to obtain a sense of belonging. Religiosity and globalization are also factors related to self-radicalization. As previously mentioned, some terrorists cite religious motives for their actions. However, Westerners that have become radicalized are not as versed in theology and may have weak religious roots. Globalization factors and cultural shifts of technology including modernization, urbanization, secularism, and displacement, may be the cause for Western radicalization, leading to a simplification of religion and all-or-nothing thought patterns. Wilner and Dubouloz argue Western Muslim youths reassert their religion or search for an identity and a sense of belonging. Instead of following a religious leader, youths pick and choose quotes or followers from the Internet and religious texts. The Internet makes attaining virtual friendships much easier, allowing individuals to find common ground regarding religion and global concerns. Religion and globalization can help create an environment priming one for radicalization, but they do not necessarily cause radicalization. The third factor of Western radicalization is the reaction to and rejection of one's native country's foreign and defense policy. Wilner and Dubouloz state that the development of government policy humiliates and angers some Western Muslims and they feel justified in seeking revenge. This may be true for non-Muslim citizens as well. Citizens may radicalize for a primary political reason, resulting in attacks on domestic soil. These politically motivated citizens may target federal buildings or controversial establishments such as Planned Parenthood. Other precursors and triggers are cited by Post, McGinnis, and Moody (2014) include death or injury to a family or friend or having a personal crisis.

Helfstein (2012) describes Doug McAdam's proposition that escalation to high-risk activism is a circular process whereby participation increases association with the activist network. This then deepens ideological socialization, furthering the development of an ideological identity, and ultimately increasing the likelihood of high-risk activism. McAdam noted there is a predisposition for joining high-risk activism at a time when young individuals assert their independence, and, as a result, see participating in these groups as a way of enhancing their status (McAdam, 1986 as cited in Helfstein, 2012).

Self-radicalization, radicalization, and recruitment

U.S. statutes associate a homegrown violent extremist or lone wolf with self-recruitment and self-radicalization (Sinai, 2008). Radicalization is a personal process in which individuals adopt extreme political, social, and/or religious ideals and aspirations, and where the attainment of particular goals justifies the use of violence (Wilner & Dubouloz, 2015). The radicalization process looks different for each individual. Some individuals actively seeking out materials whereas others fall prey to a propaganda campaign. Individuals who are considered lone wolves and who are considered to have self-radicalized do not go through a traditional external terrorism recruitment process. Rather, individuals who self-radicalize may have internal turmoil that results in an individual actively seeking out materials online to assist them in the process of carrying out a lone wolf terrorist attack. For example, Dzhokhar and Tamerlan Tsarnaev were motivated by extremist Islamic beliefs as a result of their exposure to online messages, YouTube channels, and through the online magazine *Inspire* (Weimann, 2014b, 2015).

Meloy and Yakeley (2014) reviewed findings in empirical literature and case studies to generate a list of common motivations for an individual to self-radicalize. These motivations include personal grievance and moral outrage, ideology, failure to affiliate with an extremist group, dependence on a virtual community, and occupational or educational failure. These motivations are important to consider especially in our increasingly technological world. The desire and decision to commit a terrorist act is often motivated by a combination of personal grievance and moral outrage concerning particular a historical, religious, or political event. The Internet allows individuals to engage in research and find other like-minded individuals who also like to settle these grievances. Individuals tend to commit an act of terrorism that is framed by an ideology. This means there is a specific belief system, either based on religion, political philosophy, secular commitment, that is the driving force. Although this may be the case, it has been found that the belief system is usually quite superficial and is just a cluster of chosen statements that provide a broad rationalization for the terrorist attack. Some lone wolf terrorists fail to affiliate with and may be rejected by an extremist group they are interested in joining, and this is a prelude to isolation and development of a belief system that violence is the only answer. Many lone wolves depend on a virtual community. For a lone wolf, the internet is their own virtual community where they can relate to others similar to them if they choose.

Some researchers argue that there is no such thing as self-radicalization and that terrorist groups ask for acts of lone wolf terrorism. It can be further argued that materials and propaganda on the Internet are designed to recruit and inspire acts of terrorism, and thus impossible for an individual to self-radicalize (Clark, 2014). Whether or not an individual can truly be self-radicalized, there is a recruitment and radicalization process that can be facilitated online. Simply by searching the Internet, one can find like-minded individuals or materials that support one's ideological belief system, which assists in the self-radicalization or recruitment process. The recruitment process has traditionally occurred in person at schools, prisons, and religious centers (Sparago, 2007), but with increasing advancement in technology, propaganda designed to facilitate recruitment is becoming unavoidable.

According to Weimann (2015), the interactive capabilities of the Internet, including social networking sites, video-sharing sites, and online communities, allow terrorists to also lure targeted individuals to their sites.

Individuals are inspired to join a terrorist group or carry out a terrorist attack for a number of reasons including past exposure to violence, poverty, moral obligations, desire to enter heaven, status, and to support one's family (Kruglanski & Fishman, 2009). Terrorist groups take advantage of an individual's weaknesses and tailor recruitment tactics and social media profiles to the current cultural climate and social needs of the desired audience. For example, Weimann (2006) stated websites use slogans, texts, and language designed for a specific audience whether that may be youth, teenagers, or young adults. Regardless of the audience, lone wolves have a propensity to view this material.

Gerwehr and Daly (2006) described four common tactics terrorist groups use to recruit. These techniques have primarily been studied in settings where an individual is recruited in person, but they are being adapted to cyberspace. The first tactic is the "net" recruitment style. The net involves sending a group of individuals an invitation to an event or sending every person in that group a video about their purpose. Regardless of whether they thought positively or negatively about the event or video, as long as they watched or participated they are considered primed for recruitment. The "funnel" is another recruitment tactic used to by terrorists. The funnel is used to lure potential recruits through an all-day grooming process. Some potential recruits move on to become full members while others are weeded out. The funnel tactic uses hazing rituals and group identity exercises to ensure true commitment to their cause. Online, the funnel tactic may happen on a more individual level. The *Course in the Art of Recruiting* (Al Qa'idy, 2010) written and translated by seasoned ISIS recruiters, describes the individual grooming process including identification exercises designed to test an individual's loyalty. The "infection" is another effective recruitment technique to gain recruits from within the desired population. A trusted agent enters a population with the primary goal to persuade individuals to join their cause. The infection is effective because of the degree of source credibility, social comparison, and validation, and it targets the population's appeals. The infection is a technique commonly used on the Internet. A trusted agent can enter an Internet forum or social media group to begin the recruitment process. Some desired populations are difficult to infiltrate, therefore, the "seed crystal" tactic is used. The seed crystal method involves recruiters to manipulate the environment to produce self-recruitment. On the Internet, recruiters convince individuals that their current living conditions are poor and could have a more enriched and lavish life if they convert to Islam, follow their interpretation of Quran, and join ISIS. These conversations can take place anywhere on the Internet and social media, as well as through interactive online gaming and role play games. The seed crystal is a more time-consuming approach, but this is said to be the most successful for terrorist groups and is the tactic most used on the Internet. While an individual can engage in any one of these recruitment processes, they may choose to act on their own volition with the knowledge obtained from interactions with a recruiter but without the direction of the terrorist group.

Pathology of lone wolf terrorists

Although there is no specific psychological profile of a lone wolf, there seems to be a pattern of characteristics uncovered by researchers. These characteristics may include social isolation, social alienation, living in a hostile society, disenfranchisement, and feelings of victimization of an unfair social system (Weimann, 2014b). According to Helfstein (2012), there are few people who progress to violent action in isolation, and those that do are often motivated by other forces such as mental health issues or other political grievances. Many individuals who engage in senseless acts of violence have experienced occupation or educational dissatisfaction or failure. Instead of grieving the loss of an occupational or educational experience or making an effort to move on, a lone wolf will find targets for their grievances. There are some lone wolves who kill for a political or religious cause, many of them were previously thought to suffer from a mental health disorder (McCauley, 2007), but up until now, there has been little empirical evidence concerning lone wolves or terrorist group members having a mental health diagnosis.

The existing literature did not acknowledge differences in lone actors and group actors, as well as differences in how a terrorist act was carried out (Corner, Gill, & Mason, 2016). A common suggestion is that there must be something wrong with these individuals who commit acts alone or with a group and that they must be crazy, suicidal, or psychopathological (McCauley, 2007). The roles, functions, and experiences may be fundamentally different for an individual who is a lone wolf or an individual who is a part of a terrorist group as well as potentially having different psychological factors. Recent studies suggest mental health disorders are more common in lone wolf terrorists compared to members of a terrorist group (Corner & Gill, 2015; Corner, Gill, & Mason, 2016).

As part of a larger study, Spaaij (2010) examined five cases of alleged lone wolf terrorism. The individuals studied were Theodore Kaczynski, Franz Fuchs, Yigal Amir, David Copeland, and Volkert van der Graaf. Spaaij (2010) found three of the five individuals examined had a personality disorder, one was diagnosed with obsessive-compulsive disorder, four had experienced severe depression during at least one stage in their life. These case studies presented may suggest a possible correlation between those acting as lone wolves and the co-occurrence of a psychological disturbance. However, additional research would need to be conducted to determine a causal link. Spaaij also suggests that some individuals are not members of a terrorist organization, but they sympathize or identify with their cause and ideology while others appear to be less directly influenced by terrorist groups' greater cause. All five of the individuals withdrew themselves from society and were considered unskilled socially, had few friends, and preferred to be alone.

A more recent study found that individuals who are a part of a terrorist group are unlikely to be diagnosed with a mental health disorder whereas lone wolf terrorists are more likely to have schizophrenia, depression, autism spectrum disorder, or an unspecified personality disorder (Corner, Gill, & Mason, 2016). This study found that schizophrenia is nine times more likely to exist in a lone wolf terrorist compared to terrorist group members and the general population. Autism spectrum disorder is approximately three and half times more likely to occur in a lone wolf terrorist compared to the general public. Lone wolves with characteristics of autism spectrum disorder may find comfort in online interactions. Individuals with autism spectrum disorder have deficits in social interaction, which impair an individual's ability to maintain functional relationships, but these individuals often foster intense online relationships, a characteristic of many lone wolves.

It should be of note that lone wolf terrorists with a mental health diagnosis were just as likely to engage in a range of rational attack planning behaviors as those without mental health diagnosis (Corner & Gill, 2015). These rational attack-planning behaviors could range from engaging in research on location accessibility, weapon choice, and bomb-building instructions, all of which can be discovered on the Internet. When considering a mental health diagnosis of an individual who has carried out a lone wolf terrorist attack, it is important to determine whether the terrorist attack was a by-product of the mental illness or if the mental illness is secondary.

Use of the internet

Terrorist communication and propaganda

There are several components to how terrorist groups approach communication and how they present the group. Websites contain terrorist organization's history and activities. They discuss the biographies of leaders, founders, heroes, data regarding political aims, ideology, criticisms of adversaries, and daily news. The broadcasted messages on social networks or websites owned by terrorist organizations have a specific communication structure that includes three distinct stages: propaganda, countering propaganda, and influencing propaganda (Fulea, Mircea, & Corbu, 2015). The "propaganda" conveys a lack of an alternative for the weak, poor, and helpless to fight the strong, rich, and resourceful. The "countering propaganda" rhetoric aims to counter their enemy. This is carried out on social media platforms and includes distorted and manipulated messages that say the objective of terrorism is to obtain a peaceful settlement using diplomatic negotiations. Finally, the "influencing" stage involves distributing propaganda that demonizes and questions the legitimacy of enemies. These messages are being amplified by the increasing dissemination of malicious and extremely violent videos (Fulea, Mircea, & Corbu, 2015).

Virtual training and access to materials

Today, it is not necessary for terrorists to travel to the Middle East for training; this can now be done virtually alone or with others. The use of the Internet has made it much easier to obtain coveted documents terrorist groups use. The Mujahideen Poisons Handbook, the Encyclopedia of Jihad, The Voice of Jihad, and the Art of Recruiting Handbook are just a few examples of manuals. According to Weimann (2015), there are thousands of new and updated pages of terrorist manuals, instructions, and rhetoric published on the Internet every month. Terrorist groups rely on the Internet for sharing their instructions and their training campaigns. These resources stimulate planning and tactical strategy development, regardless of expertise or skill level of an individual (Holt, Bossler, & Seigfried-Spellar, 2015).

Internet support

According to *44 Ways of Supporting Jihad* (Al-Awlaki, n.d.), an important guide for Jihad followers, the Internet has become a great medium for spreading the call of Jihad and the news of the Mujahideen. The interactive experience of the Internet, including social networking sites, video-sharing sites, and online communities allow terrorists to assume a proactive position (Weimann, 2015) meaning terrorist groups do not have to wait for people to come across their websites, they can lure targeted individuals to their sites. Online social networking provides terrorists with an ideal platform to attract and seduce, teach and train, radicalize and activate individuals all over the world (Weimann, 2015). Social media apps and file sharing platforms such as Facebook, Twitter, Instagram, Tumbler, Ask.fm, WhatsApp, PalTalk, kik, viper, and JustPaste.it are easy to use (Klausen, 2015). Terrorist groups can also establish forums, email lists to share information, post or email literature and news, and set up websites to cover specific areas of Jihad. Again, this material is designed to recruit members or to radicalize individuals. These support sites can be accessed by lone wolf terrorists to obtain information to aid in their attack or to seek help from individuals who support a similar cause. There is also a strong social component in which individuals search for social ties and validation before taking action (Helfstein, 2012).

Internet support keeps terrorist groups relevant and sustainable. Silke (2010) describes the difference between soft and hard supporters of terrorist groups. Soft supporters are accepting of terrorist groups. Soft supporters of a terrorist group will not necessarily publicly voice their approval of the terrorist group's actions or aims, and they may not provide obvious assistance to the group. Hard supporters are considered to present a further level of commitment to the group or to the cause. They try to provide practical assistance such as contributing financially or provide weapons to the group. Lone wolves may start as soft supporters. They may initially passively understand the group's cause, but not be actively involved. After a precipitating event, an individual may try to become a hard supporter and commit to a cause. Silke (2010) stated that it is unlikely for the Internet to generate many hard supporters, but the Internet will increase the amount of soft support for the groups. Internet support may come in the form of following accounts on social media, sharing posts, or participating in Internet forums.

Internet support can come in various forms. One of the most disturbing cases of lone wolf terrorism is Nicky Reilly, who attempted to carry out a suicide bomb attack in May 2008 in the United Kingdom. Reilly, a Muslim convert, researched how to make improvised explosive devices (IEDs) on the internet. He also used the internet to obtain the IED's components, as well as to investigate potential targets. Reilly was in frequent contact via the Internet with two unidentified men from Pakistan, with whom he discussed his intentions and from whom he received encouragement and information. The unidentified individuals answered his questions and directed him to bomb-making websites. The two unidentified men were never apprehended, allowing them to continue to assist lone wolves and potential recruits. However, one does not necessarily need to interact with others. With all of the propaganda on the internet, one can simply follow internet forums, email lists, and social media accounts to help facilitate their radicalization process.

Social media and video propaganda

The use of social media to develop networks and social support is critical in the formation of collective identity that can move into the real world and effect everyday life (Holt, Bossler, & Seigfried-Spellar, 2015). Social media outlets such as Facebook, Twitter, and Tumblr are popular with terrorist groups. These outlets allow terrorists to provide minute-to-minute updates that include propaganda and information about their cause. These social media websites also have mobile phone applications making it easy to sign in and retrieve updates from anywhere anytime.

Interactions on social media can increase the potential for radicalization or lone wolf terrorism. Tashfeen Malik and Syed Rizwan Farook, perpetrators of the San Bernardino County terrorist attack in 2015, advocated for Jihad in messages on social media. The government allegedly conducted three background checks on Malik prior to her emigrating from Pakistan, but they did not uncover social media postings in support of Jihad. Malik supposedly made comments under a pseudonym with strict privacy settings (Perez & Ford, 2015). Using a pseudonym, specific privacy settings, and multiple social media account identities is common for lone wolves. Most lone wolves use encryption techniques and password-protected sites for exchanging sensitive information (Kaati & Johansson, 2016).

Facebook is a powerful social media platform designed to help individuals stay connected and interact with others. Facebook is a free online social networking site that allows registered users to create a personal profile, to share status updates, to post photos and videos, and to connect with friends and other like-minded individuals. One could log onto Facebook through the website or application for the mobile device. Weimann (2014a) classifies Facebook pages with terrorist content as "official" pages and "unofficial" pages. "Official" pages are sponsored by a specific terrorist group and includes official forums whereas "unofficial" Facebook pages are created by sympathizers to share propaganda and instruction manuals (Weimann, 2014a). In May 2009, Ansar al-Mujahideen followers created a Facebook group called "Islamic Jihad Union" to connect with other Jihad supporters. On this Facebook page, Jihadists can post information on weapon maintenance, propaganda, ideology as well as post additional links to Internet forums. Members of this group were notified of the risks of having their real identity tied terrorist group social media accounts (Weimann, 2010), likely encouraging them to create fake accounts without their personal identifying information.

Twitter is a free social networking "megaphone" used to reach a large audience. Registered users are able to send and read "tweets" or messages that are limited to 140 characters. Twitter can be accessed through the Twitter website as well as its corresponding application for mobile devices. Twitter has a policy against posting acts of violence, threats, and harassment and government agencies can request to have posts removed; however, Twitter is primarily unmonitored. Twitter can also be used to quickly communicate with one another privately or openly for others to see. Weimann (2014a) stated that Twitter is terrorists' favorite Internet service and considered to be preferred over self-designed websites or Facebook. Weimann (2010) describes an incident where two terrorists communicated back and forth on Twitter about how to make a bomb. In a New York Times documentary, a young girl named "Alex," reached out to someone on Twitter claiming to be a member of ISIS to inquire why they were beheading journalists (Callimachi, 2015). This initial contact began the recruitment process. Alex regularly communicated with known terrorists and received gifts in the mail from them. Within 2 months, Alex tweeted that she had converted to Islam.

YouTube is a popular video sharing website. YouTube is used as a platform for terrorist groups and supporters. Cell phone cameras and web cameras make it easy for groups like ISIS to create training videos, refute news claims, and take credit for attacks. ISIS and Al Qaeda both disseminate videos of beheadings, prayers, and other messages they would like to get across. Terrorist groups post even more gruesome videos on a lesser known video site called LiveLeak. LiveLeak is a video sharing website based in the United Kingdom that allows reality footage and permits users to post graphic or political content. These videos are primarily used for recruitment purposes and to gain sympathy for their cause. Lone wolves and other aspiring terrorists are able to watch these videos in the privacy of their own home making it easier to become radicalized or inspire an attack. While many lone wolf terrorists have not yet been known to create videos or video manifestos, this may occur in the future. School shooters and mass murderers have moved towards creating video manifestos or video confessions such as Seung-Hui Cho, the Virginia Tech shooter and Elliot Roger, the University of California, Santa Barbara shooter. So far, Anders Breivik, a Norwegian lone wolf terrorist, has been one of the only ones to create a video of his plan prior to carrying out his attack in which he killed 77 people and injured more than 300 (Botelho, Carter, Shoichet, & Stang, 2011).

Individuals who self-radicalize or who become lone wolf terrorists may engage in conversations with individuals on social media and watch videos, but they do so to prepare for their own attack. Past lone wolves such as Ted Kaczynski, the Unabomber, had hardly any contact with like-minded individuals. Recent lone wolves maintained minimal in-person contact with individuals with similar ideologies but maintained contact with people on the Internet. The contacts made, coupled with extremist propaganda have helped perpetuate acts of terror by lone wolves.

Fangirls and female terrorists

Social media makes it easy for individuals to be constantly connected. Youth and adolescents create relationships on social media to relieve boredom and loneliness as well as provide a focus for their lives (Stever, 2009). By definition, social networks create an opportunity for terrorist recruiters to interact with and encourage youth to join and follow their cause.

Terrorist groups and organizations alike would not be able to exist without followers. The Internet makes it easier for a terrorist organization to gain followers. According to a recent George Washington University study, approximately one third of ISIS followers are young girls (Vidino & Hughes, 2015). Today's social media posts show young ISIS devotees admiring the Islamic State and their soldiers. In various countries across the world, young girls are following ISIS members because they consider them "cool." Huey and Witmer (2016) describe these young girls as fangirls. Merriam-Webster dictionary defines a "fangirl," as "a girl or woman who is an extremely or overly enthusiastic fan of someone or something." Females involved in earlier terrorist groups were well versed in the ideology of their respective organizations; today's ISIS girls and women have often little or no knowledge of Islam or ISIS's interpretation of Islam (Nacos, 2015).

Fangirls are using Tumblr, Twitter, Facebook, or other social media outlets to idolize and support ISIS fighters. Young fangirls come from diverse ethnic backgrounds and geographic locations, they were raised Muslim or recently converted to Islam, and they post about their parent's concerns regarding their recent behavior and attitudes. Although researchers do not believe every one of these fangirls are radicalized or plan to travel to join ISIS, these fans maintain an environment for recruitment to be facilitated. No one knows for sure how many Westerns girls and young women have pledged themselves to the Islamic State or have traveled to join their cause.

Researchers have not been able to determine personality traits or backgrounds of female members of ISIS, but there is a connection these girls share. There appears to be a sense of sisterhood experienced by these women on social media. They may help out one another and guide each other through the process of joining ISIS or converting to Islam. When females are ready to travel to the Middle East, this sisterhood may provide advice based on their past experience and decisions. It is important to note, men may be posing and posting under a female alias and vice versa (Klausen, 2015) suggesting men may be luring women and young girls to their cause and providing them advice.

In the United States, there have been few publicized terrorist attacks carried out by female lone wolves. One example is the 2015 San Bernardino shooting where the woman was part of a couple. In other parts of the world, females commit attacks of lone wolf terrorism, but these women are associated with a terrorist group. There have been several recent international events in the news that suggest women are becoming more involved in terrorist organizations. In August 2016, two female terrorists open fired and dropped hand-made bombs near an Istanbul police department. In September 2016, a woman was charged in connection to a terrorist attack on the Notre Dame Cathedral in Paris. All three of these women had known terrorist organization ties. While there are many women associated with terrorist groups, there are few examples of women committing acts of lone wolf terrorism similar to their male counterparts.

Internet forums

As mentioned before, there are various platforms used to disseminate propaganda. Internet forums are another tool used by terrorist groups. Torres-Soriano (2013) noted the first Jihadi Internet forum appearing in 2003. This first Internet forum, Muntada Al Ansar Al Islami, distributed materials to a website unofficially affiliated with a terrorist organization, allowing it to continue its existence in cyberspace and stock propaganda materials. Internet forums initially allowed open access, enhancing communication potential and allowing curious individuals to peruse their forums. When their forums started to get shut down, they started requiring individuals to register to view content. Over time, terrorist groups sent material to various trusted websites and forums to thwart individuals who reported their sites.

Many existing Internet forums contain the same links, videos, and propaganda, which stem from a smaller group of origin sites. These other sites and mirror sites circulate information at such a rapid pace that most of the information evades authorities' attempts to shut them down. In addition to smaller sites stemming from the origin and mirror sites, there are individuals who start forums on their own, these eventually gain enough favor from those running the origins to eventually become linked with them. Torres-Soriano (2009) noted that earlier web forums did not have the capability to upload sophisticated video and audio statements due to the forums inability to store large quantities of media. Social media outlets are now equipped to handle such files.

Torres-Soriano (2013) described how forums are set up and the function of these forums. Most forums are broken up into sections, the first of which is the "statement section," which contains direct messages and propaganda from the terrorist networks which are limited to the posts of authorized spokespersons. The "general section" is completely open to participation from all members where they can post comments, news, videos, links, or anything Jihad-related. Some other sections include history, translations, trainings on bomb-making and weapons, and discussion boards on a variety of topics. The breakdown of these sections may make it easier for a lone wolf terrorist to research the topics necessary for them to achieve their goal.

Through the creation of a large number of Internet forums linked to larger origin sites, there is a greater opportunity for people with a variety of interests to focus on and discuss specific topics. Eventually, through the conversations that take place between the forum members, similar ideas will form, counter ideas will be tamed, and the members may become more and more drawn into the websites. These groups also operate by monitoring the language and opinions of newcomers in efforts to weed out possible authorities or those who would detract from the goals of the forum. A person can become more involved in a terrorist organization or obtain information to plan their attack. Some individuals may not interact with others within the forum, but the forum provides a wide variety of resources at their disposal.

Online magazines

Dabiq and *Inspire* are two online magazines designed to motivate and encourage radicalization and initiate independent terror attacks. *Dabiq* is ISIS's official publication whose goal is to communicate to Muslims regarding factual and truthful information contrary to what the international media reports (Nacos, 2015). *Inspire* is an online magazine published by the al-Qaeda in the Arabian Peninsula. Because of the English-language publication, *Inspire* has a greater capacity for reaching and affecting readers in the United States and other English language cultures. *Inspire* features figures from the West who are wanted for crimes against Islam as well as how to handle weapons and make simple bombs (Watson, 2013). The content and format of the magazine create a sense of connectedness to the Jihadi cause and motivates individuals who seek out the digital magazine (Sivek, 2013).

The content in *Dabiq* and *Inspire* target lone wolves, particularly Muslims in the West who feel marginalized. These magazines have the power to influence individuals to commit violent attacks of terrorism. For example, the Boston Marathon Bombers Tamerlan and Dzhokhar Tsarnaev were able to obtain directions on how to make a pressure cooker bomb from *Inspire* magazine (Khan, 2013). For lone wolves, the ideologies and activities proposed by online Jihadist literature may seem a valid way to rectify other personal issues or perceived social and political injustices especially since it targets Western culture and emphasizes Western Islamophobia (Sivek, 2013).

Video games

Terrorist groups are also using video games to recruit and train young children. Video games are offered in multiple languages and are designed to target a wide range of youth. These video games are similar to the games that youth are playing every day. Terrorist groups mimic common themes in popular video games such as *Call of Duty* and *Grand Theft Auto* and market the Islamic State as a real-life version of these games. Video games developed by terrorist organizations can simulate acts of terrorism and encourage the user to engage in role-playing, by allowing the player to act as a virtual terrorist. In early 2016, ISIS introduced an Android game app that individuals are able to download. This app, like other video games, is targeted for interested individuals and promotes the use of violence against a country or prominent political figures. In these games, players are rewarded for their virtual successes.

Terrorist groups can also use computer games and video games to communicate with one another as well as potential recruits. Terrorist groups can send messages or try to communicate over fighting games, particularly role-playing games such as *World of Warcraft*. Anders Breivik, a Norwegian terrorist, reportedly trained using the online game *Call of Duty: Modern Warfare 2* (Bosco, 2013). *Second Life* is another computer game that allows players to create a virtual world used for military warfare training. In these virtual worlds, you can have a real-life conversation about terrorism training or potential attacks and act them out. These games portray simulations of war, but they can mask virtual terrorist training grounds (Bosco, 2013). Video game systems such as *PlayStation 4* and *XBox* allow users to connect directly to one another through live, interactive conversations. With these game consoles, you do not need to play the game to participate in the conversation.

Lone wolf terrorists as a pseudocommando

The term "pseudocommando" was initially used by Dietz to describe an individual who, after long deliberation and much planning, commits a mass murder (Dietz, 1986). Pseudocommandos are considered driven by fantasies of revenge and fame. They collect moments that focus on the unwanted, hated, and feared aspects of self (Knoll, 2010a, 2010b). There is a dearth of research connecting terrorists to being pseudocommandos, in fact, recent research looks at them as two separate concepts (Swanson & Felthous, 2015), despite having similar characteristics. A similar concept is that of the lone wolf being labeled a "violent true believer" (Meloy, Mohandie, Hempel, & Shiva, 2001; Meloy & Yakeley, 2014), in which a person commits an act of homicide and/or suicide to advance his or her particular political beliefs. Much of the research on pseudocommandos can certainly be applied to lone wolf terrorism as the concepts are similar, but the definition of a pseudocommando lacks the political or religious motive.

Lone wolves and their pseudo affiliation

Some individuals, like Omar Mateen who was responsible for the Pulse Nightclub Shooting, have what could be considered a pseudo affiliation. Individuals who have a pseudo affiliation may have a belief system that is not shared with a terrorist organization; rather, he or she will selectively pick and choose what to believe in. Regardless of their beliefs, pseudocommandos or individuals with a pseudo terrorism affiliation will carry out attacks and see themselves as carrying a banner for a terrorist group. Meloy and Yakeley (2014) discussed the narcissistic fantasies of having a special connection to others, potentially to a terrorist group or to a religious figurehead. Lone wolves may have a sense of admiration and respect for other individuals who have previously carried out an attack and may spend time researching them. The fantasies of belonging to a group and the fantasies of glory, combined with moral outrage and grievance and the incorporation of pieces of a belief system that encourages violence, leads these individuals on a pathway to violence (Calhoun & Weston, 2003).

In the age of the Internet, lonely isolated individuals who wish to belong and be accepted are driven to spend an inordinate amount of time online (Post, McGinnis, & Moody, 2014). Lone wolf terrorists share, at least in part, an ideological or philosophical identity with the extremist group, but may not have much interaction with the group (Weimann, 2012). For example, following the Pulse Nightclub shooting, there was evidence Omar Mateen allegedly had ties to three different terrorist organizations. This is inconsistent with a terrorist group because each terrorist group demands loyalty to their singular group. Mateen may have idealized individuals belonging to these groups and he may have felt like he was a part of them or he wanted to be a part of something bigger than himself. The pseudo affiliation gives an individual the power and courage to carry out their attack.

Social media accounts now tell us a great deal of information about who a person is, their beliefs, and what potentially led up to them committing their terrorist act. Social media allows individuals to demonstrate their dedication to a terrorist group and to share their beliefs. Social media posts can be analyzed and used to help determine whether the motivation for their act of violence was political, ideological, or something else. There is often debate in the media as to whether an act of violence should be considered a mass killing or a terrorist attack, but there is typically not enough evidence to determine a motive right away.

Individuals can follow various accounts and profiles. They may like what the terrorist group believes in, but they may not have the courage or means to fly to the Middle East. Some individuals may choose to stay in their hometown and continuing to practice their ideology. Even if they are not participating in the groups or interacting with others online, they still feel a sense of association giving them a pseudo affiliation gives them a sense of strength or power.

Conclusion

The Internet and technology have significantly changed our cultural climate. It is now easier to access information and connect with others in a way that we were unable to previously. Terrorist organizations rely on the Internet to spread their propaganda, recruit new followers, and ask followers to carry out an attack. Regardless of how a lone wolf was recruited or whether they self-radicalized, lone wolves remain a growing threat. To address the growing threat of lone wolves, counterterrorism strategies should be developed using a multidisciplinary approach. Researchers and professionals in the field are working to develop tools and techniques to answer to the online activity of terrorists.

Keeping track of terrorism propaganda and social media followers on accounts linked to terrorist groups, at least in theory, could act as a valuable means for uncovering potential lone wolf terrorists (Kaati & Johansson, 2016). When monitoring social media to detect threats towards society, the goal is to identify unconcealed communication in hopes that one can find the warnings signs to prevent an attack. Unfortunately, lone wolves may easily avoid identification and detection since most do not reveal their plans and, if they do, their plans are discussed on hard-to-find websites or secretive networks (Weimann, 2015).

References

Al-Awlaki, A. (n.d.). *44 ways of supporting Jihad.* Victorious Media.

Al Qa'idy, A. A. (2010). *A course in the art of recruiting.* Retrieved from https://ia800300.us.archive.org/32/items/ACourseInTheArtOfRecruiting-Revised-July2010/A_Course_in_the_Art_of_Recruiting_-_Revised_July2010.pdf

Botelho, G., Carter, C.J., Shoichet, C., & Stang, F. (2011). *Purported manifesto, video from Norway suspect detail war plan.* Retrieved from www.cnn.com/2011/WORLD/europe/07/24/norway.terror.manifesto

Bosco, F. (2013). Terrorist use of the internet. In U. Gurbuz (Ed.), *Capacity building in the fight against terrorism* (pp. 39–46). Amsterdam: IOS Press BV.

Calhoun, T. & Weston, S. (2003). *Contemporary threat management*. San Diego, CA: Specialized Training Services.

Callimachi, R. (2015). *ISIS and the lonely American*. Retrieved from www.nytimes.com/2015/06/28/world/americas/isis-online-recruiting-american.html

Clark, M. (2014). *There's no such thing as a "Self-radicalized" Islamic terrorist*. American Center for Law and Justice. Retrieved from https://aclj.org/jihad/self-radicalized-islamic-terrorist

Corner, E. & Gill, P. (2015). A false dichotomy? Mental illness and lone-actor terrorism. *Law and Human Behavior, 39* (1), 23–34.

Corner, E., Gill, P., & Mason, O. (2016). Mental health disorders and the terrorist: A research note probing selection effects and disorder prevalence. *Studies in Conflict & Terrorism, 39*(6), 560–568.

Dietz, P.E. (1986). Mass, serial and sensational homicides. *Bulletin of the New York Academy of Medicine, 62*(5), 477–491.

Fangirl. (n.d.). *Merriam-Webster's collegiate dictionary* (11th ed.). Retrieved from www.merriam-webster.com/dictionary/fangirl

Fulea, D.C., Mircea, C., & Corbu, M.C. (2015). Communication of terror in cyberspace. In S. Anton & I.S. Tutuianu (Eds.), *The complex and dynamic nature of the security environment*. Paper presented at International Scientific Conference Strategies XXI, "Carol I" National Defence University, Bucharest, Romania: ProQuest.

Gerwehr, S. & Daly, S. (2006). Al-Qaida: Terrorist selection and recruitment. In D.G. Kamien (Ed.), *The McGraw-Hill homeland security handbook*, (2nd ed. pp. 73–89). Dubuque, IA: McGraw-Hill. Retrieved from https://www.rand.org/content/dam/rand/pubs/reprints/2006/RAND_RP1214.pdf

Gill, P., Horgan, J., & Deckert, P. (2014). Bombing alone: Tracing the motivations and antecedent behaviors of lone-actor terrorists. *Journal of Forensic Sciences, 59*(2), 425–435. doi:10.1111/1556-4029.12312

Helfstein, S. (2012). *Edges of radicalization: Individuals, networks, and ideas in violent extremism*. Combating Terrorism Center at West Point. Retrieved from www.ctc.usma.edu/v2/wp-content/uploads/2012/06/CTC_EdgesofRadicalization.pdf

Holt, T.J., Bossler, A.M., & Seigfried-Spellar, K.C. (2015). Crime and digital forensics: An introduction. New York, NY: Routledge

Huey, L. & Witmer, E. (2016). #IS_Fangirl: Exploring a new role for women in terrorism. *Journal of Terrorism Research, 7*(1), 1–10.

Kaati, L. & Johansson, F. (2016). Countering lone actor terrorism: Weak signals and online activities. In M. Fredholm (Ed.), *Understanding lone actor terrorism: Past experience, future outlook, and response strategies* (pp. 266–279). New York: Routledge.

Khan, A. (2013). *The magazine that "Inspired" the Boston bombers*. Frontline. Retrieved from www.pbs.org/wgbh/frontline/article/the-magazine-that-inspired-the-boston-bombers/

Klausen, J. (2015). Tweeting the Jihad: Social media networks of Western foreign fighters in Syria and Iraq. *Studies in Conflict and Terrorism, 31*(1), 1–22.

Knoll, J.L. (2010a). The "Pseudocommando" mass murder: Part I, the psychology of revenge and obliteration. *Journal of American Academy of Psychiatry and the Law, 38*(1), 87–94.

Knoll, J.L. (2010b). The "Pseudocommando" mass murder: Part II, the language of revenge. *Journal of American Academy of Psychiatry and the Law, 38*(2), 263–272.

Kruglanski, A.W. & Fishman, S. (2009). Psychological factors in terrorism and counterterrorism: Individual, group, and organizational levels of analysis. *Social Issues and Policy Review, 3*(1), 1–44.

McAdam, D. (1986). Recruitment to high-risk activism: The case of freedom summer. *American Journal of Sociology, 92*(1), 64–90.

McCauley, C. (2007). Psychological issues in understanding terrorism and the response to terrorism. In B. Bongar, L.M. Brown, L.E. Beutler, J.N. Breckenridge, & P.G. Zimbardo (Eds.), *Psychology of terrorism* (pp. 13–33). New York: Oxford University Press.

Meloy, J.R. & Yakeley, J. (2014). The violent true believer as a "Lone Wolf" – Psychoanalytic perspectives on terrorism. *Behavioral Science & the Law 32*, 347–365.

Meloy, J. R., Mohandie, K., Hempel, A., & Shiva, A. (2001). The violent true believer: Homicidal and suicidal states of mind. *Journal of Threat Assessment, 1*, 1–14.

Moskalenko, S. & McCauley, C. (2011). The psychology of lone-wolf terrorism. *Counseling Psychology Quarterly, 24*(2), 115–126. doi:10.1080/09515070.2011.581835

Nacos, B.L. (2002). *Mass-mediated terrorism: The central role of the media in terrorism and counterterrorism*. Lanham, MD: Rowman & Littlefield Publishers, Inc.

Nacos, B.L. (2015). *Young Western women, fandom, and ISIS*. E-International Relations. Retrieved from www.e-ir.info/2015/05/05/young-western-women-fandom-and-isis/

Perez, E. & Ford, D. (2015). *San Bernardino shooter's social posts on jihad were obscured*. Retrieved from www.cnn.com/2015/12/14/us/san-bernardino-shooting/

Phillips, P.J. (2011). Lone wolf terrorism. *Peace Economics, Peace Science, and Public Policy, 17*(1), 1–29.

Phillips, P.J. (2012). The lone wolf terrorist: Sprees of violence. *Proceedings of the 12th Jan Tinbergen European Peace Science Conference, 18*(3), 1–11.

Post, J.M., McGinnis, C., & Moody, K. (2014). The changing face of terrorism in the 21st century: The communications revolution and the virtual community of hatred. *Behavioral Science and Law, 32*, 306–334.

Silke, A. (2010). The internet and terrorist radicalization: The psychological dimension. In H.L. Dienel, Y. Sharan, & C. Rapp (Eds.) *Terrorism and the Internet: Threats, target groups, deradicalisation strategies* pp. 27–39. doi: 10.3233/978-1-60750-537-2-27

Sinai, J. (2008). How to define terrorism. *Perspectives on Terrorism, 4*(2), 9–11.

Sivek, S.C. (2013). Packaging inspiration: Al Qaeda's digital magazine *Inspire* in the self-radicalization process. *International Journal of Communication, 7*, 584–606.

Spaaij, R. (2010). The enigma of lone wolf terrorism: An assessment. *Studies in Conflict and Terrorism, 33*(9), 854–870. doi:10.1080/1057610x.2010.501426

Sparago, M. (2007). *Terrorist recruitment: The crucial case of Al Qaeda's global Jihad terror network*. Retrieved from www.scps.nyu.edu/export/sites/scps/pdf/global-affairs/marta-sparago.pdf

Stever, G.S. (2009). Parasocial and social interaction with celebrities: Classification of media fans. *Journal of Media Psychology, 14*(3). Retrieved from http://web.calstatela.edu/faculty/sfischo/

Swanson, J.W. & Felthous, A.R. (2015). Guns, mental illness, and the law: Introduction to this issue. *Behavioral Science and the Law, 22*, 167–177.

Teich, S. (2013). Trends and developments in lone wolf terrorism in the Western world: An analysis of terrorist attacks and attempted attacks by Islamic Extremists. International Institute for Counter-Terrorism. Retrieved from www.ctcitraining.org/docs/LoneWolf_SarahTeich2013.pdf

Torres-Soriano, M.R. (2009). Maintaining the message: How Jihadists have adapted to web disruptions. *CTC Sentinel, 2*(11), 22–24.

Torres-Soriano, M.R. (2013). The dynamics of the creation, evolution, and disappearance of terrorist internet forums. *International Journal of Conflict and Violence, 7*(1), 164–178.

Vidino, L. & Hughes, S. (2015). *ISIS in American: From retweets to Raqqa.* Retrieved from https://cchs.gwu.edu/sites/cchs.gwu.edu/files/downloads/ISIS%20in%20America%20-%20Full%20Report_0.pdf

Watson, L. (2013). Al Qaeda releases guide on how to torch cars and make bombs as it names 11 figures it wants 'dead or alive' in latest edition of its glossy magazine. Daily Mail. Retrieved from www.dailymail.co.uk/news/article-2287003/Al-Qaeda-releases-guide-torch-cars-make-bombs-naming-11-public-figures-wants-dead-alive-latest-edition-glossy-magazine.html

Weimann, G. (2006). *Terror on the internet: The new Arena, the new challenges.* Washington, DC: United States Institute of Peace.

Weimann, G. (2010). Terror on Facebook, Twitter, and YouTube. *The Brown Journal of World Affairs, 16*(2), 45–54.

Weimann, G. (2012). Lone wolves in cyberspace. *Journal of Terrorism Research, 3*(2). doi:10.15664/jtr.405

Weimann, G. (2014a). *New terrorism and new media.* Commons Lab, Science and Technology Innovation Program. Washington, DC: Woodrow Wilson International Center for Scholars. Retrieved from www.wilsoncenter.org/publication/new-terrorism-and-new-media

Weimann, G. (2014b). *Virtual packs of lone wolves.* Woodrow Wilson International Center for Scholars. Retrieved from www.wilsoncenter.org/article/virtual-packs-lone-wolves

Weimann, G. (2015). *Terrorism in cyberspace: The next generation.* Washington, DC: Woodrow Wilson Center Press.

Wilner, A.S. & Dubouloz, C. (2015). Homegrown terrorism and transformative learning: An interdisciplinary approach to understanding radicalization. *Global Change, Peace and Security, 22*(1), 33–51. doi:10.1080/14781150903487956

Part IV

Scientific advancements in forensic investigations

12 Phenylketonuria (PKU) cards

An underutilized resource in forensic investigations

Scott Duncan

Introduction

Annually in the United States, authorities manage up to 90,000 reports of missing children and adults (National Missing and Unidentified Persons Fact Sheet, 2014).[1] Most of these cases are closed quickly as they are determined to be misunderstandings between family members, youths who run away and return home, or adults who were temporarily lost. For a missing person incident that remains open, law enforcement will conduct follow-up investigations to uncover clues to explain the disappearance. In some instances, detectives may even collect unique identifier evidence such as fingerprints or hair strands in hopes of resolving these cases.

Paralleling the thousands of missing person investigations, local authorities struggle to identify individuals who have died but remain unknown. A US Department of Justice study estimated that there were over 40,000 unidentified bodies that had been documented by law enforcement, medical examiners, and coroners between the early 20th Century and 2006, with about 1,000 new cases added each year (Ritter, 2007).

The volume of missing and unidentified decedents can be overwhelming. The time and expense required to investigate a cold case with no viable leads is problematic for agencies of any size, but especially for authorities in rural and/or impoverished jurisdictions where resources and personnel are limited. And despite technological advances in identification and improved conduits for communication among law enforcement and public health officials, missing persons and unidentified bodies continue to challenge communities. Nancy Ritter (2007) referred to this phenomenon as the nation's "silent mass disaster."

What can be done to assist authorities in matching missing persons to recovered unidentified bodies? Investigators and researchers use DNA to aid in this process. Saliva or blood from a personal item of the missing individual or from direct family members can now be processed, analyzed, and then submitted to searchable online databases for potential matching. One underutilized technique in obtaining an individual's DNA is the use of Phenylketonuria (PKU) test cards or Guthrie cards (also known as infant or neonatal blood spot cards).

Since the 1960s, US departments of health as well as health departments from other countries have collected neonatal PKU cards, albeit with different retention schedules. These blood spot cards are virtually untapped direct DNA sources for authorities investigating missing persons or unidentified decedents. They also contain a wealth of information for authorities responding to mass disasters or investigating unexplained deaths. This chapter explores the potential benefits of combining the emerging technologies related to DNA with a mandated health record. Specific objectives are: (1) describe the traditional function of a PKU card, (2) understand the uses of PKU cards and potential contributions to investigations, (3) gain familiarity with the varied retention schedules for these records, (4) comprehend ethical concerns and best practices related to PKU cards, and (5) recognize opportunities for future research.

Background

Since the formalization of law enforcement functions, authorities have relied on innovative techniques for identification purposes involving criminal and death cases (Swanson, Chamelin, Territo, & Taylor, 2009). If known, the names of criminal suspects and those convicted could be researched and then linked to previous crimes and criminal histories, but the process was based solely on physical recognition. In 1883, Alphonse Bertillon hypothesized that certain physical measurements for individuals were unique and could be used to identify individuals who had been previously arrested. These procedures were formalized to create the "Bertillon" system in which 11 separate measurements of an individual were recorded – ranging from cheek width to head length (2009). After a few years of use, the tediousness and the inaccuracies of the Bertillon system warranted a new approach.

The validity of dactylography, or the science of using fingerprints for identification purposes, was established by Edmond Locard and others in the early 1900s. Specifically, this approach helped investigators corroborate that an individual was present at a crime scene, held a particular item, or touched a discarded weapon. As the 20th Century progressed, dactylography was a ubiquitous and effective investigative technique in Europe and the United States (2009).

Odontology, or the scientific examination of dental records for identification purposes, also gained popularity in the early 20th Century. The matching of a deceased person's teeth to dental records of an unidentified individual was recognized as an accurate identification method. But dental records were not available for every case, and the process can be an expensive and protracted. As such, odontology could only be used in certain situations.

In fact, listing and then matching dental records vis-à-vis law enforcement databases has not been very successful. In a 1993 article in the *Journal of Forensic Sciences*, odontologist Gary Bell detailed the failure of the FBI's National Crime Information Center (NCIC) to match dental records. Bell submitted the complete dental records of four known murder victims (authorities had identified the victims using a separate system operated by the US Army) to NCIC for potential identification. None of the records were found to be a match through NCIC despite the fact that the victims' complete records were listed in the FBI's database (Halber, 2014).

Despite technological advancements since Bell's test, matching dental records using contemporary databases has not yielded impressive results for investigators either. In the 1980s, the State of Washington passed legislation requiring law enforcement agencies to collect dental records for persons missing longer than 30 days. These records could then be entered into searchable databases and used by authorities nationwide. Despite reporting the highest number of missing person records with accompanying dental records entered into NCIC, the State of Washington has not received a positive identification via NCIC online searches to date (2014).

In the late 1980s, science produced the most accurate means for personal identification: Deoxyribonucleic Acid (DNA) typing. Initially discovered in 1868, the potential of DNA for criminal investigations was not recognized until over a century later when Alex Jeffreys and colleagues (1985) described the unique properties of DNA that could be linked to an individual's identification. Police in Europe initially used forensic DNA evidence in the 1980s, and then in Florida in the early 1990s. Researchers estimate that the chance of having the same DNA pattern would be between 30 and 100 billion to 1 (Swanson et al., 2009).

In terms of personal identification, authorities now had an accurate tool for matching criminals to evidence and missing persons to unidentified decedents (2009). For intact corpses, fingerprints are still secured at the crime scene or after the body has been moved to a coroner's office. For instances that preclude investigators from obtaining fingerprints (e.g., only body parts recovered or the decedent is in an advanced state of decomposition), DNA could be used to ascertain an identity.

Despite this technological innovation, collecting DNA samples from missing persons has not become standard practice for US investigators. By policy, some law enforcement agencies will attempt to collect a DNA specimen after a certain period of time, while other agencies have no written procedures for when DNA should be collected for a missing individual. Jurisdictions where authorities with available resources, and the training in dealing with this manner of personal identification, are more likely to collect individual or family DNA specimens. In contrast, agencies with fewer resources/training might limit the search for a missing person to a more traditional approach: investigating potential locations, interviewing friends, entering physical and clothing descriptors into computer databases, and collecting photographs. These jurisdictions may not collect fingerprints or DNA evidence until years later after the case has gone cold. Without a

DNA profile for a missing person, it is difficult for authorities to match an unidentified body found, in some instances, a thousand miles away.

Further hindering the use of DNA matching was a limitation related to technology use. Prior to 2009, DNA searches were limited to databases maintained by the FBI and used by US law enforcement. Primary user groups such as medical examiners and local coroners, who are often charged with determining the identity of an unidentified decedent, did not access the FBI's system. A 2007 Bureau of Justice Statistics report indicated that 80% of the coroners and medical examiners surveyed said that they rarely or never used the NCIC missing and unidentified files as a tool for investigating unidentified bodies (Hickman, Hughes, Strom, & Ropero-Miller, 2007).

To improve the organization of identifiers such as DNA and to better connect consumers to this information, the federal government established the North American Missing and Unidentified Persons System (NamUs) at the University of North Texas (UNT). NamUs includes a database of missing persons and unidentified bodies. It was released for use by law enforcement, medical professionals and the public in 2009. NamUs provides a central repository for recording case information including personal identifiers such as tattoos, the use of a prosthesis, and DNA specimens for comparison purposes. With the advent of NamUs, all agencies investigating these types of cases have a conduit for publicizing missing persons and/or unidentified bodies, and scientists at UNT can assist in comparing DNA specimens for matches.

Still, investigators are often challenged with how to secure a DNA specimen from a missing individual. This can be problematic for a variety of reasons including if the individual has been gone for years, leads a transient lifestyle, was adopted, etc. One resource that may benefit authorities in these circumstances is the PKU card.

PKU is a rare genetic disorder discovered by Norwegian physician Ivar Asbjorn Folling in 1934. An individual diagnosed with PKU suffers from a deficiency in metabolizing the enzyme phenylalanine hydroxylase. This deficiency leads to an accumulation of the amino acid phenylalanine, which disrupts the development of the individual's brain (Wang et al., 2013). If untreated, the condition can result in problems including intellectual disability, delayed speech, seizures, and behavior abnormalities. The National Institutes of Health estimates that approximately one of every 15,000 infants born in the US suffers from this disorder ("Phenylketonuria: Screening and Management," 2000).

Fortunately, PKU is one of the few genetic disorders that can be managed by dietary practices thereby limiting the affected child's intake of phenylalanine. Accordingly, early diagnosis is essential (2000). To maximize the effect of PKU treatments, many countries now mandate testing via heel prick for the disorder in all newborns, typically within two to seven days after birth. In the US, most states began PKU testing in the 1960s. Neonatal blood spots are sent to labs for analysis of PKU, as well as other disorders (Wang et al, 2013). In the US, some home-births are missed, as well as mistakes are made in collecting or interpreting the results, nevertheless the vast majority of those inflicted are identified ("Phenylketonuria: Screening and Management," 2000).

The collection of neonatal blood specimens to test for PKU and other abnormalities is a valuable source of medical information. Over 160 polymorphisms measured from dried blood spots have been cited in epidemiological studies. Reports from these blood specimens include DNA and other biological markers, infectious agents, and potential environmental contaminants such as metals. In addition, emerging nanotechnologies allow researchers to study gene transcripts, proteins, metabolites, infectious agents, drugs, and other chemicals and pollutants in ways not previously conceived ("Michigan Neonatal Biobank: Researchers," 2015).

PKU test results are maintained by medical authorities for varying amounts of time. A study conducted by the National Center for Missing and Exploited Children determined that four states (Kansas, Missouri, Oklahoma, and South Dakota) retain PKU test information for as little as 30 days. In contrast, Florida, Maine, Michigan, New York, North Carolina, and Vermont have archived the test results indefinitely—some with patient records originating over 50 years ago (Reed, 2010).

Unfortunately, accessing a PKU record is far from an infallible process. Some cards are not submitted as mandated, while other cards are lost or destroyed. Human error can also lessen the availability of accurate information. A 1979 Netherlands study involving lead levels in one-year-olds led researchers to question the effectiveness of blood spot cards. Due to contamination in 50 of the PKU cards in their sample, researchers questioned the reliability of these records as a benchmark for understanding early lead exposure (Morgan, Hughes, & Meredith, 1979). In the 35+ years since Morgan et al.'s research, collection practices and specimen storage of neonatal blood spots have vastly improved. Still, the range of procedures for testing, processing, and securing neonatal samples in the US and in other countries could lead to problems for investigators seeking a single individual PKU card. Nevertheless, the PKU record represents a viable opportunity for investigators to obtain a DNA profile, a resource that has been scarcely used by investigators to date.

PKU cards as an investigative tool

Improvements in NCIC (e.g., promoting access to medical examiners and coroners) and the launch of NamUs have given investigators the tools to match DNA specimens. PKU cards may provide another direct specimen of an individual's DNA to facilitate matches. For identification purposes, DNA specimens are classified into two types: direct reference samples and family reference samples. Direct samples are the most accurate—referring to biological matter collected directly from the individual in question.

Sometimes, a direct sample may not be available, and family reference samples are used to compare specimens. This sample is collected from close biological relatives of the missing person (e.g., mother or father). The FBI recommends collecting samples from multiple relatives, and obtaining mitochondrial DNA (useful in examining the relatedness among species and populations) from at least one maternal relative. Further, if the individual is male, a Y-chromosome short-tandem-repeat or Y-STR analysis should be conducted on one paternal relative to elicit additional information ("Missing Person Comparison Request," 2015).

Both direct and family reference types are useful to investigators, but only a direct sample can be searched against all indexes of the FBI's Combined DNA Indexing System (CODIS). In contrast, family reference samples are only searched against records in the Unidentified Human Remains Index of CODIS (Reed, 2010). Thus, PKU cards as direct samples could be an invaluable piece of evidence. Specifically, the DNA information gleaned from neonatal PKU records has contributed to the closure of cases in three areas: missing persons and unidentified bodies, evaluating unexplained deaths, and in mass disasters.

PKU cards for missing person and unidentified body cases

The information contained on a PKU card can be useful to authorities involved in investigating missing persons or unidentified decedents. This medical record can assist authorities in identifying or ruling out whether an unidentified person is that of a missing person. With unidentified bodies, personal identification is essential. Halber (2014) characterizes "no confirmed identity" as a near zero percent chance of success for authorities in solving a case.

In the United States, data from the National Missing and Unidentified Persons System (NamUs) contains reports for more than 11,000 missing persons and over 10,000 unidentified decedents. All of these cases are assigned to law enforcement or coroners/medical examiners and are in various stages of investigation. Some are active and some are classified as cold. In 2009, authorities using resources from NamUs first used a PKU card as an investigative resource (Parmelee, 2009).

In June of 2002, a young male jumped to his death from a building in New York City. When police arrived at the scene, the decedent was not carrying any form of identification. While investigating the suicide, detectives followed the standard practices of examining recent missing person and runaway reports, consulting missing person databases, and running the decedent's fingerprints through available systems, but they were unable to find any information as to his identity. After a few months and with no new leads, the case was "tabled" to accommodate new death investigations that required immediate attention. Authorities were unable to determine the young man's name, and he was buried in an unmarked grave. It would take seven years before the mystery was solved (Pettem, 2013).

On the day before the young man's body was recovered in New York City, an officer with the Piscataway Police Department had filed a missing person report. Piscataway, New Jersey is community of about 56,000 residents approximately 40 miles southwest of New York City. The Maurer family was concerned about 17-year-old Ben Maurer, and had not had recent contact with him. Piscataway police described Ben as five-foot-eight inches in height, 135 pounds, a crewcut hairstyle, and wearing contact lenses. Authorities had no clothing description, but did learn from interviewing friends and family, that Ben had no past history of being a runaway, was well liked, had a new girlfriend, and was an average student in high school. Detectives in New Jersey established that on the evening of his disappearance, Ben was seen at local diner, later at a convenience store, and then walked to the train station at Dunellen, NJ. Investigators surmised that the missing teen might have ventured into New York City. Unfortunately, the link between Ben's disappearance and the unidentified decedent was not initially made (2013).

Investigators had submitted Ben's dental records and blood type to the National Crime Information Center database (NCIC) and had collected a mitochondrial DNA sample from his mother (i.e., a family reference sample). The family reference sample helped rule out several potential matches. One potential match could not be dismissed and was listed as "inconclusive" based on the information provided (2013).

In 2005, detectives asked the family for personal items of Ben's that could be used to collect biological matter for a direct sample. This task was obviously hindered by the three-year span since Ben's disappearance. While brainstorming for potential sources of a direct sample, Ben's mother mentioned her son's "newborn blood test." She was referring to the PKU test collected from Ben's heel when he was only days old (2013).

Navigating in unchartered waters, Detective Kevin Parmelee began researching the PKU tests and subsequent record retention by the State of New Jersey. He learned that New Jersey stores PKU results for 23 years. Parmelee used a grand jury subpoena to access the New Jersey Department of Health's PKU card for Ben Maurer. The court order for that access was approved in May of 2008, just weeks prior to Ben's 23rd birthday (2013).

With a direct sample from the missing teen and a possible match of an unidentified body in New York, investigators sent the PKU record results to the University of North Texas Center for Human Identification. Researchers analyzed the information, and entered the specifics into the Combined DNA Index System (CODIS) and the NamUs database. In June of 2009, the New York City medical examiner notified detectives in New Jersey that the decedent found in New York City and buried in an unmarked grave was that of missing teen Ben Maurer. Ben had committed suicide the day after he went missing. The Maurer family had Ben's remains exhumed and moved to a private burial in his hometown (2013).

A mother provided a suggestion to investigators about an infant blood spot card. Detectives, knowledgeable in DNA, considered an innovative approach to a missing person investigation, and then were persistent in navigating the legal environment to obtain that card. A department of health in New Jersey retained this medical information for an extended period of time. All of these are factors allowed authorities to make the identification. The PKU card, medical information that has been previously collected by every state in the US for decades, was essential in the closure of an unsolved mystery.

Surprisingly, the Maurer death investigation appears to be the only case documented in academic journals in which PKU results were used to positively match a missing person to unidentified decedent. Nevertheless, authorities still proactively pursue DNA in missing person cases. In May of 2015, the Michigan State Police and representatives from NamUs sponsored an event in Detroit to raise awareness for missing persons and to connect families affected by missing loved ones. Entitled "Missing in Michigan," the public forum allowed participants to hear about recent stories involving the identification of unidentified decedents using family reference samples (Smith, 2015).

One such case was that of Carla Tucker, reported missing from Detroit at age 14 in 1979. In 2002, workers at a Carlton, Michigan landfill found a body in a 55-gallon drum that had been encased in concrete. The female decedent was listed as unidentified, while the investigation into Carla's disappearance became stale (2015). In 2014, relatives of Carla learned of the need for family reference samples of missing persons, and submitted specimens to representatives from NamUs. A few months later, the decedent from Carlton was positively identified as that of Carla Tucker. Carla Tucker's loved ones had waited nearly four decades to learn the missing teen's fate, and authorities had unknowingly recovered her body 13 years before the confirmation was made (2015).

Building on the approach used to identify Carla Tucker, Michigan authorities, actively solicit family reference samples as an attempt to reduce a backlog of almost 300 unidentified decedent cases in that state (2015). The Michigan State Police and representatives from NamUs should be applauded for the "Missing in Michigan" event and other efforts to obtain family reference samples of DNA. Proactively soliciting and adding new evidence to cold cases, could not only assist in solving investigations in Michigan, but in other states as well where an unidentified decedent is waiting to be matched. Law enforcement officials in other states should be encouraged to pursue similar campaigns. But, what if a direct DNA sample for a missing person could also be obtained and indexed? Wouldn't that evidence be of added value?

The Michigan Department of Community Health stores PKU card results indefinitely at the Michigan Neonatal Biobank. Records have been collected since 1984. Current governmental procedures regarding this testing, give Michigan parents several options. Parents or legal guardians can opt to have extra neonatal blood specimens collected to be included for scientific research. The specimens are earmarked for research, and identified using a barcode system to ensure confidentiality. If parents opt out, blood spot cards are stored strictly for emergency purposes. The Department of Community Health also allows parents to have all blood spot records destroyed after initially screens are completed—they just need to submit a formal request to that agency. Alternatively, once the child becomes 18-years-old, he or she can formally request to have the blood spot card destroyed ("Michigan Neonatal Biobank: Researchers," 2015).

Since Carla Tucker was born in 1965, the PKU test would not have been available to detectives, but how many unsolved missing persons cases involve individuals born in 1984 or after? As of July 2015, dozens of those listed as missing in Michigan are 31 years old or younger. In addition, there are open missing children and adult cases reported in other states, where the victim's place of birth is Michigan. As such, it might behoove authorities to explore the usefulness of securing PKU cards for missing persons so that they may be entered into searchable databases.

Family reference samples are important pieces of evidence for authorities, but may not be available for every individual. What if the missing person is adopted? What if he or she is indigent and no biological relatives can be located? A direct sample could be the only hope for an investigator in obtaining a reference specimen necessary for database searches and comparisons. Again, an existing PKU card would provide this essential evidence.

Imagine that a 20-something-male with a transient background is reported missing by his friend and employer who owns a moving company. The employer states that he met the man through a local homeless shelter, and that the man had been an excellent worker at the moving company for almost two years. The employer stated the man had missed work all week. Investigators learned that the man had no regular residence, was mostly a loner, and no biological relatives could be found, but he did have a misdemeanor arrest record from several years prior. With no immediate options for a family reference sample or direct sample, authorities could submit the information gleaned from the investigation and criminal history to appropriate databases, and wait for a new lead to develop. Unfortunately, scenarios such as this are common in that social exclusion or persons who are disconnected to society are a higher risk to go missing (Kiepal, Carrington, & Dawson, 2012).

But what if a detective learns from the criminal history data that the missing man is 25 years old and reports that he was born in New York? An astute investigator might recall that the New York Department of Health retains neonatal PKU test results for 27 years ("Newborn Screening Program Retention and Screening Policies," 2015). Since the missing man was born in 1990, and New York began conducting and saving those neonatal tests in 1965, that investigator could then pursue the necessary permissions to obtain a direct reference sample of DNA for the case. Collecting that specimen could increase the chances for resolving the investigation—potentially aiding authorities in another jurisdiction that are trying to identify a recovered body.

With the Ben Mauer case or the hypothetical example of the missing transient, authorities are hampered by the lack of a national standard protocol for dealing with the missing and unidentified bodies. Though the Maurer case was close in proximity, multiple agencies from two states were involved in the investigation. Communicating within one agency can sometimes be a challenge—one that is compounded when multiple agencies are working a case. Some states have requirements for law enforcement and medical examiners, but frequently, policies are developed and followed at the county or agency level. This creates a landscape of loosely connected or even disconnected jurisdictions all with the same goal—to identify a missing person or unidentified decedent (Pettem, 2013). Conversely, innovations such as NamUs promote collaboration, and allow investigators to search cases from around the US, thereby providing DNA specimens obtained from PKU cards the necessary visibility for identification.

PKU cards for unexplained death cases

A recent article published in the *Journal of Child Health* provides additional insight into an innovative use of PKU test results (Skinner, Chong, Fawkner, Webster, & Hegde, 2004). Dr. J.R. Skinner and several colleagues documented the sudden death of a 12-year-old boy in New Zealand. Witnesses to the incident described the decedent as slowly jogging and then suddenly dropping "lifelessly" to the ground while playing hockey. An ambulance crew arrived within minutes and provided CPR and fibrillation, but the young man could not be resuscitated (2004).

Since the death was sudden and unexpected, an autopsy was conducted. The medical examiner described the decedent's heart as normal, recorded no acute problems with his other primary organs, and listed the cause of death as inconclusive. The decedent's family met with Dr. Skinner and associates to further investigate the unexplained death. There was no history of sudden death on either side of the family going back three generations. Interviews with the boy's parents and reviews of the descendant's medical records showed that he had suffered three seizures over the past five years. All of the episodes had occurred during exercise, two while swimming. After these incidents, the teen was diagnosed with epilepsy and received treatment and medications (2004).

Further study of the family's medical history indicated that the father's brother also had previously been diagnosed with epilepsy, and that the mother's heart showed signs of Long QT Syndrome (LQTS), a heart condition that can cause sudden, uncontrollable and dangerous arrhythmias during exercise or stress. The cause of this malady is unknown, but LQTS can be fatal (2004).

Dr. Skinner and associates hypothesized that the decedent may have suffered from a similar undiagnosed heart condition that led to his death. Needing blood for analysis, the researchers sought the boy's PKU card, as New Zealand requires neonatal blood spot records be collected 48 hours after birth. The 12-year-old blood spot card was located, and with parental consent, genetic tests were conducted (2004).

Dr. Skinner and associates concluded that the decedent's death was consistent with someone suffering from LQTS. The intense physical stress placed on the heart during rigorous activities like swimming and hockey had likely triggered dangerous arrhythmias and caused sudden death. This conclusion corresponded with the eyewitness descriptions of the boy's sudden and lifeless collapse. Following the investigation, the mother's own physician began actively treating her LQTS, and the mother began avoiding high-risk physical exercise. In addition, other members of the decedent's family were evaluated for LQTS, in hopes that early diagnosis and treatment might promote long-term health (2004).

Skinner and colleagues (2004) also discussed the debate in New Zealand about patient privacy and the storage of newborn PKU cards. They noted that a number of families have withdrawn neonatal blood sample records from the country's storage center. Some of these records have been secured since 1969. Despite privacy concerns, Skinner et al. argue that the diagnosis in this case would not have been possible without the genetic information obtained from the PKU card. The authors asserted that these records provide a wealth of information for health care professionals. Their example illustrated the possibility of retroactive diagnosis for individuals who had died suddenly and even decades ago while exercising or sleeping (e.g., sudden infant death syndrome or Brugada syndrome) (2004).

In 1962, Massachusetts collected the first mandatory infant PKU screenings. Since then, all other states and several countries have followed suit. The information gleaned from the cards has proven to not only be useful in evaluating neonatal health, but also for diagnosing and managing the health of older children and adults. For example, PKU cards have been used to study children suffering from deafness. They have helped medical professionals evaluate whether the cause of a hearing deficiency is related to an infection acquired during pregnancy (Green & Heffron, 2013).

The cards have also provided specimens to identify genetic disorders and Sudden Arrhythmic Death Syndrome (SADS) (2013). Similarly, Australian researchers used the blood spot cards of stillborn children to detect the presence of cytomegalovirus (CMV). Researchers are exploring connections between CMV and intrauterine deaths at less than 20 weeks gestation (Howard et al., 2009). Finally, the PKU cards have been valuable in cases where one identical twin develops leukemia. Researchers can use the genetic information collected at birth to study the likelihood that the other twin will develop leukemia as well (Green & Heffron, 2013). In sum, medical information gleaned from PKU cards can be useful for investigating unexplained deaths and other health issues.

PKU cards for mass disaster cases

Finally, neonatal PKU cards offer assistance to authorities investigating the victims of disasters. The September 11, 2001 attacks on the Unites States showed the crucial role that DNA identifications could play in a mass casualty incident. Of more than 2,000 victims at Ground Zero, fewer than 300 were intact upon recovery. As human remains were gathered, the need for identification of the deceased placed a new emphasis on DNA during such disasters (Halber, 2014).

Initially, a director from New York City's Office of the Chief Medical Examiner planned to restrict the testing of tissue and bone samples to only those larger than the size of a thumb. But after seeing the enormity of the carnage, he ordered the analysis of every fragment of human specimen recovered. Author Deborah Halber (2014) described the complexity and tediousness of this task for investigators depicting much of what was provided to scientists as barely recognizable. For instance, one researcher struggled with analyzing a tiny specimen for hours that eventually was determined to be plastic. There is no record of PKU cards being used to assist in the identification of 9/11 victims, despite New York medical authorities having collected those neonatal cards since 1965. In contrast, eight years later and over 10,000 miles away from New York City, infant blood spot cards would play an integral role in victim identifications involved in another disaster.

In February of 2009, a raging firestorm devastated over 2,700 miles in Australia's most densely populated state of Victoria. Over several days, thousands of businesses and homes were destroyed and 173 individuals perished. In addition, thousands of people were injured and/or left homeless. The country's government characterized the Victorian Bushfires Disaster as their worst national disaster (Hartman et al., 2011).

Prior to this tragedy in 2008, Australian and New Zealand scientists and other authorities involved in disaster operations implemented recommendations to strengthen collection and access to DNA in the case of a national emergency. Collection procedures of direct and family reference DNA samples for comparison purposes had been approved by the International Society for Forensic Genetics. With respect to direct samples, researchers advocated for the use of PKU cards instead of personal items collected from individuals. Researchers argued that toothbrushes, razors, hairbrushes, or similar items may be shared, leading to contaminated specimens and misidentifications. PKU cards, in contrast, are sterile and secured by medical professionals. Australia has mandatory neonatal testing within days after birth. PKU cards are created using heel pricks, analyzed for genetic disorders, and then stored in compliance with government regulations at Genetic Health Services in Victoria. These records, in general, are available for births dating back to the 1970s and can be made available to coroners for identification purposes (2011).

As part of the investigation from the Victorian Bushfires Disaster, authorities were overwhelmed with the bodies of deceased. Charred remains and partial remains made visual death or fingerprint identification impossible. Dental records could be used as identification for some, but not all of the victims had those records, and also existing dental records were not immediately available to investigators. As such, PKU cards became an essential tool to assist with individual identifications (2011).

Based on victim age (e.g., born after 1975) and place of birth, coroners eliminated 129 cases and submitted 34 requests for PKU cards to the Genetic Health Services. Health officials were able to provide 21 of the 34 records to aid in identification of the deceased. The DNA specimens obtained from the PKU cards were described as of good quality and provided authorities with full DNA genotypes. With the cards, researchers were able to directly match 11 decedents, as well as one additional identification – using the card along with other evidence to connect relations via kinship. Authorities speculate that it was the postmortem samples taken from victims of the tragedy and not specimens from the obtained PKU cards that hampered the other identifications. The disaster's extreme temperatures, in excess of 2,700 degrees Fahrenheit, limited the availability of some post mortem samples using current extraction techniques. Further, of the blood specimens obtained from the PKU cards, only three of the comparison records had available dental histories—making DNA vital to identification (2011).

The tragedy in Victoria also illustrated another complexity in DNA collection and identification. In some cases, all family members perished in the quickly spreading fire. In such instances, family reference samples were not available for identification purposes (2011). This example further demonstrates the need for access to neonatal PKU cards.

Solutions and recommendations

The use of PKU cards by investigators appears to offer a distinct advantage in identifying an unknown decedent in cases of missing persons or mass casualties, as well as assisting professionals with unexplained deaths. Protocols already exist to collect, store, and permit access to these infant blood spot records. The following recommendations may help to create the consummate environment for authorities to use these records for investigative purposes.

Record place of birth for all missing persons

Currently, the collection of information regarding missing persons varies widely by agency. Documenting a missing person's date of birth is standard practice, but listing the individual's place of birth may not be collected by every reporting agency. In order to determine the usefulness of a PKU record investigators must establish the date of birth and place of birth. For instance, if an individual who was born in Michigan in 1991 is reported missing in that state, authorities could potentially obtain the person's infant blood spot card as a source for a direct DNA profile as Michigan currently retains PKU records indefinitely. Further, if police in Wisconsin are investigating a missing man who was born in 1995 and they learn that his place of birth was Detroit, detectives could attempt to obtain the infant blood spot card from Michigan's Neonatal Biobank.

Obtain infant PKU cards sooner rather than later

When the investigation dictates, authorities should follow established protocols for accessing specimens of the infant blood spot cards and immediately obtain those records. In cases of missing persons, there are no guarantees that the PKU card will be available at some later time. In Texas and Minnesota, cards were destroyed as per court orders (2009 and 2012, respectively) (Tarini, 2011; Schultz, 2014). Elsewhere, historical records may be unavailable for other reasons. For instance, in early 2000, severe flooding to health record storage facilities in Vermont destroyed thousands of infant PKU cards. As a result, the neonatal information is available only for births in 2002 or after (Pettem, 2013). As such, it is imperative that authorities obtain consent from parents or legal guardians or use court orders to access this potentially valuable form of personal identification as early as possible in the investigation.

Develop collaborative relationships with departments of health

Partnerships with medical professionals may provide opportunities for authorities in missing person investigations. Understanding what resources, records, and storage practices are used by Departments of Health can provide investigators with information about what specimens may be available for DNA identification or historical comparisons. As previously illustrated in the death case of Ben Mauer in New Jersey, investigators had not considered PKU cards as a source for DNA until a mother's comment serendipitously reached an astute and diligent detective's ears.

Further, collaborative relationships may help authorities advocate for the preservation of health-related information, whether agencies hold the information indefinitely or for shorter periods of time. With advances in DNA, these records hold current value, but also likely have potential value as technology improves. As with the legal actions in Texas and Minnesota that led to the destruction of millions of historical neonatal PKU cards, judges reviewing the cases did not speak to or realize the potential benefits these records could hold for authorities in certain circumstances. In the future, it would behoove authorities to be more vocal as to the investigative value of the PKU records.

Recognize the sensitivity involved and support innovative alternatives for retention

Through collaborative relationships, investigators and medical professionals may improve public trust in the use of medical records. A study published in *Academic Pediatrics* provides insights into building trust via communication in relation to parental perspectives of neonatal PKU cards (Hendrix, Meshlin, Carroll, & Downs, 2013). Dr. Kristin Hendrix and colleagues surveyed 506 low-income parents with at least one child aged 17 or younger born in Indiana. Researchers studied the impact of certain factors on parental support for the continued use of neonatal blood spot cards after the required initial screenings were complete. Specifically, the study explored parental attitudes about infant blood spot card research based on whether a child's identity would be linked to the blood specimen, who would be conducting the research, and whether and how often parental consent would be sought prior to research being initiated. Results showed that consent for each instance of research was the most important factor of those surveyed. The authors observed that this finding was consistent with previous quantitative and qualitative research conducted on the use and collection of infant blood spot records (Tarini et al., 2010)—that parents and legal guardians desire input into the research process. Data also showed that parents prefer anonymity for their children in terms of these medical records, and that research be conducted by universities (Hendrix et al., 2013).

Authorities should be sensitive to citizens' privacy concerns about retaining the PKU test cards. In Texas and Minnesota for example, the long-term storage of PKU cards by government health officials has been denied by recent court rulings. As such, authorities in these states should support innovative methods in securing the information contained on infant blood spot cards.

To address the privacy concerns of citizens regarding PKU cards, Virginia officials recently implemented an innovative program. Specifically, government-held PKU samples are collected, tested, and destroyed after a short period of time. As an alternative approach that balances citizens' privacy needs with the need for medical research, parents or legal guardians have the option of taking custody of the infants' blood specimens ("Child ID Program," 2015). As of 2012, Virginia law requires hospitals to provide parents or legal guardians the newborn blood spot records. The parents then have the option of maintaining that record for future use if DNA analysis is necessary to help identify a child in the event of an emergency such us a disaster or abduction. By giving the infant blood spot record to parents, government officials abdicate responsibility for storage and security. Future access to the record would require researchers or officials to obtain written informed consent from parents or legal guardians (2015). For example, an investigator working a missing person case involving a child born in Virginia in 2012 could obtain a DNA sample from a neonatal PKU card in the parent's possession.

Ethical issues

With confidential health records like that of the PKU card, accessing the information can be challenging. Fortunately, the US and several other countries (e.g., Australia, United Kingdom) have prescribed policies for the collection, use, and retention of neonatal blood spot records. Following established protocols, investigators can argue the need for the information and a legal authority can decide whether to approve the request. Still, ethical issues related to PKU card information remain controversial for at least two reasons: patient confidentiality and the responsibilities of authorities.

First, the confidentiality of patient medical records is a concern. Use of infant blood spot cards to screen for various medical conditions is common practice in developed countries. Conversely, protocols for obtaining and storing the DNA of all newborns born in the US have been the subject of recent controversy. In the 1960s, when many health departments instituted newborn screening programs, collecting blood samples was not seen as intrusive by most parents. It was viewed as necessary and important to ensure patient safety. In the ensuing decades, advances in DNA technology have created a more difficult landscape for officials to navigate. Parents may ask relevant questions including: Why does the government need to hold a baby's blood sample for more than a year? What safety protocols are present to protect this private medical record? How much access do scientific researchers have to this information? If abnormalities are identified in a baby's genes, could that potentially be used to deny services or opportunities in the future?

Two recent legal challenges have explored some of these questions. In 2009, several families filed a civil suit against the State of Texas. Allegations included that members of Texas' health department had provided the mandatory blood spot card information to the US military for a forensic investigations database without parental consent. At that time, health officials were saving Texas PKU cards indefinitely. The plaintiffs triumphed in the legal action and, as a result, Texas officials were forced to destroy millions of infant blood spot cards that had been collected since the 1960s. A revised state health policy restricts Texas officials, in general, from retaining PKU records and related information for more than 24 months (Tarini, 2011).

A similar civil court case questioning the necessity of retaining infant blood spot cards occurred in 2011. In *Bearder v. State of Minnesota*, the Minnesota Supreme Court ruled that infant blood samples were regulated by the 2006 Genetic Information Act (GIA). This mandate governs the storage, dissemination, or use of blood samples. The GIA requires that written informed consent be obtained from parents or legal guardians, something that had not been performed consistently over the decades of collecting this information (Wadman, 2012). As a result of the ruling, Minnesota officials destroyed historical infant PKU cards from 2011 to the original collection year of 1965. Under a revised policy, state health officials collect and retain infant blood samples for 71 days unless abnormalities are present. For newborns affected by a heritable or congenital screening disorder, medical officials seek informed consent from parents or guardians to store the records longer. Unless there is an approved exception, all other collected specimens are destroyed after two years (Schultz, 2014).

Dr. Beth Tarini (2011) stated that parents should be encouraged to participate in the decision-making process regarding the storage and use of infant blood information. Tarini argued that deceptive practices by the government and researchers create an environment of distrust. Regarding the previously mentioned Texas civil case, she asserted: "Paradoxically, it is likely that allowing parents the opportunity to say 'no' may actually get them to say 'yes.'" Tarini (2011) added that Andrea Beleno, one of the plaintiffs in the Texas lawsuit, said: "And if they'd asked me if I would consent for this blood to be used for specific medical research… I would have probably said 'yes'" (p. 620).

Patient confidentiality is also being debated in California (Ghorayshi, 2015). In March of 2015, California legislators began discussing infant blood collection and retention practices. Specifically, proposed legislation entitled the "Newborn Blood Sample Privacy Bill" is intended to protect the rights of infants and parents from government intrusion. Supporters of the bill argue that such protections are necessary to prevent the potential abuses related to state-created DNA databases that would house records of neonatal testing required by California and other states. Legislators also alleged that the government's current practices are confusing at best. New parents are bombarded with information and neonatal waivers regarding infant blood tests that can easily be misunderstood or forgotten. In addition, the California bill would require all research facilities to obtain consent from parents before using newborn blood spot records (Ghorayshi, 2015).

In Texas, Minnesota, and California, the discussion is about the potential problems of retaining PKU cards for extended time periods, including the lack of parental and patient protections of private medical information. If legislation like that described in California is passed, an unintended consequence may be that parents prematurely remove PKU cards from government storage or that state governments adopt restrictive retention schedules for health departments—both of which may hamper investigations that could benefit from the use of these neonatal blood spot records. The challenge for authorities is to defend the retention of neonatal PKU cards by highlighting the positives, like that of personal identification and assistance in death investigations.

A second category of ethical challenges with PKU cards involves the responsibilities of authorities who handle the cases. Policing and criminal investigations in the US are primarily performed at the local level. Personnel who are responsible for investigations have a range of experience and competencies. An inexperienced investigator for example, may not know the potential value of a PKU card. And since agencies often work autonomously, it may be difficult for federal and state officials to offer assistance without being seen as overstepping territorial boundaries.

Dissimilarities in US law enforcement practices are not the only issue, but the differing backgrounds among medical examiners and coroners in the United States also contribute to varied investigations. In a study by the Bureau of Justice Statistics, medical examiner/coroner offices serving more than 100,000 residents had an average of 20 employees, while some smaller offices did not have any full-time workers (Hickman et al., 2007). Despite having fewer employees, Hickman et al. (2007) found medical examiners/coroners in smaller jurisdictions (under 250,000 in population) conducted death scene investigations at a much higher rate (72%) as compared to equivalent medical professionals from larger jurisdictions. Further, the authors determined that smaller jurisdictions were less likely to have record retention policies for unidentified bodies. Specifically, only 38% of agencies surveyed serving populations between 10,000 and 24,999 residents had retention policies (as compared to 95% for jurisdictions of 1,000,000 or more residents). As such, with no retention schedules, case files for unidentified bodies are destroyed leaving nothing to compare information gleaned from PKU cards. Disturbingly, the authors surmise that the current reported total of 13,486 unidentified bodies nationwide is grossly underestimated as the records for some cases are simply no longer available (Hickman et al., 2007).

Hickman et al. also described substantial differences in the policies and resources of large and small coroners/medical examiners. In the offices surveyed, the mean large agency annual budget was over $1.1 million, while the average small office worked with only $41,000 per year. Further, 30% of coroner/medical examiner offices serving populations of less than 2,500 people even had established policies governing unidentified bodies (2007). Understandably, this leads to challenges when no entity has oversight over the autonomous practices of these agencies that vary widely in size, access to resources, and levels of expertise.

Investigators are vested in protecting the rights of citizens. If it is necessary to access the neonatal PKU card of a missing teen born in Rhode Island, then investigators should follow established health department protocols and obtain the necessary consent by the parent or legal guardian. If the missing person is age 18 or older, then investigators should pursue a court order for the PKU card. To assure patient privacy, authorities should describe to the magistrate how personal information from that record would be removed from NCIC and other databases if the missing individual were located.

Raising awareness of the use of PKU cards through scientific journals, trade publications, and traditional media coverage would be beneficial. Further, partnering with professional medical and law enforcement associations would provide opportunities to educate practitioners as to the potential benefits of this underutilized record. Through a proactive and comprehensive approach, the value of infant PKU records could be trumpeted and thereby circumvent the problem of lack of control over individual agency practices. In sum, this knowledge should be promoted to all authorities.

Applications in diverse forensic settings

The advantage to using infant blood spot records as a source of identification or medical history is that the resource can be used in diverse forensic settings—including instances of criminal activity or if involvement in a crime is unknown. As described previously, investigators may find the card beneficial in an unidentified body case, missing person, or if examining an unexplained death. Jurisdictions already have protocols in obtaining comparison samples from a decedent and the infant PKU card represents another professionally obtained specimen viable for comparison.

Future research directions

The use of PKU cards in the three scenarios described (e.g., missing persons, unidentified bodies, and undetermined cause of death investigations) has not been well studied. The following suggested areas of research may offer direction for not only academicians exploring this topic, but also for practitioners as they include information from infant blood spot cards in their investigations.

Cataloging of retention methods

The most comprehensive list of PKU retention practices by health departments in the US was conducted under the direction of Pamela Reed by the National Center for Missing and Exploited Children in 2010. Though outdated, the list is still being disseminated (Pettem, 2013). For instance, health officials in Texas and Minnesota appear on the 2010 list as maintaining infant blood spot cards indefinitely, despite those states having reduced retention schedules to that of months. Further, reducing retention schedules for PKU records is the subject of contested debate in places like California. It is essential for authorities to have a current list of PKU card retention schedules to know how this information can assist in a missing person, unidentified body, or unexplained death case.

Knowledge and use of retained records

Over 18,000 law enforcement agencies exist in the US with varying responsibilities. Law enforcement officers' knowledge about PKU cards as an investigative resource is currently unknown. Does an investigator in Mississippi working a case know that a DNA specimen via an infant blood spot card may exist for a missing 18 year old who was born in New York? Certainly, opportunities for assessment that can be used for informational programs exist and would benefit society. Further, as with law enforcement, medical examiners and coroners vary in job requirements and responsibilities. Again, the knowledge of PKU cards by these professionals is not currently known.

In addition, little is known about the use of PKU cards by authorities. For example, personal communications with a representative from the Michigan Neonatal Biobank revealed that their agency has released infant blood spot records to law enforcement after parent authorization about ten times in the last several years and twice to the National Center for Missing and Exploited Children (C. Langbo, personal communication, July 16, 2015). Overall, it is unknown how frequently law-enforcement and medical authorities access PKU cards, or in what capacity they have aided investigations.

Best practices for retaining specimens

Historically, PKU cards were secured by health department personnel wherever practical. Under what conditions should these records be stored that best allow researchers to elicit information? Some states, now store the PKU records under strict protocols. For instance, the Michigan Neonatal Biobank has been storing infant blood spot cards at −20 degrees Celsius since 2009—a temperature that is said to better preserve the specimens when used in advanced medical research ("Michigan Neonatal Biobank: Researchers," 2015). Research opportunities exist regarding best practices for retaining, storing, requesting, and using PKU cards for forensic purposes.

Conclusion

Obtaining DNA profiles can be an essential part of investigating cases of missing persons, unidentified bodies, or unexplained deaths. Though family reference DNA specimens are useful to authorities, a direct specimen from the individual involved is of even greater value. For years, health departments across the United States as well as medical authorities from around the globe have collected neonatal PKU records. These blood spots contain the direct DNA sought by investigators in cases of identification. They also contain valuable information for authorities examining the potential causes in unexplained deaths.

Nevertheless the retention of PKU cards is a controversial topic insofar as it warrants sensitivity to juvenile and adult patients' confidentiality. In response to court orders, health officials in two states have already destroyed millions of infant blood spot cards. Accordingly, investigators should act quickly in cases where PKU cards could be accessed, as those records may not be available in the future as was discussed in the context of recent court decisions in Texas and Minnesota. Obtaining the "place of birth" for an individual being investigated is also imperative in determining whether or not an infant blood spot record is available.

Further, additional research is necessary in determining the frequency of PKU card use, as well as current storage practices and retention schedules in the US and abroad. Only a single case exists in the literature detailing the use of a PKU card to solve a cold case death investigation. Also, no updated and publically available retention schedule exists that specifically describes the varying management practices by health departments on neonatal PKU cards. Consequently, educating investigators as to the availability and potential uses of infant PKU cards, while advocating to the public for government or alternative retention of these records will allow greater use and preservation when sources of DNA are being sought.

As emerging technologies continue to advance DNA applications, researchers will be able to use new techniques to aid in investigations. The collection and continued use of infant blood spots has the potential to play a pivotal role in supporting this growth. The value of a PKU card continues to be an underutilized and unrealized resource for law enforcement and medical authorities.

Key terms and definitions

CODIS (mp): Stands for the Combined DNA Index System for Missing Persons or also referred to as the National Missing Person DNA Database. Established by the FBI in 2000, this searchable database contains information on DNA obtained from unidentified remains, reference samples from the individual being investigated, and from relatives of missing persons.

Direct Reference Sample: Biological material of an individual that is collected and then used for DNA comparisons to establish identity. Sample specimens would include a tooth, saliva from a toothbrush, biopsy matter, etc.

Family Reference Sample: Specimens of blood, saliva, or other biological material collected from an individual's biological relatives (e.g., mother or father). The samples are then used for DNA comparisons to establish identity.

NamUs: Released in 2009, the National Missing and Unidentified Persons System is an online searchable database for law enforcement, medical examiners, coroners, researchers, and members of the public examining missing persons or unidentified decedents. Based at the University of North Texas, this federally funded initiative also provides case management and biometric support services to assist with investigations.

National Crime Information Center (NCIC): The FBI's primary crime database dedicated to supporting and serving local, state, and federal law enforcement agencies.

Phenylketonuria (PKU): A genetic disorder caused by an identifiable enzyme deficiency. Tests are used for neonatal identification, and specialized diets are successful in managing the affliction. If untreated, it can cause intellectual disability.

Phenylketonuria (PKU) Test Card/Guthrie Card/Neonatal Bloodspots: A medical record containing the blood sample of an infant that is used to test for the presence of Phenylketonuria and other disorders. Specimens are obtained through a neonatal heel prick, and are typically collected between 24 hours and 7 days after birth. Retention of these records varies by jurisdiction.

Note

1 I would like to thank Dr. Mary Katherine Duncan for her valuable feedback and direction in the development of this project.

References

Bell, G.L. (1993). Testing of the national crime information center missing/unidentified persons computer comparison routine. *Journal of Forensic Sciences*, *38*(1), 13–22.

Child ID Program. (2015). Retrieved July 7, 2015, from the Virginia Hospital and Healthcare Association website: www.vhha.com/childidprogram.html.

Ghorayshi, A. (2015). *Most parents don't know their babies' blood is given to scientists — But that may change*. BuzzFeed News, 3/17/2015. Retrieved July 8, 2015, from www.buzzfeed.com/azeenghorayshi/most-parents-dont-know-their-babys-blood-is-given-to-scienti#.bs9vEL4pq.

Green, A. & Heffron, M. (2013). *Newborn screening bloodspot cards*. Report: Royal College of Physicians of Ireland, 1–26. Retrieved July 9, 2015, from www.rcpi.ie/content/docs/000001/1619_5_media.pdf?1392207018.

Halber, D. (2014). *The Skeleton Crew*. New York: Simon and Schuster, 1–284.

Hartman, D., Benton, L., Morenos, L., Beyer, J., Spiden, M., & Stock, A. (2011). The importance of Guthrie cards and other medical samples for the direct matching of disaster victims using DNA profiling. *Forensic Science International*, *205*(1–3), 59–63.

Hendrix, K., Meshlin, E., Carroll, A., & Downs, S. (2013). Attitudes about the use of newborn dried blood spots for research: A survey of underrepresented parents. *Academic Pediatrics*, *13*, 451–457.

Hickman, M., Hughes, K., Strom, K., & Ropero-Miller, J. (2007). Medical examiners and coroners' offices, 2004. *Bureau of Justice Statistics, Special Report, June, NCJ216756*, 1–7.

Howard, J., Hall, B., Brennan, L.E., Arbuckle, S., Craig, M., Graf, N., & Rawlinson, W. (2009). Utility of newborn screening cards for detecting CMV infection in cases of stillbirth. *Journal of Clinical Virology*, *44*, 215–218.

Kiepal, L., Carrington, P., & Dawson, M. (2012). Missing persons and social exclusion. *Canadian Journal of Sociology*, *37*(2), 137–168.

Michigan Neonatal Biobank: Researchers. (n.d.). Retrieved July 13, 2015, from www.mnbb.org/researchers.

Missing Person Comparison Request. (n.d.). Retrieved June 28, 2015, from www.fbi.gov/about-us/lab/biometric-analysis/codis/missing-person-comparison-request.

Morgan, M., Hughes, M., & Meredith, P. (1979). Assessment of the PKU card as a retrospective index of neonatal blood lead status. *Toxicology*, *12*(3), 307–12.

National Missing and Unidentified Persons Fact Sheet. (2014). Retrieved July 27, 2015, from www.findthemissing.org/documents/NamUs_Fact_Sheet.pdf.

Newborn Screening Program Retention and Screening Policies. (2015). Retrieved July 31, 2015, from the New York State Department of Health website: www.wadsworth.org/newborn-screening/nbs-specimen-retention.

Parmelee, K. (2011). PKU card: A new tool in the search for missing and unidentified individuals. *Journal of Forensic Identification, 61*, 1.

Pettem, S. (2013). *Cold Case Research Resources for Unidentified, Missing, and Cold Homicide Cases*. Baton Rouge, FL: CRC Press, Taylor & Francis Group, 1–301.

Phenylketonuria: Screening and Management. (2000). Retrieved June 29, 2015, from the National Institutes of Health website: www.nichd.nih.gov/publications/pubs/pku/pages/sub3.aspx.

Reed, P. (2010). Direct DNA references for missing persons. *UNT Health Science Center, Forensic Services Unit Bulletin*, July.

Ritter, N. (2007). Missing persons and unidentified remains: The nation's silent mass disaster. *National Institute of Justice Journal*, 256. Retrieved July 15, 2015, from www.nij.gov/journals/256/Pages/missing-persons.aspx.

Schultz, D. (2014, January 13). *Lawsuit settlement allows newborn screening program to move forward*. Minnesota Department of Health, Press Release. Retrieved July 16, 2017, from www.health.state.mn.us/news/pressrel/2014/newbornscreening011314.html.

Skinner, J.R., Chong, B., Fawkner, M., & Hegde, M. (2004). Use of the newborn screening card to define cause of death in a 12-year-old diagnosed with epilepsy. *Journal of Paediatrics and Child Health, 40*(11), 651–653.

Smith, Kyla. (2015). *State police to collect information on missing people*. Detroit Free Press, 5/15/2015. Retrieved June 16, 2015, from www.detroitnews.com/story/news/local/detroit-city/2015/05/15/police-collect-information-missing-people/27409319/.

Swanson, C., Chamelin, N., Territo, L., & Taylor, R. (2009). *Criminal Investigation*, 10th Edition. Boston, MA: McGraw Hill, 13.

Tarini, B.A. (2011). Storage and use of residual newborn screening blood spots: A public policy emergency. *Genetics in Medicine: Official Journal of the American College of Medical Genetics, 13*(7), 619–620.

Tarini, B., Goldenberg, A., Singer, D., Clark, S., Butchart, A., & Davis, M. (2010). Not without my permission: Parents' willingness to permit use of newborn screening samples for research. *Public Health Genomics, 13*, 125–130.

Wadman, M. (2012). Minnesota starts to destroy stored blood spots. *Nature: International Weekly Journal of Science*, 02/03/2012. Retrieved December 16, 2017, from: www.nature.com/news/minnesota-starts-to-destroy-stored-blood-spots-1.9971.

Wang, C., Zhu, H., Cai, Z., Song, F., Liu, Z., & Liu, S. (2013). Newborn screening of phenylketonuria using direct analysis in real time (DART) mass spectrometry. *Analytical and Bioanalytical Chemistry, 405*, 3159–3164.

13 Detection of impairing drugs in human breath

Aid to cannabis-impaired driving enforcement in the form of a portable breathalyzer

Nicholas P. Lovrich, Herbert H. Hill, Jessica A. Tufariello, and Nichole R. Lovrich

Preface

The advent of a wide range of laws of varying scope, range and degree of state regulation in 23 U.S. states permitting the production, sale and use of medical marijuana has led in due course to even more dramatic change in the cannabis use landscape of the United States. The subsequent enactment of even more permissive laws in Colorado, Washington, Oregon, Alaska and Washington, DC allowing for the state-regulated production, sale and use of recreational marijuana – with additional states (including California) poised to vote on similar legislation in 2016 – raises the specter of **a serious challenge to traffic safety and the enforcement of impaired driving laws** (Svrakic et al., 2012). The federal government retains the right to preempt all state-level lawmaking in the area of substances such as cannabis listed on Schedule I of the *Controlled Substances Act* (21 U.S.C. § 903 and § 821)(2011). It is unclear how the courts will rule on the issue of preemption given the differing court opinions across the nation. As far as Colorado and Washington are concerned, in what is perhaps the only statement by a federal court relating to preemption of the Colorado and Washington laws, in *In re: Rent-Rite Super Kegs West LTD, 484 B.R. 799 (Dec 19, 2012),* a bankruptcy court noted (in what was clearly *dicta*) that

> conflict preemption is not an issue here. Colorado constitutional amendments for both medical marijuana, and the more recent amendment legalizing marijuana possession and usage generally, both make it clear that their provisions apply to state law only. Absent from either enactment is any effort to impede the enforcement of federal law.

In addressing the preemption potential present, the Office of the U.S. Attorney General has directed that federal law enforcement take no action against private persons and state officials who are operating under color of state statutes *so long as certain conditions are being met in said states*. The most critical of those conditions involve the strict regulation and control over distribution (such as those pertaining to alcohol and tobacco) to minors, transfer to other states and beyond, and the prevention of penetration by organized crime into the legal marijuana business (Memorandum for U.S. Attorneys from James M. Cole, Deputy Attorney General, *Guidance Regarding Marijuana Enforcement*, August 29, 2013 (hereinafter Cole 2013 Memorandum)). In addition, the states are required to assure the **effective enforcement of impaired driving laws** with respect to cannabis-impaired driving.

The impaired driving enforcement effort has long been highly focused on alcohol-impaired driving, and this is for very good reason. The danger of driving while under the influence of alcohol is well known, the incidence of this offense has been and continues to be great, and the "tools" available to law enforcement are relatively well-developed and have been incorporated into the forensic evidence mainstream of court operations throughout the country. Patrol officers are trained to perform a standardized field sobriety test (SFST), they are equipped with a portable breath test device (PBT), and calibrated stationary breath testing devises are available for securing a scientifically validated measurement of blood alcohol content level (BAC) from breath samples taken under controlled conditions. The correspondence between breath alcohol content and blood alcohol content is well established in scientific research, and the technologies for sample collection and laboratory analysis of biological evidence in urine, saliva, blood and breath are all highly refined. In the case of the PBT, it is important to note that *no judicial search warrant is required for securing the breath samples involved*. The PBT is an important tool for determining probable cause of alcohol impairment, and the results of its use add importantly to the effective handling and fair treatment of persons suspected of impaired driving. In the U.S. states where blood evidence is required, the PBT provides the arresting officer with the probable cause to secure a search warrant for securing a blood sample for entry into the chain of evidence in a case of a suspected *Driving While Intoxicated* (DWI) or *Driving Under the Influence* (DUI) offense.

Unfortunately, as the incidence of alcohol-impaired driving has been declining (a very good thing) the **incidence of drug-impaired driving has been rapidly on the rise** – particularly in those U.S. states with poorly regulated "leaky" medical marijuana laws (e.g., permit home growth, allow easy access to medical referrals, do not require a state registry) and in states permitting the legal sale of recreational cannabis (Sewell, Poling & Sofuogli, 2009; Garvey & Yeh, 2014; Whitehall, Rivara & Moreno, 2014). Even in states which have not enacted either medical or recreational marijuana statutes the incidence of drugged driving has increased dramatically (SAMHSA, 2013), and even in nations which adhere to the international treaties and conventions on drugs which treat marijuana as a prohibited substance vis-à-vis possession, use, production, processing, and sale invariably the most common impairing substance after alcohol found in fatal traffic collisions and impaired driving convictions is MARIJUANA (EMCDDA, 2009).

Law enforcement's principal weapon for dealing with drug-impaired driving are the **DRE** (Drug Recognition Expert) and the **ARIDE** (Advanced Roadside Impaired Driving Enforcement) programs of the National Highway Traffic Safety Administration. The latter program is an initiative of the White House Office of Drug Policy and will result in the training of 10,000+ patrol officers to bridge the gap between the SFST training received by most officers and the highly specialized training of the relatively few (~6,000) DRE officers present in each state (DuPont et al., 2012). Experience has taught the DRE officers that rapidly metabolizing drugs such as cannabis require *point-of-contact evidence of impairment*; in the case of cannabis-impaired driving, the detection of Δ in the case of cannabisTHC) at the time of initial police contact is essential to effective prosecution. The combination of ARIDE-trained officers, DRE officers on call, and a THC breathalyzer available for point-of-contact documentation of presence of the drug would permit much more effective cannabis-impaired driving enforcement. State-supported efforts are under way in both Colorado and Washington, the first two states to enact recreational marijuana statutes via citizen initiatives, to develop such a hand-held portable field device for law enforcement (and other applications for school, workplace and transportation systems safety). Early laboratory results on reliable detection, human subjects testing in controlled laboratory settings, breath to blood concentration level correlations, and field trials of prototype devices are ongoing to establish proof of concept and determine the best course toward commercialization and scaled up production.

This chapter provides the historical setting for the problem, sets forth the legal process within which the collection of evidence for impaired driving takes place in most U.S. states, and describes in brief the type of science – *ion mobility spectrometry* (IMS) – that provides the most likely path toward creating an analog to the PBT for law-enforcement use in cannabis-impaired driving. In due course, impairing drugs other than cannabis can be detected and documented from human breath samples in the same manner using this particular technology. The IMS detection technology is similar to that used in airports for the detection of explosives, in field ambient scanning in hazardous materials settings, and in cross-border illicit drug commerce (Phillips, 2015). The authors have been working as a team on this THC breathalyzer project since 2010 in a collaborative effort of the Hill Laboratory in the Department of Chemistry at Washington State University and the Division of Governmental Studies and Services, both units of the WSU College of Arts and Sciences. This research has been supported by grants from the State of Washington (Alcohol and Drug Addiction Research Program), by Washington State University, and by the Chemring Detection Systems Company of Charlotte, North Carolina.

The historical setting

One major consequence of the legalization of marijuana, either for medicinal use or for recreational purposes, and of the progressive decriminalization of its possession in small amounts and personal use virtually everywhere in the country, is that *the incidence of cannabis-impaired driving will increase substantially.* Even before the dramatic voter-approved changes in drug laws in the Western states of Colorado and Washington in the November elections of 2012, in both cases permitting the state-regulated production and retail sale of marijuana to adults, the problem of "drugged driving" (i.e., impaired driving caused by one or more drugs **not including alcohol**) was growing as a major challenge for law enforcement, traffic safety planners, and public policy makers. The drug-impaired driving problem is indeed widespread and growing throughout the country, and is of substantial concern to the several federal agencies for which the combination of drug law enforcement and transportation safety policies come into close contact. In reflection of these concerns, the Office of National Drug Control Policy in the Executive Office of the President in 2014 set out as one of its national drug control strategy goals for improving the public health and public safety of Americans "the reduction of the prevalence of drugged driving by 10 percent" (ONDCP, 2014, p. 4).

In like manner the member states of the European Union have all experienced the same phenomenon of the progressive displacement of alcohol by drugs among their impaired drivers. Both illicit (controlled substances) and prescription drugs are involved, and just as we are doing in the United States, these European nations are each considering varying legal approaches to this major transportation safety and public health challenge. These nations, acting in concert, have reported recently on findings derived from a comprehensive review of the available research conducted by European scientists to better understand and manage the worsening problem on their respective national and collective European Union highways and roadways (DRUID, 2012).

Extent of growing problem

A story circulated by the *Associated Press* penned by reporter Joan Lowry enjoyed wide distribution in U.S. newspapers on February 6, 2015 under the title "Federal Report: Fewer Drivers Drinking; More Using Drugs" (Lowry, 2015). The author of that story noted that the U.S. National Highway Traffic Safety Administration (NHTSA) in the U.S. Department of Transportation (USDOT) has sponsored a long-running series of anonymous surveys of large numbers of randomly selected drivers a total of **five times over the past 40 years**, providing a rather revealing and well-documented picture of the changing landscape of drug-impaired driving in the United States over that substantial time period. That large-scale field research has been carried out episodically following the same fundamental sampling and research protocol; most recently the survey was conducted by the highly regarded *Pacific Institute for Research and Evaluation* (PIRE). The survey process in question entails a very careful attempt to secure a random selection of drivers from the normal traffic stream in dozens of locations across the country, with each person who is recruited for participation in the study being interviewed by a trained interviewer about their driving habits and patterns of use of alcohol and drugs, and then being tested for the presence of impairing drugs if they are willing to provide breath and biological samples (saliva and blood). The PIRE weekend "roadside surveys" are carried out at three separate periods of the day on Fridays and Saturdays, days of the week when most fatal and serious injury traffic collisions tend to occur on American roadways. High visibility signage calling for voluntary and compensated participation in the survey is positioned in high traffic volume areas featuring ready access to a suitably large parking lot where 12–16 trained and experienced PIRE survey crews conduct their data collection on Friday afternoon (1:30–3:30 pm), Friday Evening (10:00 pm to midnight), and the early hours of Saturday morning (1:00–3:00 am). The second day of data collection occurs Saturday afternoon, Saturday evening, and during the early hours of Sunday morning. Each PIRE data-collection session occurs at a different location within the cities wherein the roadside survey is being conducted in a given year to maximize the likelihood of the collection of a representative cross-section of the driver population.

The precise physical locations for the data-collection activity are agreed to in advance by local law enforcement, municipal engineering and transportation departments, traffic safety advocates, and relevant property owners. Those locations and the exact times of data collection are held in strict confidence by all parties concerned until the absolute latest possible moment in order to keep the problem of self-selection of volunteers to a minimum. The survey volunteer participants in the roadside sampling process are instructed not to contact their friends and/or relatives about taking part in the survey; the temptation to do so is most certainly present because the compensation for participation is rather ample, particularly if participants agree to provide saliva and permit a phlebotomist to extract a blood sample at the end of the interview process. The compensation for full participation is in the range of $50–$60.

To further guard against self-selection, filter questions are present in the survey to weed out persons who may have been contacted by phone or through social media and alerted to the weekend roadside survey location. Local law-enforcement agencies are involved in both site selection and traffic-management planning, and they provide security support backup for the survey crew; however, **law enforcement is not in visible presence** among the PIRE crew collecting survey data, securing saliva swabs, collecting breath samples for alcohol detection, and obtaining blood samples for laboratory analysis and documentation of drugs present.

If the PBT results reveal a reading of 0.08 or higher – thereby indicating intoxication – the survey participant is promptly offered transportation home or to a place of their choice, but the person in question is not permitted to drive away from the location for purposes of public safety. Importantly, no traffic citation is issued and no fine is imposed on such persons, and provision is made for the safe and prompt return of the vehicle to the operator in due course at the expense of PIRE. The local news media and press in each roadside survey city are given advanced notice of the weekend roadside survey, and local news media leaders are asked to cooperate faithfully with the important goal of maintaining anonymity to survey participants by not recording any images of faces, of distinguishing images of human bodies or unique clothing, or of license plate numbers in their news coverage of the survey event.

The roadside survey volunteer participants are informed of the purpose of the survey, they are likewise informed of their rights as human subjects of research as to anonymity and the requirement of *informed consent*, and are told about the level of compensation associated with the survey, the PBT, the saliva sample, and the blood sample (collected in that order). If agreement to continue is given, all survey volunteers are then asked to complete a short questionnaire administered by PIRE staff featuring demographic background items and some questions on attitudes about drug use and driving habits. Next, the participants in the survey blow into a PBT device to measure BAC, and they provide a saliva sample on a cheek swab which is securely stored and coded to the participant's study ID number. Finally, many survey volunteers move on to a professional phlebotomist who extracts a blood sample for laboratory analysis. Generally, 60%–75% of survey volunteers agree to all of the aspects of the survey, including the blood sample. Both the saliva and blood samples are sent to a forensic laboratory for analysis regarding the presence of impairing drugs.

The findings from the latest roadside survey conducted (2013–2014) indicated that among American weekend nighttime drivers the prevalence of alcohol-involved drivers was down by about 30% from the level documented in the previous survey administered in 2007, and down by 75% since the first roadside survey conducted in 1973. However, *in excess of 15% of drivers tested positive for at least one illegal drug*, an increase from 12% seen in 2007. The percentage of nighttime drivers with marijuana in their blood stream grew by nearly 50% over that time period – 8.6% in 2007 as compared to 12.6% in 2014. **It is noteworthy that the one drug demonstrating the largest increase over time in weekend nighttime prevalence was THC.** These estimates of prevalence are indeed *conservative* ones, of course; it is highly likely that many drivers who were knowingly high from cannabis use at the time of the roadside survey in their community would not be inclined to volunteer for the weekend roadside survey. Upon viewing the survey roadside signage recruiting volunteers, they would likely pass up the opportunity for compensation for fear of arrest, fine or being detained; this would particularly be the case in states without either legal medical marijuana or recreational cannabis laws in force. Even though such outcomes would not occur given the research subject protections in place for observing the rights and interests of study participants built in to the study protocol, people in the general driving public are not aware of those protections.

Comparable prevalence figures reported for the United States have been documented in European countries as well. The DRUID study (*Driving While Under Influence of Drugs, Alcohol and Medicines*), a research project amply funded by the European Commission for the European Union and its several member states, constitutes the most far-reaching study of impaired driving undertaken to date. The study featured both original data collection on a vast scale, and two meta-analyses covering virtually all available drug-impaired driving research done by European scientists over the course of the past two decades. The DRUID project required five years to complete (2006–2011), and its voluminous findings were published in 2012.

Reliable data on the prevalence of psychoactive substances among European drivers were generated from roadside surveys conducted between January of 2007 and July of 2009. All of these surveys were of very similar design to that described above for the PIRE weekend roadside surveys carried out in the United States; they were conducted in the thirteen European countries that agreed to follow a uniform sampling process and study design template. Saliva and blood samples taken from nearly 50,000 randomly selected drivers were analyzed, and with regard to cannabis the authors of the report noted the following noteworthy observation:

> Cannabis seemed to be a weekend drug mainly used by young male drivers. There was a significant difference in the prevalence of cannabis in different time periods, most prevalent in weekend days and least prevalent in weekend mornings. However, cannabis was found during all days and hours of the week in most countries.
>
> (DRUID, p. 22)

The cut off levels used for the documentation of a positive TCH concentration were 1.0 ng/mL in blood and 27 ng/mL in saliva. **The highest drug prevalence noted among illicit drugs was that of THC**, followed by cocaine and amphetamines in distant second and third positions, respectively. For all nations combined alcohol was found in 3.48% of the drivers, illicit drugs were observed in 1.90% of motorists, medicinal impairing drugs were documented in 1.36%, and combinations of two or more drugs and drugs consumed in combination with alcohol were found in another 67% of randomly selected drivers (DRUID, p. 17). These are high figures indeed given the ubiquitous availability of mass transit options for movement about cities in Europe.

The extent of the problem of cannabis consumption and driving impairment is estimated in part from prevalence documented in roadside surveys;

this is indeed one extremely important form of evidence to be quite sure. However, a second key type of evidence is also very important in forming an informed estimation of the scale of the drugged driving problem. That second type of evidence comes in the form of official collision and fatal injury archival record data mining and relevant analysis. In the United States, the FARS (Fatality Analysis Reporting System) data are constructed, updated and dutifully maintained by each of the fifty U.S. states in close conformity with federal standards of reporting and classification; the ever-expanding FARS dataset provides an accurate longitudinal record of alcohol and drug involvement in fatal and serious injury collisions across the entire United States. The FARS record-keeping standards in question are set by NHTSA, and reports for the nation and the individual states are developed and disseminated periodically by the Federal Highway Administration, which is also responsible for determining the *vehicle miles traveled* (VMT) each year to serve as a base denominator for the calculation of comparable fatalities/VMT ratios over time. All official reports emanating from the FARS data archive are prepared by the National Center for Statistics and Analysis within USDOT's NHTSA. These periodic reports provide a rather sobering account of the steady increase in the incidence of drug-involved fatal collisions in the United States, particularly among novice drivers and young adult male drivers (National Center for Statistics and Analysis, 2015, January).

Not surprisingly, the European public health and public safety researchers associated with the DRUID study documented the same unfortunate pattern of disproportionate youth and young adult drug-impaired involvement in their review of research on automobile crashes involving motorist hospitalizations for injuries and/or fatalities. For persons hospitalized and killed in traffic crashes in the thirteen European countries taking part in the DRUID study, persons aged 25–34 (predominantly males) were most likely to test positive for impairing drugs, and among those impairing drugs **cannabis is the most often detected controlled substance** (2012, p. 23).

Likely future patterns of cannabis use as judged from youth attitudes & behaviors

A high visibility nationwide effort to dissuade American youth from indulging in the use of cannabis was actively waged for a substantial period on a truly grand scale. Marijuana was broadly portrayed as representing the major "gateway drug" to serious drug dependence, the use of which would result in likely further drug experimentation and near certain progression on to much more harmful banned substances. This active campaign to dissuade youth from marijuana use is perhaps best described as a classic "lost cause" phenomenon. Despite some fairly solid evidence that cannabis use in adolescence is likely associated with higher risk of driving collisions, drug dependence and adverse psychosocial outcomes and mental health problems in adulthood (Hall, 2014; Filbey et al., 2014), American youth by-and-large *tuned out that message*.

The great many D.A.R.E. (Drug Abuse Resistance Education) and related school-based programs in the United States (and replicated to some extent in 54 nations across the globe) were specifically designed to dissuade youth from involvement with drugs (Miller, 2001). These programs have been submitted to multiple short-term and long-term outcome assessments alike. Systematic meta-analyses carried out on many of those systematic program evaluation studies uniformly conclude that little to no beneficial effect can be attributed to this type of formal instructional intervention by specially trained police officers; this appears to be the case with both primary and secondary school students (see the research summarized in West & O'Neal, 2004 and Pan & Bai, 2009). The core message of the D.A.R.E. training in question was that of *absolute prohibition*; themes stressing responsible use and/or reduction in harm approaches intentionally were not part of the original curriculum, program conceptualization, or teaching pedagogy.

An estimated 80% of public school districts across the country **continue** to maintain their D.A.R.E. programs and similar cops-in-schools initiatives which they implemented in the early days of the "war on drugs" despite this rather compelling evidence of few if any beneficial outcomes vis-à-vis youth desistence from drug use in high school and early adulthood. Administrators of these school districts tend to give reasoned justifications for maintaining their D.A.R.E. and D.A.R.E.-like spin-off programs, reasons which are quite worthy of note here. These justifications have much more to do with building and sustaining better relations between schools and the police, and between the police and youth in their respective communities, than any suspected beneficial effect on the likelihood of drug experimentation in high school and during early adulthood (Miller, 2001; Birkeland, Murphy-Graham & Weiss, 2005).

A reliable and trusted source of direct evidence of the lack of overall impact of this prohibitionist/abstinence approach to dissuading youth from an abiding inclination to experiment with "grown up" behaviors and activities is found in the series of annual surveys conducted by the Survey Research Center at the University of Michigan for the longstanding *Monitoring the Future* program. These ongoing, yearly national surveys involve random samples of 8th, 10th and 12th grade students (weighted Ns ranging between 2,000 and 3,600 for each grade level) over the period 1975 through 2014. The de-identified individual-level data from these annual surveys are available to researchers and the general public alike through the long-established *Interuniversity Consortium for Political and Social Research* (ICPSR). These particular datasets are generated under the federally supported National Addiction & HIV Data Archive Program (NAHDAP).

The survey items featured in the rather lengthy *Monitoring the Future* study questionnaire cover a wide range of drugs of abuse, including the drugs LSD, cocaine, ecstasy, crack, heroin, amphetamines, crystal methamphetamines (ice), bath salts (synthetic stimulants), the pain medications most often used recreationally (e.g., OxyContin, Percocet), and **marijuana**. In the survey, American middle school and high school students are asked about their customary patterns of drug use – with response options ranging from yearly use, monthly use, to daily use. Responses to these items permit a careful monitoring of increases and decreases in the rates of reported use of each of the drug categories over an extended time period. Equally valuable, additionally, are the questions relating to the *perception of harm* resulting from use of these controlled substances. The U.S. middle school and high school students from across the entire country are asked to respond to the following question in this regard: ***How much do you think people are harming themselves (physically or in other ways), if they...***

For each of the categories of drugs of abuse included in the survey the students are asked about how much harm they ascribe to trying the drug "once or twice," to making "occasional use" of the drug, and to making "regular use" of the substance in question. Figure 13.1 sets forth findings for 12th grade students on perception of the harm associated with **smoking marijuana regularly**.

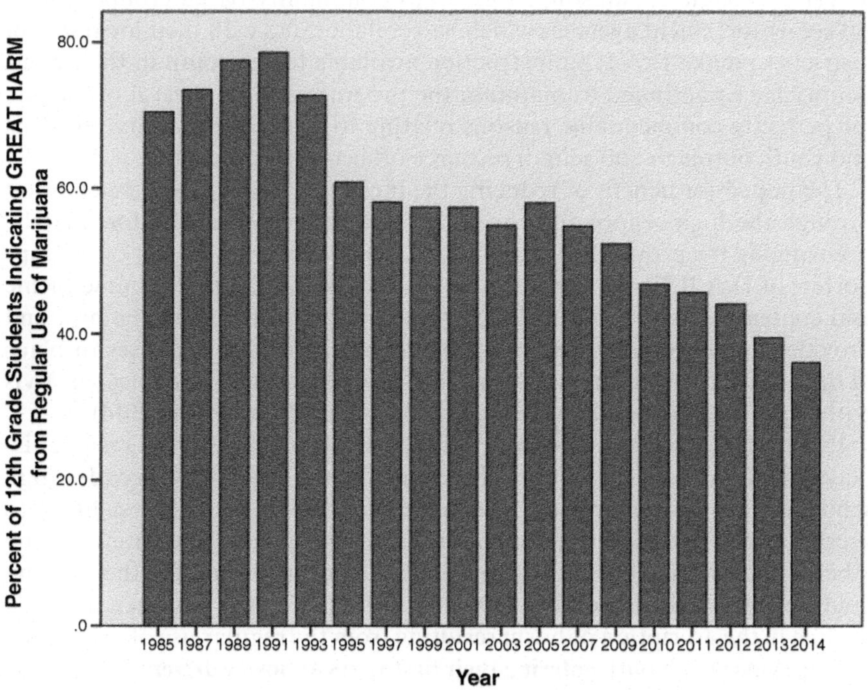

Figure 13.1 Declining rates of ascription of harm to regular use of marijuana, 1985–2014, among 12th grade students in the United States.

Two specific observations are called for at this point before characterizing the trend in annual U.S. youth assessments of the potential harm of regular use of marijuana. First, the pattern of *ascribed harm* depicted in Figure 13.1 for the nation's 12th grade students is substantially the same for its 8th and for 10th grade students. Secondly, with respect to patterns of *reported use* the same observation applies – that is, the patterns over time for U.S. 8th grade students and for U.S. 10th grade students are very similar to those of U.S. high school seniors. As for reported daily use, the frequency has remained steady between 2% and 8% for the entire run of the surveys for 12th grade students. Monthly use has likewise remained nearly constant, fluctuating between just under and slightly over one-in-five (20%). Finally, lifetime use for U.S. 12th grade students has held rather steady in the mid-40% range over the entire course of the annual *Monitoring the Future* survey series.

This evidence of failure to reduce the rate of reported use in the middle and high school student population contributed mightily to the ultimate judgment made on the part of the preponderance of American criminal justice and criminology researchers that the longstanding and ubiquitous D.A.R.E. program was not achieving the goal of desistence from drug experimentation among youth hoped for by its many promoters, by at-risk youth advocacy groups, and by its Los Angeles Police Department originators (Rosenbaum & Hanson, 1998). Many community-based groups and local law-enforcement agencies which had collaborated with their local school districts to make D.A.R.E. instruction available to the youth in their community have continued to maintain the program and its several off-shoots for perfectly commendable reasons relating to police/community relations and youth outreach and school resource officer-related benefits.

The hoped for benefit of reducing the incidence of drug experimentation through the high school and the early adult years has not, unfortunately, been among the proven benefits of the D.A.R.E. program. Many of the supporters of D.A.R.E. are hopeful that the new "keepin' it real" course theme and content, which replaced the original abstinence theme of the past and provides for the substitution of problem-based learning exercises in place of drug fact-laden lectures, will produce more favorable outcomes vis-à-vis patterns of drug use and abuse over the life course (Nordrum, 2014).

It is evident from the dramatic *Monitoring the Future* survey results displayed in Figure 13.1 that over the course of the past three decades high school seniors in the United States have become increasingly less likely to ascribe harmful effects to the regular use of marijuana. The gateway drug label pinned to marijuana may have rung somewhat true for the nation's youth of the 1980s and early 1990s, but after 2005 there has been a steady decline in the ascription of harm resulting from the regular use of marijuana among American youth **entering their first years as novice drivers**. The strong abstinence message once characterizing D.A.R.E. was quite likely landing on increasingly deaf ears in more recent years. It is hoped that the responsible use and reduction in harm approaches now featured in the D.A.R.E.

training will lead to better traffic safety and public health outcomes as U.S. states move toward ever-greater liberalization of marijuana laws.

In their analysis of the *Monitoring the Future* data collected over the decade period of 2001–2011, researchers O'Malley and Johnson (2013) show how this declining "perception of harm" ascribed to marijuana may sadly translate directly into high risk driving behavior on the part of American youth. These researchers duly note that high school seniors in the United States are reporting increasingly lower levels of likelihood of driving after drinking over this time period, **but higher likelihood of driving after using marijuana**. O'Malley and Johnson opine that given this evidence stronger efforts are needed to combat adolescent drugged driving; it might be added that such efforts are important in those states having legalized cannabis and in those states potentially moving toward the legalization of recreational marijuana where no persons under 21 are allowed to possess or use marijuana (legal or illicit), let alone drive after doing so. It should be noted that a recent report entitled "Daily Marijuana Drug Use Among U.S. College Students Highest Since 1980" issued by the University of Michigan news service observed that a survey of college students taken in 2014 (survey is 41st in an annual series) revealed that the following: "Daily marijuana use among the nation's college students is on the rise, surpassing daily cigarette smoking for the first time in 2014" (Wadley & Bronson, 2015).

Research evidence on the driving impairment consequences of cannabis

There is some degree of disagreement among researchers on how impairing marijuana use is on one's ability to operate a motor vehicle (Ingraham, 2014; NORML, 2014). Much of the early research on marijuana use and driving performance was carried out in strictly controlled laboratory settings involving driving simulators. These research studies consistently documented progressively adverse effects at increased doses, and indicated that cannabis consumption impairs the psychomotor skills known to be necessary for safe driving (see summaries of the studies set forth in Iversen, 2003; Ramaekers et al., 2004; Hartman & Huestis, 2013). However, research studies done outside of the laboratory and simulator context in the more "naturalistic" settings of everyday life (i.e., people driving under normal circumstances) have not produced such clear findings of cannabis-impaired driving. Three types of such naturalistic research studies are found in the published grant-supported research literature; cross-sectional studies, cohort studies, and case-control studies are the types of studies in question.

Cross-sectional studies involve the analysis of archival data on injured drivers and fatally injured drivers. In these types of studies cannabis is consistently shown to be one of the most often documented psychoactive substances present (though **always second to alcohol**), and it is well established that individuals who drive within 2 hours of using marijuana have raised rates of collision (Asbridge, Poulin & Donato, 2005). Based on a comprehensive meta-analysis published by Elvik (2013), it can be concluded that the same dose response relationship demonstrated in lab studies with driving simulators is found in the naturalistic setting offered by official archival data on injury and fatal crashes. Likewise, the driving impairment impact of THC consumption is in some evidence, although the impact is **far less** than is evident for alcohol consumption. Similar findings regarding dose response and comparisons to alcohol impairment were reported in the DRUID meta-analysis based on European epidemiological studies (p. 31).

Fewer studies have been reported of the cohort and case-control (e.g., culpability studies such as Longo et al., 2000) variety where more controlled (hence more revealing) comparisons are possible, but in the few studies done the findings are definitely mixed – some report an impairing consequence to marijuana use (e.g., Mann et al., 2007), some find little effect (e.g., Lachenmeier & Rehm, 2015), and some even report a beneficial effect (Ronen et al., 2008). In virtually every case of a direct comparison of impairing consequences in such studies; however, alcohol is always found to be more seriously impairing than cannabis by a wide margin. This is equally the case in U.S.-based studies and in the research done in Europe and summarized in the meta-analyses conducted for the massive DRUID study (Asbridge, Hayden & Cartwright, 2012).

Whatever might be the actual risk to driving impairment present from the consumption of cannabis, it is truly noteworthy that the research literature appears to be virtually *rock solid* on **adverse additive effects** occurring when cannabis and alcohol are used simultaneously. During concurrent use whatever supposed beneficial effects might exist for cannabis over alcohol, such as driving more slowly and being more deliberate in decisions while behind the wheel, these alleged beneficial effects are not at play and the level of driving impairment is commonly marked (Marks & MacAvoy, 1989; Ramaekers et al., 2004; Sewell, Poling & Sofuogli, 2009; Terry-McElrath, O'Malley & Johnson, 2014).

Unfortunately, the incidence of mixing alcohol and marijuana in the pursuit of recreational intoxication is all too commonplace in the United States, and likewise in the European Union nations where the highest prevalence of mixing the two was found among the 25–34 young adult age population segment (DRUID, p. 22). Research suggests that the same young adult-concentrated incidence of conjoint use of cannabis and alcohol is present in the United States (Bingham, Shope & Zhu, 2008). In this regard, Logan, Mohr, and Talpins conclude their recent study of oral fluids testing approaches to the documentation of cannabis consumption in connection with driving with this most pertinent observation:

> ...the data confirm reports in other related populations with respect to prevalence of combined alcohol and drug use on the impaired driving

population. Policies that exclude drivers with blood or BrAC concentrations above the alcohol per se limit are missing substantial numbers of drivers with co-morbid drug and alcohol problems – in this cohort as high as 53% of all drug using drivers.

(2014, p. 6)

This sobering observation leads directly to the question of how is law enforcement responding to this challenge of multiple impairing drug cases, and the question of how much value would there be in having officers equipped with BOTH a PBT and a THC breathalyzer device in carrying out impaired driving enforcement?

Law enforcement response to the challenge of cannabis-impaired driving enforcement

As the prospects for continued liberalization of marijuana laws appeared to be looming large for the nation's law enforcement, the leadership of the International Association of Chiefs of Police (IACP) placed the topic of drug-impaired driving on their agenda for the 119th annual conference held in San Diego, California in October of 2012. Upon robust discussion and due consideration on the part of the law-enforcement executives organization's membership, and upon broad recognition of the imminent passage of state-level legislation permitting state-regulated recreational marijuana to be sold to adults in the states of Colorado and Washington, the following Official Resolution drafted by the IACP's *Narcotics and Dangerous Drugs Committee* was duly adopted by the entire organization.

Combating the Dramatic Increase in Drug-Impaired Driving Offenses

October 3, 2012

WHEREAS, the International Association of Chiefs of Police ("IACP") recognizes that drug-impaired driving constitutes a significant law enforcement and societal problem; and

WHEREAS, according to the "Drugged Driving Research: A White Paper," prepared for the National Institute of Drug Abuse by the Institute for Behavior and Health, Inc., within the United States drugs other than alcohol are involved in approximately 18 percent of motor vehicle driver deaths; and

WHEREAS, the 2012 National Drug Control Strategy outlined a policy focus for a 10 percent reduction in drugged driving by 2015; and

WHEREAS, an estimated $59.9 billion in costs are attributable to drugged driving; and

> **WHEREAS, according to the National Highway Traffic Safety Administration marijuana accounted for 70 percent of illicit drugs used by drivers**; and
>
> **WHEREAS, studies by the U.S. Department of Transportation and the Dutch Ministry of Transport concluded that the effects of THC, the active ingredient in marijuana, significantly impairs drivers and makes them more likely to fall asleep at the wheel**; and (emphasis added)
>
> WHEREAS, preventing citizens from operating motor vehicles while under the influence of drugs is critical to public safety; however, there is no consistent method of identifying drug impairment and the presence of drugs in the body; and
>
> WHEREAS, drug-impaired drivers are less frequently detected, prosecuted, or referred to treatment than drunk drivers because few police officers are trained to detect drug impairment and prosecutors lack a clear legal standard under which to prove drugged driving cases; and
>
> WHEREAS, the "Policy Focus Reducing Drugged Driving" section of the 2012 National Drug Control Strategy recommends five strategies to address this growing problem: 1) encourage states to apply the per se standard used for commercial drivers to drivers impaired by illegal drugs and the impairment standard used for intoxicated drivers to other drug-impaired drivers; 2) collect further data through more consistent use of the Fatality Analysis Reporting System ("FARS") and more frequently conducted National Roadside Surveys; 3) educating communities and professionals – particularly new drivers, drivers on prescription drugs, and medical professionals – about drugged driving risks and legal consequences; 4) implementing the Drug Evaluation and Classification ("DRE") program across jurisdictions so that law enforcement is uniformly trained to detect drugged drivers; and 5) developing standard laboratory methodologies and further researching oral fluid testing to determine if it constitutes a reliable and widely-available roadside test; now, therefore, be it
>
> RESOLVED, that the IACP recommends adopting the strategies outlined in the 2012 National Drug Control Strategy to address this significant public safety issue.

It should be recalled, of course, that vigorous enforcement and prosecution of impaired driving under the influence of marijuana was explicitly included in the list of conditions which must be observed to avoid federal intervention (pre-emption) under the *Controlled Substances Act of 1971*. Attorney General Holder declared on August 29, 2013 that in those states adopting laws permitting recreational and medical marijuana the absolute ban against driving while impaired by marijuana must be strictly observed.

The traffic safety forces of the several U.S. states have all followed the recommendation of the IACP resolution to implement "the Drug Evaluation and Classification ("DRE") program across all jurisdictions so that law enforcement is uniformly trained to detect drugged drivers..." The DRE officers in the state and local law-enforcement agencies across the country can be considered the veritable *first line of defense* against drug-impaired driving. The program was originally designed to address a problem of *occasional occurrence* wherein a driver is detained for suspicion of impaired driving and a PBT reveals that there is no alcohol present to account for the observed signs of impairment. The officer responsible for the initial contact with such a suspect motorist then issues a request through police agency dispatch that a DRE-trained officer be directed to the location of the traffic stop to conduct a detailed, 12-step drug influence evaluation.

That DRE field assessment entails a considerably enhanced standard field sobriety test, featuring a series of interrogations concerning events leading up the traffic stop (including drug use), multiple pulse rate notations, pupil size estimation, vertical nystagmous assessment, stimulus tracking, eye convergence, eyelip droopiness notation, the Rhomberg balance test, a walk and turn test, a one-legged stand test, internal clock test, nose touch, a hippus test, rebound dilation, reaction to light, a check for needle marks, blood pressure and temperature notations. The DRE 12-step assessment is sanctioned by the joint action of the IACP and NHTSA, and provides the basis of expertise-based testimony in contested cases of drug-impaired driving.

There are two basic theories by which most state courts determine whether DRE testimony is admissible – the *Frye* standard and/or the *Daubert* standard. The legal standard expressed in *Frye v. United States*, commonly referred to as the "general acceptance" test, dictates that scientific evidence is admissible at trial only if the methodology or scientific principle upon which the evidence is based is sufficiently established in the research community to have gained general acceptance in the particular field in which it belongs [*Frye v. United States*, 293 Fed. at 1014 (D.C.Cir.1923)]. The *Frye* admissibility test concerns itself almost exclusively with the methodology that is being employed rather than the actual opinion which is drawn from it, and the test also requires courts to wait for the scientific community to accept a methodology as being of proven worth and reliability prior to the admission of evidence at trial reflecting the methodology in question.

Daubert employs the *Frye* general acceptance test as only one factor in considering the admissibility of scientific evidence. Under *Daubert*, scientific evidence can be validated in court even before it has reached a state of acceptance in the relevant scientific community. On the other hand, under *Daubert*, an acceptable piece of scientific evidence can be rejected if it is misleading or impairs the fact-finding process. The *Daubert* court stated in this regard that the admissibility of scientific evidence should be subject to a variety of considerations. First, the subject of the testimony must be "scientific knowledge rooted 'in the methods and procedures of science'" [*Daubert v. Merrell Dow Pharmaceuticals, Inc.*, 509 U.S. 579 (1993)]. *Daubert* then set out four nonexclusive factors which can be considered in determining admissibility: (1) whether the theory or technique can be, and has been, tested; (2) whether the theory has been subjected to peer review and achieved publication; (3) the known or potential rates of error have been identified; and (4) whether it is generally accepted in the relevant scientific community. *Id.* Next, the court must determine that the evidence is relevant to the facts in the case. Finally, the court should consider the impact of other rules of evidence; for example, it should consider whether the evidence is more prejudicial than probative, a determination which could render the evidence inadmissible. *Id.*

In *People v. Kelly*, 17 Cal. 3d 24, 130 Cal. Rptr. 144, 549 P.2d 1240 (1976), the State of California adopted a modified version of the *Frye* test. The court created a three-part test for the admissibility of novel scientific evidence. First, the reliability of the method in question must be established, usually by experts who can demonstrate that the method is generally accepted within the relevant scientific community. Next, the testifying expert must be properly qualified to take part in court proceedings. Finally, the proponent must show that the correct scientific procedures were used in the specific case before the court. This admissibility test became known as the *Kelly/Frye* test (Lustre, 2001).

Each state applies some variation of these standards when considering the admissibility of DRE testimony in criminal cases. Federally, *U.S. v. Everett*, 972 F. Supp. 1313 (1997) stands for the proposition that DRE testimony is not governed by *Daubert* because it is not scientific in nature. The Court found that upon the appropriate foundation being laid, the Drug Recognition Evaluation protocol conducted by DRE officer Ranger Bates, together with his conclusions drawn there from, shall be admitted into evidence to the extent that the DRE can testify to the probabilities, based upon his or her observations and clinical findings, but cannot testify, by way of scientific opinion, that the conclusion is an established fact by any reasonable scientific standard. In other words, the otherwise qualified DRE cannot testify as to scientific knowledge, but can provide testimony as to specialized knowledge which will appreciably assist the trier of fact to understand the evidence.

In *State v. Baity*, 991 P.2d 1151 (Wash. 2000) the DRE Protocol is considered a novel scientific test and must be analyzed under *Frye*. The court in that case determined the protocol and the testimony of the officer were admissible, provided that all 12 steps of the protocol are used. In summary, after analyzing the DRE protocol and the approach of other courts to its admissibility, the court held the DRE protocol and the chart used to classify the behavioral patterns associated with seven categories of drugs both have scientific elements meriting evaluation under *Frye*. The court found that the protocol was accepted in the relevant scientific communities. The justices' opinion was confined to situations where all 12 steps of the protocol have been carried out by the DRE-trained officer. Moreover, the opinion holds that *a DRE officer may not testify in a fashion that casts an aura of scientific certainty to testimony given.* The officer also may not predict the specific level of drugs present in a suspect. The DRE officer, if properly qualified, trained, and certified, may express an opinion that a suspect's behavior and physical attributes are or are not consistent with the behavioral signs associated with certain categories of drugs known to impair driving.

In addition to conducting this evaluation, the DRE officer must designate which drug category (or categories) he or she believes to be the cause of the impairment documented in the 12-step assessment from the following list:

1 Central nervous system depressant
2 Central nervous system stimulant
3 Hallucinogen
4 PCP
5 Narcotic analgesic
6 Inhalant
7 **Cannabis**

______ **Medical Rule Out (Confounding medical condition precludes testing)**

Under current law in the State of Washington and numerous other U.S. states, if the DRE officer feels that he or she has probable cause to suspect impairment the officer must request a judicial search warrant for the taking of blood evidence. If the warrant request is granted, the DRE officer completes a narrative report on the matter and transports the person suspected of a DWI to a medical facility where a blood sample is taken. The DRE officer then sends the blood evidence through the chain of custody to the Washington State Crime Lab for testing and the generation of an official toxicology report. That report is subsequently attached to the DRE field assessment and a narrative account (developed primarily for use by prosecutors); the DRE program coordinator receives these three reports for each DRE assessment completed and is responsible for monitoring officer performance statewide. DRE officers in good standing must conduct a minimum number of assessments each year, and they must demonstrate a 70% "hit rate" in their assessments overall – that is, in at least seven-in-ten evaluations they must have identified at least one drug category that produced a positive verification in the associated toxicology report. Veteran DRE officers participate in the training and field assessment of new DRE officers in training.

It should be noted that the DRE assessment process in Washington State had long relied upon the premise of an *implied consent* to extract blood from drivers suspected of drug-impaired driving. It was reasoned that the privilege of having a driver's license obligated the possessor of that license, as a reasonable condition of enjoyment of that privilege, to submit to a blood test if suspected of impaired driving. However, the U.S. Supreme Court issued its judgment in ***Missouri v. McNeely*** (2013) holding that the implied consent premise long-established in impaired driving enforcement was *no longer permissible*. In rejecting the legal argument customarily made in impaired driving cases that the requirement of a judicial warrant for blood draws raises a high risk of lost evidence, the court ruled in favor of privacy rights and raised the barrier to routine timely collection of blood evidence in marijuana-impaired driving cases. Unfortunately, marijuana is a drug which metabolizes fairly quickly – often within two hours after smoking – and much of the psychoactive element in cannabis *THC* metabolizes into other compounds (principally carboxy-THC) which are NOT impairing.

Peak THC concentration occurs very rapidly with smoked cannabis, but then dissipates quickly; in the case of ingestion of marijuana in infused foods or drinks, the peak high comes on more slowly and is generally lower than with smoked cannabis. The THC will be metabolized just as quickly after the THC enters the bloodstream, but the assimilation into the blood occurs over a longer period of time. As noted by Wong, Brady and Li (2014),

> THC is extremely lipid soluble and is widely distributed in the body to tissues at rates dependent on blood flow. THC in blood rapidly decreases over time, typically declining…within 3–4 hours. Therefore, delayed blood collection often results in an inaccurate THC measurement, which may not be reflective of a driver's level of impairment while driving.
>
> (Wong, Brady & Li, 2014, p. 4)

Because of this rapid rate of metabolizing of cannabis, the added requirement of a judicial search warrant for the securing of a blood sample means that some cannabis-impaired drivers will likely escape prosecution for lack of a blood draw being taken in time to detect THC in the blood sample submitted in legal evidence. Clearly, the need for point-of-contact detection and documentation of THC presence for marijuana-impaired driving enforcement is evident; interest in the feasibility of a THC breathalyzer arises from these conditions of cannabis-impaired driving enforcement.

Once the threshold of marijuana legalization was crossed with the passage of the Colorado and Washington initiatives in 2012 it was clear to the law-enforcement community across the country that their respective state's cadres of DRE officers were not going to provide sufficient coverage for drug-impaired driving in their states. The approximately 6,000 DRE certified officers spread across the country would be no match for the scale of the problem lying ahead (ONDCP, 2010). It was decided within NHTSA and IACP circles to enhance law-enforcement readiness for the pending challenge by carrying out training in the ARIDE (Advanced Roadside Impaired Driving Enforcement) program on an emergency action schedule. The course resulted from the recognition by those organizations of the rising problem of drug-impaired driving across the nation and the need for many patrol officers – not just DRE specialists – to be aware of the issue and to be capable of dealing with rising numbers of drug-impaired motorists. Nationally, over 8,500 officers are currently certified as graduates of ARIDE training (Bill O'Leary, NHTSA, USDOT via personal communication, 02/04/2014), with 10,000 officers trained being the immediate goal for the start of 2016.

Obstacles to effective enforcement of marijuana-impaired driving

Since the research literature on alcohol-impaired driving is so vast, thoroughly researched, and well-developed in case law, there are highly standardized practices for dealing with alcohol-impaired drivers and clear, well-established blood alcohol content limits and methods of measurement upon which to rely in prosecuting alcohol-impaired driving cases. The same cannot be said of marijuana-impaired driving, however (Barcotte & Scherer, 2015). The levels of THC required for the legal presumption of driver impairment vary from state to state, the physical evidence required to prosecute differs likewise, and the treatment of TCH and its metabolites differs as well (Grotenhermen et al., 2007).

It is the case in most all local jurisdictions across the United States that many cannabis-impaired drivers also have alcohol blood levels above 0.08 and are *prosecuted for alcohol impairment* ***alone*** given the added difficulties associated with drug-impaired driving under the influence of cannabis. According to researchers Brady and Li (2013), approximately 25% of drivers injured in automobile crashes test positive for two or more drugs, with cannabis and alcohol being *the most common combination*. Recently, Wilson, Stimpson and Pagan (2014) reported that 54.9% of marijuana-positive drivers in the United States had elevated blood alcohol levels. To complicate matters further, polydrug use is fairly commonplace and the DREs' ability to sort out the multiple physiological effects of each drug is limited. Wolff and Johnson (2014) report in this regard that in Europe between 20% and 30% of marijuana use among drivers was combined with other drugs, with marijuana being the most frequently used drug in combination with both cocaine and benzodiazepines. It is in circumstances of multiple drug use wherein marijuana is one of the drugs that a THC breathalyzer in the hands of the DRE and ARIDE-trained patrol officer would be of major law-enforcement utility. Probable cause of impairment could be established quickly, and if required by state law a judicial search warrant for blood could be secured for timely verification of drugged driving.

Currently the detection of THC in impaired drivers can entail use of urine, oral fluids (saliva), or blood. Blood sampling is the most reliable medium, but it is also the most invasive. Even in the case of blood there is no standardization as of yet regarding how the blood sample is to be secured with respect to time of collection, location of collection, or levels of THC presumed to be impairing. At this point seven U.S. states have set marijuana metabolite limits in statute (Colorado, Iowa, Montana, Nevada, Ohio, Pennsylvania and Washington), with all but Iowa using blood level (Iowa using urine). Sixteen countries in Europe have set legal non-zero THC concentrations, and all of these countries use either blood or blood serum as the tested medium. The U.S. states are far from uniform in their standards and permissible limits, and the same can be said of the European nations which have yet to reach agreement on levels and optimal mediums for evidence collection.

In terms of approaches to enforcement of impaired driving laws on marijuana impairment, there are four discernible principal paths taken by U.S. states and European nations alike. The first approach might best be characterized as *impairment-based*. Under this approach a prosecutor must prove that the drug in question impaired the driver's ability to operate his or her vehicle. Since a uniformly accepted definition of impairment does not exist for different drugs (including cannabis in most states), in comparison to cases of drunk driving traffic stop cases involving drugged driving are relatively rarely prosecuted in the United States (Stecker, 2015). Eleven European countries employ this very challenging impairment-based approach. In light of the many difficulties associated with the enforcement of drugged driving prohibitions under the effect-based approach, seven U.S. states and five European countries have adopted *per se* laws which set a threshold concentration as a legal limit, and exceeding that limit serves as legal proof of impairment (Lacey, Brainard and Snitow, 2010). **The *per se* level approach for THC (set at around 5 ng/mL of blood) is advocated in the ICAP resolution with respect to preferred public policies to combat drugged driving.**

In yet another approach to drug-impaired driving some U.S. states and European countries have taken a *zero tolerance* approach whereby the discovery of any minimum reliably detectable level of a drug which has been shown to impair driving (such as marijuana) makes one subject to prosecution for drugged driving. Most interestingly, nine European countries have adopted a *two-tier* approach which combines the impairment-based approach and the *per se* approach to penalize a driver who has any amount of an illicit drug in their bloodstream with a noncriminal or misdemeanor fine, and more severely punishes drivers who are shown to be impaired at the time of arrest via a SFST or DRE-like assessment process by any substance known to be impairing. Marilyn Huestis, who heads up the Chemistry and Drug Metabolism Intramural Research Program at the National Institute on Drug Addiction (NIDA) in the National Institutes of Health endorses this two-tier approach enthusiastically (Huestis, 2015).

The timeliness of a THC breathalyzer

The difficulty in detecting and documenting cannabis-impaired driving is apparent even for the well-trained and experienced DRE officers, let alone the officers prepared only with the newly devised and disseminated ARIDE training, as useful as that formal and targeted training likely is for managing the increased volume of cannabis-involved and other drugged-driving cases. Relevant research has shown that the impact of marijuana smoking (the most common form of cannabis use among marijuana-impaired drivers) differs widely among people. A recent meta-analysis of more than 120 independently conducted studies demonstrated that regular marijuana users were considerably less impaired than occasional users ingesting comparable THC concentrations (Wolff & Johnson, 2014). There would appear to be both physiological tolerance and learned compensatory driving behavior at play. In the case of the 23 U.S. states where medical marijuana is legalized, the likelihood of this complication being a commonplace aspect of marijuana-impaired driving enforcement is particularly high.

Added to this scenario of difficulty of assessment for the DRE officer and his or her ARIDE-trained associates are the substantial limitations placed upon police arising from the ruling of the U.S. Supreme Court in the ***Missouri v. McNeely*** (2013) case with respect to the requirement of a judicial search warrant for routine police blood draws. In the process of setting aside the implied consent presumption commonly employed in the past, this court holding adds further to the urgent need for "point-of-contact" collection of physical evidence for documenting THC presence at the time of the traffic stop. As noted, this type of evidence collection "is important for detecting rapidly metabolized drugs and being able to relate observed driving performance to a toxicological result" (Logan, Mohr & Talpins, 2014, p. 1).

Human sweat is one area where point-of-contact evidence collection has been explored by some researchers. The hands and feet are covered with a natural secretion produced by the eccrine glands, and that secretion – sweat – represents a mixture of water, salts, and other trace compounds such as THC. Christian Elsner and Bernard Abel from the Leibniz Institute of Surface Modification and the W. Oslwold Institute for Physical and Theoretical Chemistry, with funding from the German Science Foundation, make use of laser desorption mass spectrometry imaging to record three dimensional finger prints – with the third dimension permitting the determination of the chemical composition of finger pore secretions (Elsner & Abel, 2014). This is a promising area of work, particularly since sweat testing has the great advantage of being noninvasive; unfortunately, research on the presence of THC in sweat is still quite limited (de la Torre & Pichini, 2004; Huestis et al., 2008). Preliminary findings in the available literature indicate that the amount of THC in sweat is quite low, hence THC is going to be difficult to detect and document reliably in this medium.

In this important area of research on THC detection a number of researchers in the United States and Europe alike are focusing on the collection and on-site analysis of oral fluids. A report prepared for the American Academy of Forensic Sciences by Maggitti, Logan, and McMullin describes in excellent outline form the work being done with oral fluids for the detection of impairing drugs generally, and THC in particular (2012). Field testing is ongoing of four prototype oral fluids devices in a study being carried out by PIRE for NHTSA in jail booking area settings in California as this chapter is being written. Similarly, comparable work is being reported on studies being conducted in the European countries represented in the DRUID study (Chu et al., 2012).

The use of oral fluids to document a driver's THC level has gained relatively wide use in the United States, Europe and Australia in part because it is a relatively noninvasive source and THC concentrations in oral fluids correlate strongly with blood concentrations on the population level. However, its use remains problematic for the estimation of blood THC concentrations based on the oral fluids samples of **individuals** (Huestis et al., 2011; Wong, Brady & Li, 2014). For use in impaired driving enforcement it is clearly essential that accurate documentation of THC levels in specific individuals take place.

While this work on oral fluids is indeed important, serious concerns have been raised by bio-ethicists and privacy rights advocates concerned with the ever-expanding reach of the state in regard to DNA profiles (Rosen, 2003). Oral fluids carry human DNA information along with potential evidence of drug consumption, and advocacy groups such as the ACLU are very much opposed to the collection and retention of DNA profiles for persons who have not been convicted of any serious crime (Maschke, 2008). Similar concerns for privacy have been raised in both the United Kingdom and in the European Community (Shellberg, 2003). In 2008 the European Court for Human Rights in Strasbourg ruled that the maintenance of DNA records of persons innocent of any crimes constitutes a violation of Article Eight of the U.N. Human Rights Convention with regard to 'the right to respect for private and family life."

In the United States, the California ACLU challenged California's Proposition 69 (2004) which was voted into law by a wide margin permitting the expansion of "DNA Fingerprinting" databases to include arrestees, not just certain categories of convicted persons. Similarly, working with funding provided by the National Human Genome Research Institute of the National Institute of Health, attorneys Simoncelli and Seinhardt (2005) contributed a timely article to the *DNA Fingerprinting and Civil Liberties Project* of the ***American Society for Law, Medicine and Ethics*** which provides an excellent overview of the issues leading to ACLU concerns.

In reflection of such serious concerns, some researchers have turned to the possibility of detecting the presence of THC in breath samples. Wong, Brady and Li (2014) have observed the following in this regard:

> Some studies have examined the possible detection of THC in breath, a noninvasive and easily observed drug screening method. Himes et al. (2013) found that both chronic and occasional marijuana smokers who smoked a single marijuana cigarette experienced significant

> decreases in THC breath concentration after controlled smoking. The window for detecting cannabinoids in breath ranged from 0.5 to 2 hours and coincided with impairment. THC-COOH (carboxy-THC) was also undetected in both groups of marijuana smokers, indicating that marijuana metabolite limits should reflect concentrations of THC, not THC-COOH, when a breath test is conducted. The short detection window for THC indicates that breath may be a viable alternative to oral fluids for detecting marijuana use, but only when a driver is tested <2 hours following smoking.
>
> (Wong, Brady & Li, 2014, pp. 5–6)

Reflecting this interest in breath testing for cannabis-impaired driving enforcement, the Colorado state legislature in 2014 allocated $250,000 to the state's Department of Economic Development to provide grant funding to a Colorado-based business (Lifeloc Technologies) which manufactures alcohol breathalyzers to work on a comparable field device for detecting THC. Likewise, a Canadian corporation by the name of *Cannabix* is working on the development of such a field breath testing device, as are analytic chemistry researchers in the Hill Lab at Washington State University in Pullman. The Hill Lab researchers are involved in a longstanding public/private partnership with Chemring Detection Systems of North Carolina to apply the science of differential ion mobility spectrometry to the task of point-of-contact breath testing for THC. Their long-term plan is to make the same device capable of detecting and documenting many drugs known to be impairing, both illicit substances and prescription medications.

The analytical chemistry science – *differential mobility spectrometry* (DMS) – that provides perhaps the most likely path toward creating an analog to the PBT for law-enforcement use in cannabis-impaired driving (Sonnberg et al., 2015) is being pushed into an accelerated pace by a recent award of funds from the State of Washington. The detection process in question involves the capture of THC suspended in breath resulting either from the inhalation of marijuana smoke or the oral ingestion of cannabis-infused drinks or solid foods. The IMS detection technology is the same one used in airports for the detection of explosives, in field ambient scanning in hazardous materials settings, and in cross-border illicit drug commerce investigations.

In addition, ambient mass spectrometry is also being investigated for drug detection in the field. Recently a grant was awarded under the *Applied Research and Development in Forensic Science for Criminal Justice Purposes* program in the Office of Justice Programs in the U.S. Department of Justice entitled "Assessing the Probative Value of Physical Evidence at Crime Scenes with Ambient Mass Spectrometry and Portable Instrumentation" [Award No. 2011-DN-BX-K552] to Professors Christopher Mulligan and Adam O'Leary at Illinois State University (Mulligan & O'Leary, 2015). Although miniature and ambient mass spectrometers have been proposed as potential field screening devices for drugs and other hazardous and/or dangerous compounds, the complex nature of this instrumentation has limited successful application in the field.

The authors of this chapter and a group of faculty, post-doc, and doctoral student researchers have been working as a team on this THC breathalyzer project for the past five years. Their work represents a long term, ongoing collaborative effort joining faculty and graduate students in the Department of Chemistry and the Department of Criminal Justice and Criminology at Washington State University. The WSU team is engaged in active collaboration with scientists and engineers at the Chemring Corporation. This research endeavor has been supported by grants from the State of Washington, Washington State University central administration, the College of Arts & Sciences, and the Chemring Detection Systems Company of Charlotte, North Carolina.

The work of this team has taken the form of the development of a prototype device which is capable of reliable detection of THC in the breath of volunteer subjects (both chronic smokers and occasional smokers of varying body mass index ratings) smoking marijuana with known THC concentration under controlled conditions. The THC can be captured on filters which are capable of long-term storage as *a form of time- and place-stamped physical evidence*, and which can be heated to release the stored THC for rapid detection in a hand-held analytical device at the point-of-contact in the field. In due course, impairing drugs other than cannabis can be detected and documented from human breath samples in precisely the same manner using this particular technology.

Upon completion of laboratory testing and additional human volunteer breath sample testing under controlled conditions, the next phases of development work entail breath to blood concentration correlations and then prototype testing under field setting conditions experienced by law enforcement. In this regard, the DRE training sessions conducted twice annually by the Impaired Driver Unit of the Washington State Patrol are a likely location for such preliminary testing. The Washington State Patrol, the Washington Traffic Safety Commission, and the Washington Association of Sheriffs and Police Chiefs have all agreed to collaborate in the testing to take place beyond the confines of the Hill Laboratory. NHTSA Region 10 administrators have been of critical importance in facilitating the timely dissemination of information to the Hill Lab team, and have expressed an ongoing interest in and provided support for advancing the work of that team.

Conclusion

In closing, it is quite apparent that the challenge posed to law enforcement by marijuana-impaired driving is a very serious one, but it is also clear that potent countermeasures such as ARIDE training and DRE officers are in place and other technology-based tools are being developed that will help law enforcement meet the present challenge and prepare for the future. The numerous problems associated with amassing the necessary scientific evidence of dose response outcomes to sustain *per se* legal limits must be addressed in time, and the much more effective gathering of point-of-contact physical and biological evidence must take place for the police to keep up with the problem of driving while high. While the social institution of the *designated driver* is now a well-established one among alcohol drinkers, there is evidence that drivers who use cannabis are inclined to believe that it is safe to drive after either smoking cannabis or consuming marijuana-infused foods or drinks (O'Malley & Johnson, 2013). While cannabis may even be substantially less impairing than alcohol, it is still the case that prudence would dictate that the designated driver concept be promoted among marijuana users recreating in groups.

Perhaps the not too distant future will bring a day when the designated driver norm is adopted by cannabis users and the patrol officers on duty will have ARIDE training certification and have at their disposal breathalyzers for BOTH alcohol and drug detection for the collection of point-of-contact evidence in cases of suspected impaired driving. That scenario could lead to both greater deterrence of drug-impaired driving and more effective arrest and prosecution of offenders. This would be a major accomplishment by the public health and traffic safety communities as they struggle to meet the NHTSA goal of *Target Zero* – namely, the avoidance of any traffic fatalities by 2030 through a combination of safer vehicles, more effective road engineering, more efficacious traffic law enforcement, and greater compliance by drivers with existing laws and regulations.

References

Asbridge, M., Hayden, J.A., and Cartwright, J.L. 2012. "Acute Cannabis Consumption and Motor Vehicle Collision Risk: Systematic Review of Observational Studies and Meta-Analysis." *British Medical Journal*. 344: e536.

Asbridge, M., Poulin, C., and Donato, A. 2005. "Driving Under the Influence of Cannabis and Motor Vehicle Collision Risk: Evidence from Adolescents in Atlantic Canada." *Accident Analysis and Prevention*. 37: 1025–1034.

Barcotte, B. and Scherer, M. 2015. "The Great Pot Experiment: Legalization Keeps Rolling Ahead, but Because of Years of Government Roadblocks on Research We Don't Know Nearly Enough about the Dangers of Marijuana – Or the Benefits." *Time*. 25: 38–45.

Bingham, C.R., Shope, J.T., and Zhu, J. 2008. "Substance-Involved Driving: Predicting Driving After Using Alcohol, Marijuana, and Other Drugs." *Traffic Injury Prevention*. 9(6): 515–526.

Birkeland, S., Murphy-Graham, E., and Weiss, C. 2005. "Good Reasons for Ignoring Good Evaluation: The Case of the Drug Abuse Resistance Education (D.A.R.E.) Program." *Evaluation and Program Planning*. 28: 247–256.

Brady, J.E. and Li, G. 2013. "Prevalence of Alcohol and Other Drugs in Fatally Injured Drivers." *Addiction*. 108(1): 104–114.

Chu, M., Gerostamoulos, D., Beyer, J., Rodda, L., Boorman, M., and Drummer, O.H. 2012. "The Incidence of Drugs of Impairment in Oral Fluid from Random Roadside Testing." *Forensic Science International*. 215: 28–31.

Daubert v. Merrell Dow Pharmaceuticals, Inc., 509 U.S. 579 (1993).

De la Torre, R. and Pichini, S. 2004. "Usefulness of Sweat Testing for the Detection of Cannabis Smoke." *Clinical Chemistry*. 50(11): 1961–1962.

DRUID (Driving Under Influence of Drugs, Alcohol and Medicines). 2012. *Final Report: Work Performed, Main Results, and Recommendations*. Berlin: European Union [Principal research group leaders Horst Schulze, Markus Schumacher, Raschid Urmeew, Kerstin Auerback, Anja Knoche, Inger-Marie Bernhoft, Marjan Hagenzieker, Javier Alvarez, Monika Pilgerstorfer, Bojan Zlender, and Han de Gier].

DuPont, R., Voas, R., Walsh, J.M., Shea, C., Talpins, S.K., and Neill, M.M. 2012. "The Need for Drugged Driving Per Se Laws: A Commentary." *Traffic Injury Prevention*. 13(1): 31–42.

Elsner, C. and Abel, B. 2014. "Ultrafast High-Resolution Mass Spectrometric Finger Pore Imaging in Latent Finger Prints." *Scientific Reports*. 4, 6905. doi:10.1038/srep06905.

Elvik, R. 2013. "Risk of Road Accident Associated with the Use of Drugs: A Systematic Review and Meta-Analysis of Evidence from Epidemiological Studies." *Accident Analysis and Prevention*. 60: 254–267.

EMCDDA (European Monitoring Centre for Drugs and Drug Addiction). 2009. *Responding to Drug Driving in Europe*. Lisbon: EMCDDA.

Filbey, F., Asian, S., Calhoun, V.D., Spence, J.S., Damaraju, E., Caprihan, A., and Segall, J. 2014. "Long-term Effects of Marijuana Use on the Brain." *PNAS*. 111(47): 16913–16918.

Frye v. United States, 293 Fed. at 1014 (D.C.Cir.1923).

Garvey, T. and Yeh, B. 2014, January 13th. *State Legalization of Recreational Marijuana: Selected Legal Issues. Report Prepared for Members and Committees of Congress*. Washington, DC: Congressional Research Service.

Grotenhermen, F., Leson, G., Berghaus, G., Drummer, O.H., Kruger, H., Longo, M., Moskowitz, H., Perrine, B., Ramaekers, J.G., Smiley, A., and Tunbridge, R. 2007. "Developing Limits for Driving Under Cannabis." *Addiction*. 102(12): 1910–1917.

Hall, W. 2014. "What Has Research over the Past Two Decades Revealed about the Adverse Health Effects of Recreational Cannabis Use?" *Addiction Monograph*. doi:10.1111/add.12703.

Hartman, R. and Huestis, M. 2013. "Cannabis Effects on Driving." *Clinical Chemistry*. 59(3): 478–492.

Himes, S.K., Scheidweiler, K.B., Beck, O., Gorelick, D.A., Desrosiers, N.A., Huestis, M.A. 2013. "Cannabinoids in Exhaled Breath Following Controlled Administration of Smoked Cannabis." *Clinical Chemistry*. 59(12): 1789–1789.

Huestis, M. 2015. "Deterring Driving under the Influence of Cannabis: An Editorial." *Addiction*. doi:10.1111/add.13041.

Huestis, M., Scheidweiler, K., Saito, T., Fortner, N., Gustafson, R.A., and Smith, M. 2008. "Excretion of Delta 9-tetrahydrocannabinol in Sweat." *Forensic Science International*. 174(2–3): 173–177.

Huestis, M., Verstraete, A. Kwong, T., Morland, J., Vincent, M.J., and de la Torre, R. 2011. *Clinical Chemistry*. 57(6): 805–810.

In re: Rent-Rite Super Kegs West LTD, 484 B.R. 799 (Dec 19, 2012).

Ingraham, C. 2014, February 9th· "Stoned Drivers Are a Lot Safer than Drunk Ones, New Federal Data Show." *The Washington Post.*

Iversen, L. 2003. "Cannabis and the Brain." *Brain.* 126: 1250–1270.

Lacey, J., Brainard, K., and Snitow, S. 2010. *Drug Per Se Laws: A Review of Their Use in States.* Washington, DC: National Highway Traffic Safety Administration.

Lachenmeier, D.W. and Rehm, J. 2015. "Comparative Risk Assessment of Alcohol, Tobacco, Cannabis and Other Illicit Drugs Using the Margin of Exposure Approach." *Scientific Reports.* 5, 8126. doi:10.1038/srep08126 (2015).

Logan, B.K., Mohr, A., and Talpins, S.K. 2014. "Detection and Prevalence of Drug Use in Arrested Drivers Using the Drager Drug Test 5000 and Affinition Drug-Wipe Oral Fluid Drug Screening Devices." *Journal of Analytical Toxicology.* 1–7. doi:10.1093/jat/bku050.

Longo, M.C., Hunter, C.E., Lokan, R.J., and White, M.A. 2000. "The Prevalence of Alcohol, Cannaboids, Benzodiazepines and Stimulants amongst Injured Drivers and Their Role in Driver Culpability: Part II: The Relationship between Drug Prevalence and Drug Concentration, and Driver Culpability." *Accident Analysis and Prevention.* 32(5): 623–632.

Lowry, J. 2015, February 6th. "Federal Report: Fewer Drivers Drinking; More Using Drugs." *Denver Post.*

Lustre, A.B. 2001. "Post-Daubert Standards for Admissibility of Scientific and Other Expert Evidence in State Courts." *American Law Review.* 453: 90.

Maggitti, A., Logan, B.K., and McMullin, M. 2012. "Tools, Techniques and Findings for the Qualitative Analysis of delta-9-tetrahydrocannabinoids in Oral Fluid." *Proceedings: American Academy of Forensic Sciences.* XIX, 474–475.

Mann, R.E., Adlaf, E., Zhao, J., Ialomiteanu, A., Smart, R.G., and Asbridge, M. 2007. "Cannabis Use and Self-Reported Collisions in a Representative Sample of Adult Drivers." *Journal of Safety Research.* 38: 669–674.

Marks, D. and MacAvoy, M. 1989. "Divided Attention Performance in Cannabis Users and Non-Users Following Alcohol and Cannabis Separately and in Combination." *Psychopharmacology.* 99: 397–401.

Maschke, K.J. 2008. "DNA and Law Enforcement." In *From Birth to Death and Bench to Clinic: The Hastings Center Bioethics Briefing Book for Journalists, Policymakers, and Campaigns,* ed. Mary Crowley. Garrison, NY: The Hastings Center, pp. 45–50.

Miller, D. 2001. "D.A.R.E. Reinvents Itself – With Help from Its Social Science Critics." *Chronicle of Higher Education.* October 19.

Missouri v. McNeeley 2013. State of Missouri v. Tyler Gabriel McNeeley. No. 11–1425. United States Supreme Court.

Mulligan, C.C. and O'Leary, A.E. 2015. *Accessing the Probative Value of Physical Evidence at Crime Scenes with Ambient Mass Spectrometry and Portable Instrumentation: Final Report.* Normal, IL: Department of Chemistry, Illinois State University. Report submitted to the Applied Research and Development in Forensic Science for Criminal Justice Purposes Program of the Office of Justice Programs, National Institute of Justice, U.S. Department of Justice.

National Center for Statistics and Analysis. 2015, January. *Early Estimate of Motor Vehicle Traffic Fatalities for the First Half (Jan–Jun) of 2014. (Crashes*Stats Brief Statistical Summary. Report No. DOT HS 812 093).* Washington, DC: National Highway Traffic Safety Administration.

NORML (National Organization for the Reform of Marijuana Laws). 2014. *Drugged Driving.* Washington, DC. http://norml.org/drugged-driving. Accessed July 14, 2014.

Nordrum, A. 2014, September. "The New D.A.R.E. Program: This One Works." *Scientific American.*

O'Leary, B. 2014. Personal communication by author Nicholas Lovrich with National Highway Traffic Safety Administration office providing oversight of ARIDE program, February 4th.

O'Malley, P.M. and Johnson, L.D. 2013. "Driving after Drug or Alcohol Use by U.S. High School Seniors, 2001–2011." *American Journal of Public Health.* 103(11): 2027–2034.

ONDCP (Office of National Drug Control Policy). 2010. *National Drug Control Strategy, 2010.* Washington, DC: Executive Office of the President.

ONDCP (Office of National Drug Control Policy). 2014. *National Drug Control Strategy, 2014.* Washington, DC: Executive Office of the President.

Pan, W. and Bai, H. 2009. "A Multivariate Approach to a Meta-analytic Review of the Effectiveness of the D.A.R.E. Program." *International Journal of Environmental Research and Public Health.* 6(1): 267–277.

People v. Kelly, 17 Cal. 3d 24, 130 Cal. Rptr. 144, 549 P.2d 1240 (1976).

Phillips, R. 2015. "The Ion Investigators." *Washington State Magazine.* 14(4): 30–36.

Ramaekers, J.G., Berghaus, G., van Laar, M., and Drummer, O.H. 2004. "Dose-Related Risk of Motor Vehicle Crashes after Cannabis Use." *Drug and Alcohol Dependence.* 73(2): 109–119.

Ronen, A., Gershon, P., Drobiner, H., Rabinovich, A., Bar-Hamburger, R., Mechoulam, R., Cassuto, Y., and Shinar, D. 2008. "Effects of THC on Driving Performance, Physiological State and Subjective Feelings Relative to Alcohol." *Accident Analysis and Prevention.* 40: 926–934.

Rosen, C. 2003. "Liberty, Privacy, and DNA Databases." *The New Atlantis: A Journal of Technology and Society.* 1: 37–52.

Rosenbaum, D.P. and Hanson, G.S. 1998. "Assessing the Effects School-based Drug Education: A Six-Year Multilevel Analysis of Project D.A.R.E." *Journal of Research in Crime and Delinquency.* 35(4): 381–412.

SAMHSA (Substance Abuse and Mental Health Services Administration). 2013. *Results from the 2012 National Survey on Drug Use and Health: Summary of National Findings.* Rockville, MD: U.S. Department of Health and Human Services.

Sewell, R.A., Poling, J., and Sofuogli. 2009. "The Effect of Cannabis Compared with Alcohol on Driving." *American Journal of Addiction.* 18(3): 185–195.

Shellberg, T. 2003. "DNA Justice Speaks." *Presentation for the American Prosecutors Research Institute,* November 20th.

Simoncelli, T. and Seinhardt, B. 2005. "California's Proposition 69: A Dangerous Precedent for Criminal DNA Databases." *DNA Fingerprinting and Civil Liberties: Second in a Series of Articles.* Boston, MA: American Society for Law, Medicine and Ethics.

Sonnberg, S., Armenta, S., Garrigues, S., and de la Guardia, M. 2015. "Detection of Tetrahydro-cannabinol Residues on Hands by Ion-Mobility Spectrometry (IMS). Correlation of IMS Data with Saliva Analysis." *Analytical and Bioanalytical Chemistry.* 407: 5999–6008. doi:10.1007/s00216-015-8784-1.

State v. Baity 2002. State of Washington v. Jasper Leeroy Baity, No. 26328-9-11 (Wash. Ct. App. Feb. 20, 2002).

Stecker, K. 2015. "The Significance of a Drug Recognition Expert Prosecutor in the Courtroom." *Between the Lines.* 23(4): 1–2 [Publication of the National Traffic Law Center, National District Attorneys Association].

Svrakic, D., Lustman, P., Mallya, A. Lynn, T.A., Finney, R., and Svrakic, M. 2012. "Legalization, Decriminalization and Medical Use of Cannabis: A Scientific Public Health Perspective." *Modern Medicine.* 109(2): 90–98.

Terry-McElrath, Y.M., O'Malley, P.M., and Johnson, L.D. 2014. "Alcohol and Marijuana Use Patterns Associated with Unsafe Driving Among U.S. High School Seniors: High Use Frequency, Concurrent Use, and Simultaneous Use." *Journal of Studies on Alcohol and Drugs.* 75(3): 378–389.

U.S. v. Everett, 972 F. Supp. 1313 (1997).

Wadley, J. and Bronson, A. 2015. "Daily Marijuana Use among U.S. College Students Highest Since 1980." *Michigan News* (August 31st).

West, S.L. and O'Neal, K.K. 2004. "Project D.A.R.E. Outcome Effectiveness Revisited." *American Journal of Public Health.* 94(6): 1027–1029.

Whitehall, J.M., Rivara, F., and Moreno, M. 2014. "Marijuana-Using Drivers, Alcohol-Using Drivers, and Their Passengers: Prevalence and Risk Factors among Underage College Students." *Journal of the American Medical Association, Pediatrics.* 168(1–2): 342–354.

Wong, K., Brady, J.E., and Li, G. 2014. "Establishing Legal Limits for Driving Under the Influence of Marijuana." *Injury Epidemiology.* 1: 26.

Wilson, F.A., Stimpson, J.P., and Pagan, J.A. 2014. "Fatal Crashes from Drivers Testing Positive for Drugs in the U.S., 1993–2010." *Public Health Report.* 129(4): 342–350.

Wolff, K. and Johnson, A. 2014. "Cannabis Use: A Perspective in Relation to the Proposed UK Drug-Driving Legislation." *Drug Test Analysis.* 6(1–2): 143–154.

Part V

Ethical concerns in forensic science

14 The ethical considerations of forensic science management

The case for state oversight and accreditation

Wendy L. Hicks

Introduction

As the rapid expansion of forensic science continues, concerns pertaining to analytic error, insufficient laboratory protocol, and misrepresented results have served to generate concern and skepticism in the field. While such problems have been identified in labs nationwide, in 2002 the Houston Police Department Crime Laboratory was the focus of much scrutiny. After irregularities were discovered following a series of investigative reports by a local television station, government officials began an official inquiry into laboratory processes and protocols. According to Goldstein (2011) problems affecting the Houston laboratory included the fabrication of analytic results, a DNA laboratory supervisor without any prior experience performing DNA analysis, and a leaking roof that allowed water to backup and contaminate evidence. Upon completion of an investigation, officials determined that multiple units of the Houston facility failed to meet accepted standard practices and protocols, "posing major risks of contributing to miscarriages of justice in extremely significant cases, including death penalty cases" (p. 225).

The problems affecting the Houston laboratory are unfortunately not unique. Laboratories nationwide have faced scrutiny after laboratory error or analyst negligence has resulted in postconviction exonerations (DiFonzo, 2005; Goldstein, 2011; Hansen, 2013). At times poor regulation and inadequate supervision have allowed such problems to go unrecognized. When laboratory inconsistencies have been identified, such occurrences have often been attributed to a lack of state oversight, certification, or accreditation.

Those working within policy development have a duty to ensure forensic analyses are conducted with ethical considerations to laboratory protocol and scientific accuracy. As Real Science Radio host Bob Enyart stated in 2005, "It is not a Justice System. It is just a system". His statement sums up many quandaries currently affecting the field of forensic science. Too many within the field view their positions as undertaking scientific analyses within the U.S. Justice System. Unfortunately, our justice system is simply a system, similar in many ways to any governmental or bureaucratic system operating in other facets of our society. It is subject to the same issues, problems, and foibles as any other system. It is just a system and as such should be regulated out of consideration for those whose lives are affected by the products produced within the various forensic laboratories operating across the nation. The argument for increased oversight of forensic science labs has been couched in terms of professional ethics, scientific accountability, state bureaucratic obligations, and human rights (Donnelly, 2012).

There is a sound argument to be made for increased state oversight, accreditation, and certification of both forensic laboratories and the personnel working within such facilities. In 2009, the National Research Council proposed accreditation, certification, and increased state oversight as possible remedies for the ills plaguing the field of forensic science in their report on Strengthening Forensic Science in the United States. Such strategies could serve to ensure adherence to adequate scientific protocol, increase accountability, and maintain rigorous quality standards for both laboratories and personnel.

The argument can be framed in terms of ethical constructs, terms of the social contract, or basic human rights. In whatever form the argument is provided, there is a need for enhanced oversight of forensic personnel and the laboratories in which they work. It is the purpose of this chapter to examine issues currently affecting forensic science nationwide and formulate an argument for accreditation, state oversight, and certification that are currently being debated by government officials and professionals within the field.

Historical demand for forensic science

As is often the case, a good amount of interest and growth within the area of forensic science began with popular culture. There are those today within the disciplines of criminal justice, criminology, and forensic science that attribute much of the growth in the field to the early literary work of Sir Arthur Conan Doyle (O'Brien, 2013; Saferstein, 2011). His portrayal of Sherlock Holmes as an intelligent crime fighter intent on uncovering the identity of various nefarious characters through a more scientific approach was ground breaking for its time.

Holmes was renowned for his use of the latest in finger print technology, handwriting analysis, ciphers, footprint identification, firearms identification, and serology. In the Adventure of the Norwood Builder, Sherlock Holmes states very emphatically;

> That it was written in a train. The good writing represents stations, the bad writing movement, and the very bad writing passing over points. A scientific expert would pronounce at once that this was drawn upon a suburban line, since nowhere save in the immediate vicinity of a great city could there be so quick a succession of points.
>
> (Sir Arthur Conan Doyle, 1905, p. 468)

These skills were compounded by his capabilities in chemistry. Conan Doyle imbued his detective with skillfulness in the use of dyes, coal tar, poisons, acetones, and gemstones to help solve the latest caper. In The Naval Treaty, Holmes proclaims after conducting a basic litmus test that, "If this paper remains blue, all is well. If it turns red, it means a man's life" (Sir Arthur Conan Doyle, 1905, p. 306). In A Study in Scarlet, Homes notes, "I've found it. I've found it. I have found a re-agent which is precipitated by haemoglobin and by nothing else" (Shreffler, 1989, p. 98). The ability of Sherlock Holmes to make use of scientific methodology to solve crime was groundbreaking for its era. Not only did it provide readers with a glimpse into new techniques for criminal identification and profiling, it also forced people to look toward the future and set the stage for the development of a viable field of inquiry that was to provide many important tools in the fight against criminal offending.

What is forensic science?

Throughout the years researchers and criminal justice practitioners have utilized a variety of categorization techniques in an effort to define the field of forensic science. The single biggest mistake made by many outside of formal criminal justice channels centers on the fact that the field of forensic science is not comprised of a single scientific methodology or technique designed to assist analysts in solving crime. Forensic science is comprised of a variety of different disciplines each designed to assist in criminal investigations. According to the National Institute of Justice the following classification is used to help distinguish between the varied disciplines comprising the general field of forensic science (National Research Council, 2009):

1. General toxicology
2. Firearms/toolmarks
3. Questioned documents
4. Trace evidence
5. Controlled substances
6. Biological/serology screening (including DNA analysis)
7. Fire debris/arson analysis
8. Impression evidence
9. Blood pattern analysis
10. Crime scene investigation
11. Medicolegal death investigation
12. Digital evidence.

As stated by the National Research Council (2009), "The forensic science disciplines exhibit wide variability with regard to techniques, methodologies, reliability, level of error, research, general acceptability, and published material" (p. 38). Some of the disciplines comprising the greater forensic science field are laboratory based while others are based on expert analysis and interpretation of observed patterns such as toolmarks, fingerprint, or dental impressions. Select disciplines under the umbrella of forensic science require the analytical expertise and education of trained scientists such as chemists, physicists, or biologists. While some analyses are conducted by those working in law enforcement, others require the special skills of a medical professional.

One of the biggest hurdles facing the overall field of forensic science is the fact that many of the techniques employed by the various disciplines are empirical applications of generalized scientific principles. Or, as noted by the National Research Council (2009), "they are not based on a body of knowledge that recognizes the underlying limitations of the scientific principles and methodologies used for problem solving and discovery" (p. 38). Some practices and analytical techniques within forensics are not grounded in scientific methodology and do not have a well-developed knowledge base refined through stringent replicable research. It is precisely due to this nuance that it is imperative that those working with criminal justice, academia, and forensic science concentration on methods to "improve, systematize, and monitor the activities and practices in the forensic science disciplines and related areas of inquiry" (National Research Council, 2009, p. 39).

Growth and controversy in modern scientific criminal investigation

Great strides in the field of forensic science have been made in the modern era of criminal investigation and detective work. Rather than remaining the fancy of literature, forensic science began to be utilized in efforts to unravel complex crimes. In the 20th century substantial growth occurred in the area of forensics, bringing the rigor of scientific inquiry into detective and police work. In 1923, August Vollmer established the first and oldest crime laboratory, housed within the Los Angeles Police Department (Saferstein, 2011). Vollmer would later go on to found the nation's first university program in criminology and criminalistics at the University of California, Berkeley (Johnson, 1988).

In the nascent stages of development in the field of criminalistics, law enforcement made use of laboratories to assist with finger print identification, handwriting comparisons, tool mark identification, ballistics, drug analysis, blood testing, and trace evidence. According to Saferstein (2011), the development of crime laboratories within the United States has been "characterized by rapid growth accompanied by a lack of national and regional planning and coordination" (p. 10). This rapid expansion has created many challenges and questions now facing governmental authorities.

However, a host of Supreme Court decisions in the 1960s served to change the emphasis of law enforcement agencies to securing more scientifically valid evidence from crime scenes. In Breithaupt v. Abram (1957), the Court wrote "Modern community living requires modern scientific methods of crime detection lest the public go unprotected" (p. 439). Again, in Escobedo v. Illinois, the Court ruled,

> We have learned the lesson of history, ancient and modern, that a system of criminal law enforcement which comes to depend on the 'confession' will, in the long run, be less reliable and more subject to abuses than a system which depends on extrinsic evidence independently secured through skillful investigation.
>
> (1964, p. 488)

The Court would rule again in 1966 in the case of Schmerber v. California, (1966) that the prohibition against compulsory self-incrimination applied only to testimonial evidence, not to physical evidence. This paved the way for the insertion of blood alcohol analysis testing that was in full compliance with 5th Amendment constitutional mandate protecting the suspect against self-incrimination. Justice Brennan's opinion held promise for the increased use of scientifically procured evidence as the Schmerber decision incorporated handwriting exemplars, fingerprints, voice exemplars, and other biological samples that could be obtained from a suspect without violating 5th Amendment requirements.

Again, in 1967, Justice Brennan would draft the opinion in Warden v. Hayden, in which the Court overturned prior cases prohibiting the seizure of mere evidence. According to Wigmore and McNaughton (1961) mere evidence is defined as anything "which is neither contraband nor tools [or more commonly, instrumentalities] nor fruits of crime, but which consists of private documents or other chattels of the defendant wanted by the government solely for its evidentiary value" (p. 45). Thereafter, according to the courts, scientific evidence such as blood, hair, gunshot residue, and blood spatter was no longer considered mere evidence.

After the early debates surrounding the admissibility of scientific evidence, controversy then erupted with regard to voice identification by spectrographic examination of speech patterns, typically referred to as voiceprint documentation (Bolt, Cooper, David, Dens, Pickett, & Stevens, 1970). Those in law enforcement often used this technique during investigations into kidnapping, extortion, and other crimes involving the use of telephoned threats or demands. A report drafted by the National Academy of Sciences (1979) brought into question the scientific basis for voiceprint recognition techniques stating,

> Estimates of error rates now available pertain to only a few of the many combinations of conditions encountered in real-life situations. These estimates do not constitute a generally adequate basis for a judicial or legislative body to use in making judgments concerning the reliability and acceptability of aural-visual voice identification in forensic applications.
>
> (p. 60)

The report did not go so far as to render voiceprint evidence in admissible in all jurisdictions, however (State v. Coon, 1999).

In the early years of forensic science the courts required some legal test to determine the admissibility of increasingly sophisticated scientific evidence. In the 1980s and 1990s law enforcement agencies began to make use of DNA analytic techniques adopted from Dr. Alec Jeffreys from the University of Leicester in England (Newton, 2008). By 1990 DNA evidence had been utilized in 185 cases in 38 states and the U.S. military (Newton, 2008). By the end of the century DNA evidence and analytic techniques were admissible in every state and federal circuit in the nation (Newton, 2008). In People v. Wesley (1988) the court affirmed that DNA analysis was "the single greatest advance in the search for truth since the advent of *cross-examination*" [italics in original] (p. 643).

Despite the promises of DNA analysis, the technique did not come without its critics. In People v. Castro (1989), the court ruled that the test results applied to Castro were inadmissible, faulting the specific manner in which the analysts conducted the test. Soon thereafter Eric Lander, a scientists at MIT's Whitehead Institute who had testified for the defense in Castro along with FBI expert Bruce Budowle, characterized early controversies over lab protocols:

> The initial outcry over DNA typing standards concerned laboratory problems: poorly defined rules for declaring a match; experiments without controls; contaminated probes and samples; and sloppy interpretation of autoradiograms. Although there is no evidence that these technical failings resulted in any wrongful convictions, the lack of standards seemed to be a recipe for trouble.
>
> (Lander & Budowle, 1994, p. 735)

The issues brought up in the Castro decision highlighted many of the salient problems plaguing the field of forensic science today. In 1992, the National Academy of Science's National Research Council drafted a report recommending stringent written laboratory protocols, both objective and quantitative for identifying patterns, declaring matches, and methods for the identification of potential artifacts (National Research Council, 1992). The reported went further to note the absence of mandatory proficiency testing for laboratory analysts in its statement, "No laboratory should let its results with a new DNA typing method be used in court unless it has undergone such proficiency testing via blind trials" (p. 55).

The controversies surrounding various laboratory procedures were not quelled by the 1992 National Research Council's report. Instead, critics declared the

> report did not eliminate all controversy. Indeed, in propounding what the committee regarded as a moderate position [for dealing with the statistical

> issues] the report itself became the target of criticism from scientists and lawyers on both sides of the debate on DNA evidence in the courts.
>
> (National Research Council, 1996, p. 1)

The National Academy of Science commissioned a second report in 1996 in an attempt to deal more effectively with the myriad of issues identified in their previous report.

One issue that has plagued the field of forensic science centered around the question of the admissibility of evidence in court. Courts have, throughout the years, utilized three competing standards in efforts to address the admissibility of forensic evidence into court proceedings. The first is the general acceptance standard initially applied in Frye v. United States (1923). Under the Frye standard,

> courts do not inquire directly into scientific consensus that the scientific truth, but ascertain whether the scientific community has reached the consensus that the scientific procedure in question rests on a valid theory and generates reliable results when properly applied.
>
> (Kaye, 1993, p. 104)

Therefore, Frye imposes a burden upon the court to determine whether a specific technique is considered generally accepted by the scientific community.

The second admissibility standard adopted by the U.S. Supreme Court in 1993 has come to be referred to as the reliability test (Newton, 2008). This standard has as its foundation the case of Daubert v. Merrell Dow Pharmaceuticals, Inc. (1993), which pertained to the admissibility of expert epidemiological testimony in a civil suit concerning whether the antinausea drug Bendectin caused birth defects. In Daubert, the Court unanimously ruled that existing federal evidentiary rules "implicitly" reject the Frye standard (Kaye, 1993, p. 105). The Court based their decision upon rule 702 of the Uniform Rules of Evidence which states that,

> If scientific, technical, or other specialized knowledge will assist the trier of fact to understand the evidence or to determine a fact in issue, a witness qualified as an expert by knowledge, skill, experience, training, or education, may testify thereto in the form of an opinion or otherwise if (1) the testimony is based upon sufficient facts or data, (2) the testimony is the product of reliable principles and methods, and (3) the witness has applied the principles and methods reliably to the facts of the case.
>
> (United States Code, 2000, p. 893)

While the Daubert reliability standard acts as a statutory case it is not binding at the state level. Individual states are free to follow the Frye standard in determining the admissibility of scientific evidence (Newton, 2008; Goodwin & Gurulé, 2009).

The last admissibility standard, the relevancy test, is considered by many to be an alternative approach to the application of Frye (Giannelli & Imwinkelried, 2002). According to the relevancy test, evidence is admitted into court based on whether "its probativeness, materiality, and reliability outweigh its tendency to mislead, prejudice, or confuse the jury" (United States v. Jakobetz, 1990; Goodwin & Gurulé, 2009). The relevancy test is subject to rule 403 of the Federal Rule of Evidence which states that

> The court may exclude relevant evidence if its probative value is substantially outweighed by a danger of one or more of the following: unfair prejudice, confusing the issues, misleading the jury, undue delay, wasting time, or needlessly presenting cumulative evidence.
>
> (Ingram, 2015, p. 209)

Noted legal scholar Charles McCormick, stated his support for the relevancy test.

> Any relevant conclusions which are supported by a qualified expert witness should be received unless there are other reasons for exclusion. Particularly, its probative value may be overborne by the familiar dangers of prejudicing or misleading the jury, unfair surprise and undue consumption of time.
>
> (McCormick, 1954, p. 363)

Slapdash protocol

In the 1980s, the applicability of DNA and other assorted forensic evidence in solving crime became commonplace. No longer was the use of scientifically analyzed evidence the fanciful imaginings of crime authors and fictional sleuths. However, the use and admissibility of forensic evidence, most notably DNA material, did not come without some reservations. In 1990, the Office of Technology Assessment was called upon by the Senate Committee on Labor and Human Resources to evaluate

> the scientific, legal, and ethical issues surrounding forensic applications of DNA tests: the validity and reliability of DNA tests for forensic casework, quality assurance and standards for DNA analysis by forensic laboratories, the legal basis for the admissibility of such tests in courts of law, privacy and civil liberties concerns about collecting, using, and storing genetic information and material, and criminal justice interest in employing DNA tests at the Federal, State, and local level.
>
> (p. 3)

Despite any legal debates surrounding the admissibility of forensic science in the nation's courts, there are numerous examples of haphazard adherence to scientific laboratory technique, insufficient laboratory protocol, and inadequate analyst training. As stated by DiFonzo (2005), "The record is littered with slapdash forensic analyses often performed by untrained, underpaid, overworked forensic technicians operating in crime labs whose workings reflect gross incompetence or rampant corruption" (p. 2).

It remains an unfortunate fact of reality that the lamentations of DiFonzo remain very much in the consciousness of policy makers and professionals within the field of forensics. Due to the nature of the analyses performed by those in forensic science, it is vital for staff working in the nation's laboratories to maintain proper and strict scientific protocol. This is where laboratory accreditation and personnel certification would serve to remedy many of the problems currently existing in the field. One of the most salient arguments for accreditation and certification centers on the fact that the scientific basis of forensic science has the potential to mislead naive jurors into believing that scientific evidence, especially DNA, negates the need to examine other corroborating or contradictory evidence (DiFonzo, 2005).

The current cultural phenomenon euphemistically referred to as the CSI Effect has led juries to accept forensic evidence and testimony, often with little or no scrutiny. Properly defined, the CSI Effect is "the perception of the near-infallibility of forensic science in response to the TV show" (Pyrek, 2007, p. 306). The potential for people to blindly accept seemingly credible scientific evidence in court proceedings has led to attorneys to *voir dire* prospective jurors on their television viewing habits.

The potential for contamination of forensic evidence is especially salient when considering legal cases based on DNA evidence; in spite of the fact that DNA evidence is considered by many to be nearly infallible if handled and processed with due care, the reality surrounding DNA paints a much different picture. Experts in the field have noted, "DNA samples recovered from crime scenes are often so small and in such disintegrated condition that they are easy to mishandle or manipulate" (McDonald, 1998). Thompson, Ford, Doom, Raymer, and Krane (2003) concluded that the criminal justice system as a whole "does a poor job of distinguishing unassailably powerful DNA evidence from weak, misleading DNA evidence" (p. 1).

Due to the fact that forensic evidence is often utilized by prosecuting attorneys in legal cases dealing with capital crimes it seems logical that analysts would be compelled to take due care with the storage, handling, and analytic techniques employed. However, a Chicago study by Possley, Mills, and McRoberts (2004) demonstrated that of 200 death row exoneration cases since 1986 more than a quarter involved faulty or inadequate crime lab work or testimony. In 1997 Giannelli noted that "major abuses in the use of scientific evidence have surfaced, including perjury by expert witnesses, faked laboratory reports, and testimony based on unproven techniques" (p. 441).

To illustrate a small sampling of the extent of the problem with failed protocol, a lack of analyst integrity, and questionable ethics occurring in laboratories nationwide five high profile cases came to light between 1994 and

2005. In 1994, FBI Supervisory Special Agent Frederic Whitehurst complained to the Office of the Inspector General and the Department's Criminal Division concerning laboratory irregularities. Agent Whitehurst, who had previously conducted chemical analyses of explosives and associated residue, questioned the reliability of various laboratory procedures, analyst integrity, and the veracity of analyst testimony during court proceedings (Office of Inspector General, 2014). Whitehurst alleged, "Lab examiners improperly testified outside their expertise, presented unsupportable scientific conclusions, committed perjury, fabricated evidence, and failed to follow appropriate procedures" (Office of Inspector General, 2014, p. 2). The ensuing investigation by the FBI revealed, "significant instances of testimonial errors, substandard analytical work, and deficient practices by many lab examiners" (Office of Inspector General, 2014, p. 3).

In 2004 evidence was discovered indicating an analyst at the FBI laboratory, Jacqueline Blake, had failed to follow established procedures for 2 years, forcing FBI officials to notify relevant parties involved in affected cases so evidence could be reanalyzed and defense attorneys could have the opportunity to challenge existing analyses and take corrective action (Solomon, 2003). The Inspector General stated that "Blake's misconduct, and the Laboratory's failure to detect it for a period exceeding two years, has damaged intangibly the credibility of the FBI Laboratory" (Office of the Inspector General, 2014, p. 8). In May 2004, Jacqueline Blake pled guilty to a misdemeanor criminal charge of making false statements about following protocol in 100 DNA analysis reports and received 2-year probation (Possley et al., 2004). In another episode, again occurring in 2004, after an FBI affidavit determined a fingerprint match was a "100 percent positive identification", American attorney Brandon Mayfield was arrested and jailed in connection to the Madrid terrorist train bombing. Mayfield was subsequently released after two weeks behind bars after the FBI acknowledged that the laboratory had erred (Feige, 2004).

In 2012, media attention was once again levelled at the reliability of forensic analyses and laboratory procedures as the convictions of Donald E. Gates, Santae A. Tribble, and Kirk L. Odom were brought into question (Hsu, 2012). According to documents obtained from the Office of the Inspector General, each individual had served sentences in excess of 21 years, based partly upon hair analyses and testimony that was subsequently proven erroneous. The three defendants were later exonerated. However, another defendant, Benjamin H. Boyle, whose conviction had rested in part on FBI hair and fiber analysis that was later determined to be flawed, had been executed (Hsu, 2012).

In 2011, Annie Dookhan, an analyst in a state run drug lab in Boston, was indicted on 27 counts of obstructing justice, tampering with evidence, perjury and other charges stemming from her work on assorted drug cases during her nine years of work (Hansen, 2013). Dookhan was accused of "faking test results, intentionally contaminating and [adding suspected drug samples, forging co-workers' signatures on lab reports, and falsely claiming to have a master's degree in chemistry" (p. 2). The investigation into Dookhan's work led to the closing of the lab, the release of numerous drug offenders, and the termination of one lab official, and the resignation of Public Health Commissioner John Auerbach (Hansen, 2013).

As the examples provided earlier indicate, forensic laboratories have a need for improved protocol and oversight. According to the Innocence Project since 1989, there have been a total of 330 postconviction exonerations due to faulty forensic analysis (Innocence Project, 2015). Since 2000 there have been 263 postconviction exonerations due primarily to faulty or inadequate forensic analysis (Innocence Project, 2015). Unvalidated or improper forensic analysis has played a role in 47% of the postconviction exonerations (Innocence Project, 2015).

One of the biggest difficulties facing forensic analysis centers on the fact that many techniques employed by laboratories across the nation have no scientific basis or existing standards. Forensic techniques developed out of a need to solve various crimes. No scientific standards existed for many techniques for tool mark identification, fingerprint detection, voice recognition analysis, and bite mark matching. While some forensic techniques might be grounded in scientific methodology, there nonetheless, remains a dearth of effective baseline standards as the techniques have evolved beyond the traditional scientific laboratory.

A second problem plaguing forensic science is the fact that many analysts are woefully under educated and lack proper training. Often legal testimony of forensic analysts goes further than the professional capability of the individual would ordinarily allow. Many laboratories only require a bachelor's degree for employment as a forensic analyst. The lack of advanced education in the hard sciences can be an indicator that many analysts have not experienced serious scientific research methodology and laboratory protocols. In addition, despite the fact that many forensic techniques have been employed for decades, there remains a serious lack of scientific research to support the reliability and validity of these analytic procedures.

Forensic science in a human rights framework

Forensic science has become an integral, some might say essential, aspect of law enforcement. It has become intertwined with culture and the popular media. There is no escaping the influence of forensic science. Forensic analysts hold in their hands the very lives of criminal defendants. One would hope with such power came the acknowledgement of great responsibility and due care. However, it is an unfortunate fact of reality that far too many instances of analyst error, unethical behavior, failure to follow established protocol, deception, and fabrication of results have tainted the façade of the field.

The increasing reach of forensic science into our lives, coupled with the power of the field raises important ethical questions. Paralleling the scientific and legal developments of forensic science is a growing public concern over privacy issues and the potential for governmental infringement of personal liberties. These concerns pave the way for a valid argument for the accreditation of forensic laboratories and the certification of its analysts. Proper accreditation and certification would bring a standardization or normalcy to the field. Standards of ethics and protocol would provide a measure of consistency and regulation to the field.

As framed by Donnelly (2012), forensic science is viewed by many within the field as value neutral. Attention to proper scientific methodology allows one to engage in scientific inquiry in an unbiased, detached manner. Science is "considered uncontaminated with political values, and therefore an objective arbiter of truth" (McCartney, 2006, p. xv). Even for a forensic analyst, if established scientific protocols are followed results are obtained, even in the most volatile of legal cases, in an unbiased fashion. According to Koppl (2005), the essential function of forensic science is to facilitate the extraction of the truth.

However, by contrast, criminal justice is very much value laden. The administration of justice is colored by "competing political, ethical and financial objectives" (Donnelly, 2012, p. 94). Law enforcement officials and prosecuting attorneys work in conjunction in an adversarial justice system to ensure criminal convictions. The power wielded by practitioners within forensic science can be substantial. It is, after all, their results, which can lead to a successful prosecution or see a defendant go free.

The need for adherence to proper scientific methodology and laboratory protocol can be addressed through certification and accreditation. In addition, the importance of ethical behavior among forensic practitioners could be stressed through certification requirements for analysts. In a study by Donnelly (2012) it was suggested that forensic science "needs to find its voice if it is to engage and inform the issues that lie at the intersection of law and science" (p. 96). It is precisely because of the fact that forensic science lies at the intersection of such prominent, competing forces that certification and accreditation should be entertained.

Contemporary crime labs

In 2009, a total of 411 publicly funded crime laboratories existed within the United States (U.S. Department of Justice, 2012). This is an increase from 2002, in which 305 publicly funded laboratories were in existence nationwide (U.S. Department of Justice, 2005). Figure 14.1 provides a graphical illustration of the growth in publicly funded laboratories operating nationwide.

Along with an increase in labs comes the accompanying growth in personnel. In 2009 laboratories within the U.S. employed 13,100 full-time personnel, compared to 11,900 in 2005 and 11,000 in 2002 (U.S. Department of Justice, 2012). Growth in laboratory personnel across jurisdiction can be seen in Table 14.1.

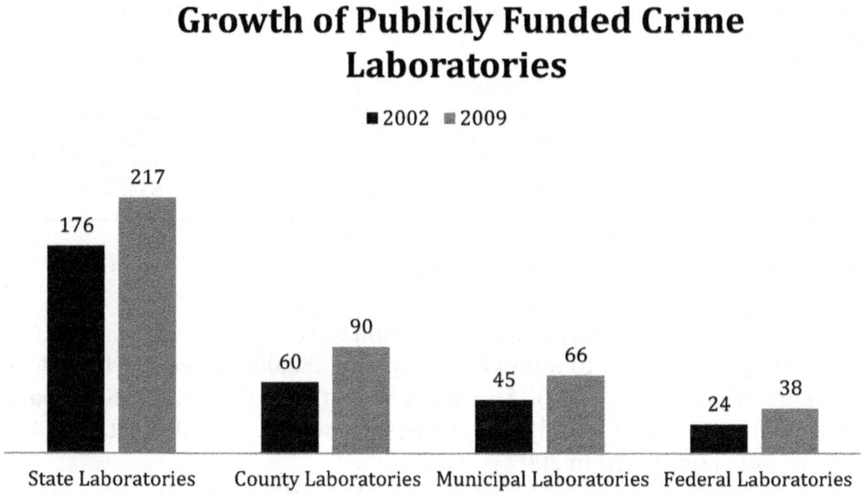

Figure 14.1 Growth of publicly funded crime laboratories 2002–2009.
Source: Bureau of Justice Statistics (2005, 2012).

Table 14.1 Full-time laboratory personnel across jurisdiction 2002–2009

Type of jurisdiction	*2002*	*2005*	*2009*
All laboratories	11,000	11,900	13,100
State	5,200	5,600	6,100
County	1,900	2,200	2,500
Municipal	1,900	1,900	2,200
Federal	2,000	2,100	2,300

Source: Bureau of Justice Statistics (2005, 2008, 2012).

As the total number of laboratories and personnel increased annual operating budgets also saw rapid growth. In 2002, the total operating budget for a state laboratory was $454 million (U.S. Department of Justice, 2005). From 2002 to 2009, the operating budget for state laboratories increased from $454 million and $680 million (U.S. Department of Justice, 2005, 2012). Table 14.2 provides a breakdown of annual operating budgets for laboratories across jurisdictions.

Despite the increasing number of laboratories and accompanying personnel in existence across the nation the requests for service have been growing far in excess of the capabilities of facilities and scientists. In 2002, U.S. labs received 2,695,269 requests to analyze evidence obtained from various cases (U.S. Department of Justice, 2005). In 2009, the requests for service increased to 4,120,000 (U.S. Department of Justice, 2012). Of the total requests

Table 14.2 Annual operating budget for crime laboratories across jurisdictions

Type of jurisdiction	*2002*	*2005*	*2009*
All laboratories	$1,036,000,000	$1,155,000,000	$1,562,000,000
State	$454,000,000	$529,000,000	$680,000,000
County	$172,000,000	$236,000,000	$314,000,000
Municipal	$112,000,000	$130,000,000	$219,000,000
Federal	$299,000,000	*	$348,000,000

Source: Bureau of Justice Statistics (2005, 2008, 2012).
* Data unavailable due to insufficient response rate.

for service in 2009, approximately 2.4 million were sent to state labs, 820,000 to county labs, 650,000 to municipal labs, and 280,000 to federal labs (U.S. Department of Justice, 2012). According to statistics from the U.S. Department of Justice (2012), 3,905,000 requests for service out of a total of 4.12 million were completed in 2009.

The increase in requests for service coupled with demands from law enforcement agencies for rapid turnaround of laboratory results only serves to exacerbate an already overburdened system. This growth served only to create a surge in the backlog of cases for laboratories from 262,637 in 2002 to 1,193,800 in 2009 (U.S. Department of Justice, 2003, 2012). According to the U.S. Department of Justice, a case is considered backlogged if it "has been submitted to a crime lab, but has not yet been examined and reported to the submitting agency within 30 days" (U.S. Department of Justice, 2012, p. 4).

Many of the problems pertaining to inadequate laboratory protocol, oversight, and analyst training can be alleviated through proper accreditation. Unfortunately, formal accreditation has not been quick to develop. For example, in 2005 82% of all labs nationwide were accredited. However, in 2009 the percentage of accredited labs rose to only 83% (U.S. Department of Justice, 2012). Ironically, it is state laboratories that lead accreditation with 92% of those labs having either certification by the American Society of Crime Laboratory Directors (ASCLD)/Laboratory Accreditation Board, Forensic Quality Services, or American Association for Laboratory Accreditation (U.S. Department of Justice, 2012). Figure 14.2 provides detailed information on accreditation for state, county, municipal, and federal laboratories.

Early suggestions for strengthening the forensic sciences

According to the Office of Technology Assessment, early concerns about the use of DNA evidence included, whether the detection methods were scientifically valid and whether DNA analysis of forensic samples is reliable (1990). Suggestions for enhanced quality control measures included quality assurance protocols implemented by federal and state government, membership of forensic scientists in professional organizations, and the facilitation of

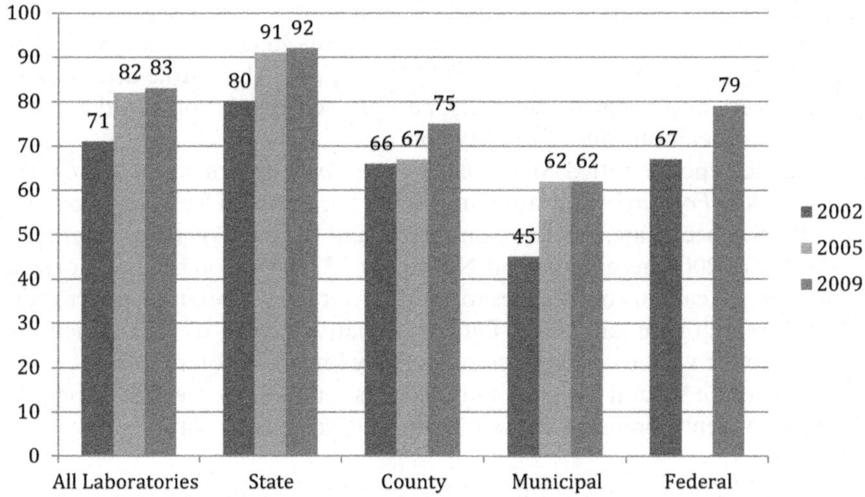

Figure 14.2 Percent of publicly funded forensic crime labs accredited by a professional forensic science organization, by type of jurisdiction, 2002, 2005, and 2009.

Source: Bureau of Justice Statistics (2012).

*Data not presented for Federal labs in 2009 due to low response rate to BJS survey

laboratory self-regulation. It was believed that proper self-regulatory mechanisms would enable forensic scientists and laboratories to engage in best practices to enhance reliability and validity of forensic techniques without adding a layer of governmental regulatory organization to complicate matters.

The 1990 report by the Office of Technology Assessment suggested membership for analysts in one of several professional organizations devoted to discussion relevant to the field. Organizations such as the American Academy of Forensic Sciences, ASCLD, California Association of Criminalists, and California Association of Crime Laboratory Directors were hailed as potential sources of support and assistance for analysts and crime laboratories searching for effective means of enhancing quality, reliability, and validity in the field. Appendix 14.1 provides a listing of the various professional organizations devoted to the various disciplines comprising the field of forensic science.

Additional nonregulatory means of quality control included the authorization of ongoing scientific research, directed at cross disciplinary endeavors designed to apply basic scientific methodologies to more grounded forensic casework. Federal initiatives designed to encourage the use of conferences to facilitate enhanced quality control mechanisms were also suggested. The use of nonregulatory mechanisms designed to develop improved the quality, reliability, and validity of forensic science were favored over the more stringent, rigid federal mandated regulatory procedures. The issue was one of whether "high quality necessarily flows from mandatory regulation" (Office of Technology Assessment, 1990, p. 12). Rather than stifle attempts at self-discovery and scientific curiosity, perhaps self-regulation could accomplish more efficient means of quality control.

A second report entitled *Strengthening the Forensic Science in the United States: A Path Forward* was drafted in 2009 by the National Research Council. After the Science, State, Justice, Commerce, and Related Agencies Appropriations Act of 2006 became law on November 22, 2005, Congress authorized the National Academy of Sciences to conduct a study on forensic science (National Research Council, 2009). Due to a relative dearth of valid, scientific insight into the various disciplines collectively known as forensic science, the study was given several concrete mandates as set forth by the Senate report. The independent Forensic Science Committee was tasked with:

1 assess the present and future resource needs of the forensic science community, to include State and local crime labs, medical examiners, and coroners;
2 make recommendations for maximizing the use of forensic technologies and techniques to solve crimes, investigate deaths, and protect the public;
3 identify potential scientific advances that may assist law enforcement in using forensic technologies and techniques to protect the public;
4 make recommendations for programs that will increase the number of qualified forensic scientists and medical examiners available to work in public crime laboratories;
5 disseminate best practices and guidelines concerning the collection and analysis of forensic evidence to help ensure quality and consistency in the use of forensic technologies and techniques to solve crimes, investigate deaths, and protect the public;
6 examine the role of the forensic community in the homeland security mission;
7 [examine] interoperability of Automated Fingerprint Information Systems [AFIS]; and
8 examine additional issues pertaining to forensic science as determined by the Committee (National Research Council, 2009, p. 2).

Within the report were contained 13 various recommendations on how to strengthen forensic science. First among these recommendations was the suggested establishment of an independent federally funded entity, the National Institute of Forensic Science (NIFS). According to the report, the NIFS

> should have a full-time administrator and an advisory board with expertise in research and education, the forensic science disciplines,

physical and life sciences, forensic pathology, engineering, information technology, measurements and standards, testing and evaluation, law, national security, and public policy.

(National Research Council, 2009, p. 19)

Most notable of the Council's recommendations was the first. Contained within the verbiage of the this recommendation were suggestions for establishing and enforcing best practices for forensic analysts and laboratories, mandatory laboratory accreditation, the promotion of scholarly competitive peer-reviewed research, the development of improvements in forensic science education, the allocation of funding to support forensic science methodologies, the establishment of a code of ethics, and assessing the introduction of new technologies into the forensic science disciplines (National Research Council, 2009). Many of the valued hallmarks of traditional academe were suggested for improvements within forensic science. The full recommendations outlined in the 2009 National Research Council's report can be seen in the following text.

Recommendations of the National Research Council, 2009

1 To promote the development of forensic science into a mature field of multidisciplinary research and practice, founded on the systematic collection and analysis of relevant data, Congress should establish and appropriate funds for an independent federal entity, the NIFS. NIFS should have a full-time administrator and an advisory board with expertise in research and education, the forensic science disciplines, physical and life sciences, forensic pathology, engineering, information technology, measurements and standards, testing and evaluation, law, national security, and public policy.

2 The NIFS, after reviewing established standards such as ISO 17025, and in consultation with its advisory board, should establish standard terminology to be used in reporting on and testifying about the results of forensic science investigations. Similarly, it should establish model laboratory reports for different forensic science disciplines and specify the minimum information that should be included.

3 Research is needed to address issues of accuracy, reliability, and validity in the forensic science disciplines. The NIFS should competitively fund peer-reviewed research in the following areas: (a) studies establishing the scientific bases demonstrating the validity of forensic methods. (b) The development and establishment of quantifiable measures of the reliability and accuracy of forensic analyses. (c) The development of quantifiable measures of uncertainty in the conclusions of forensic analyses. (d) Automated techniques capable of enhancing forensic technologies.

4 To improve the scientific bases of forensic science examinations and to maximize independence from or autonomy within the law enforcement

community, Congress should authorize and appropriate incentive funds to the NIFS for allocation to state and local jurisdictions for the purpose of removing all public forensic laboratories and facilities from the administrative control of law enforcement agencies or prosecutors' offices.

5 The NIFS should encourage research programs on human observer bias and sources of human error in forensic examinations.

6 To facilitate the work of the NIFS, Congress should authorize and appropriate funds to NIFS to work with the National Institute of Standards and Technology, in conjunction with government laboratories, universities, and private laboratories, and in consultation with Scientific Working Groups (SWGs), to develop tools for advancing measurement, validation, reliability, information sharing, and proficiency testing in forensic science and to establish protocols for forensic examinations, methods, and practices.

7 Laboratory accreditation and individual certification of forensic science professionals should be mandatory, and all forensic science professionals should have access to a certification process.

8 Forensic laboratories should establish routine quality assurance and quality control procedures to ensure the accuracy of forensic analyses and the work of forensic practitioners.

9 The NIFS, in consultation with its advisory board, should establish a national code of ethics for all forensic science disciplines and encourage individual societies to incorporate this national code as part of their professional code of ethics. Additionally, NIFS should explore mechanisms of enforcement for those forensic scientists who commit serious ethical violations. Such a code could be enforced through a certification process for forensic scientists.

10 To attract students in the physical and life sciences to pursue graduate studies in multidisciplinary fields critical to forensic science practice, Congress should authorize and appropriate funds to the NIFS to work with appropriate organizations and educational institutions to improve and develop graduate education programs designed to cut across organizational, programmatic, and disciplinary boundaries.

11 To improve medicolegal death investigation: (a) Congress should authorize and appropriate incentive funds to the NIFS for allocation to states and jurisdictions to establish medical examiner systems, with the goal of replacing and eventually eliminating existing coroner systems. (b) Congress should appropriate resources to the National Institutes of Health (NIH) and NIFS, jointly, to support research, education, and training in forensic pathology. (c) NIFS, in collaboration with NIH, the National Association of Medical Examiners, the American Board of Medicolegal Death Investigators, and other appropriate professional organizations, should establish a SWG for forensic pathology and medicolegal death investigation. (d) All medical examiner offices should be

accredited pursuant to NIFS-endorsed standards within a timeframe to be established by NIFS. (e) All federal funding should be restricted to accredited offices that meet NIFS-endorsed standards or that demonstrate significant and measurable progress in achieving accreditation within prescribed deadlines. (f) All medicolegal autopsies should be performed or supervised by a board certified forensic pathologist.

12 Congress should authorize and appropriate funds for the NIFS to launch a new broad-based effort to achieve nationwide fingerprint data interoperability.

13 Congress should provide funding to the NIFS to prepare, in conjunction with the Centers for Disease Control and Prevention and the Federal Bureau of Investigation, forensic scientists and crime scene investigators for their potential roles in managing and analyzing evidence from events that affect homeland security, so that maximum evidentiary value is preserved from these unusual circumstances and the safety of these personnel is guarded.

Source: National Research Council (2009).

The need for quality, reliability, and validity for the nation's forensic science laboratories had been addressed previously by Congress. In 1988, Clinical Laboratory Improvement Amendments was passed that subjected forensic laboratories to requirements such as the qualifications of directors, standards for the supervision of lab testing, qualifications for technical personnel, management standards, and quality control programs (Office of Technology Assessment, 1990). Shortly thereafter, in 1990 the Health Care Financing Administration (HCFA) of the U.S. Department of Health and Human Services obtained regulatory authority over forensic laboratories. The HCFA established standards for employment and maintenance for medical laboratories nationwide. The HCFA also developed a comprehensive program to monitor each lab and impose sanctions in an effort to establish quality control procedures.

Strengthening the forensic sciences 2014

Laboratory accreditation

In answer to the 2009 report from the National Research Council, the National Science and Technology Council Committee on Science Subcommittee on Forensic Science drafted the report *Strengthening Forensic Science in the United States: A Path Forward* in 2014 in an effort to outline the current state of the field of forensics. Topics such as laboratory accreditation, certification of forensic examiners and medicolegal personnel, proficiency testing of forensic examiners, and ethics were contained within the detailed report.

The Subcommittee on Forensic Science was chartered in 2009 and completed the work on the final report in 2012. For 3 years, subcommittee members "actively pursued the investigation and analysis of critical issues that can inform a coordinated and meaningful response to the NAS report" (National Science and Technology Council, 2014, p. 1). This report endeavored to pursue remedies to the ills plaguing the field of forensic science in a more rigorous manner than ever before. The 2014 report was intended to address matters that serve to enhance forensic science policy, research, and practice.

The report was the culmination of "policy makers in government as well as forensic scientists, laboratory managers, and other practitioners working together to strengthen the forensic sciences" (National Science and Technology Council, 2014, p. ii). The National Science and Technology Council is the primary mechanism for the coordination of science and technology policy across the various components of the federal research and development initiatives and is well positioned to begin a serious conversation about the merits of accreditation and certification for those involved in the forensic sciences.

The Subcommittee on Forensic Science was comprised of approximately 200 subject matter experts across a variety of levels of government spanning a total of 23 federal departments and agencies. In an effort to better address the original 13 recommendations laid out by the original 2009 report, 4 interagency working groups were created, each tasked with "receiving information from a multitude of stakeholders including state and local advisory members, national associations and organizations representing the forensic science and medicolegal death investigation communities, academia, private industry, and other entities" (National Science and Technology Council, 2014, p. 1). In addition, the Outreach and Communications Interagency Working Group was established to "facilitate the process of gaining input from the public and private sectors" and communicate issues under study by the other groups to interested stakeholders (National Science and Technology Council, 2014, p. 2). The interagency working groups consisted of:

1 Accreditation, Certification Interagency Working Group
2 Standards, Practices, and Protocols Interagency Working Group
3 Education, Ethics, and Terminology Interagency Working Group
4 Research, Development, Testing, and Evaluation Interagency Working Group.

In the contemporary arena of forensic science accreditation can vary widely. In a 2009 report by the Bureau of Justice Statistics, 79% of federal and 92% of state laboratories were accredited (U.S. Department of Justice, 2012). However, only 75% of county, 62% of municipal were accredited (U.S. Department of Justice, 2012). Accreditation rates are also relatively low for medical examiner and coroner offices. According to the National Science and Technology Council (2014), accreditation "attests that an office has a functional governing code; adequate staff, equipment, and training; and a suitable physical facility, and produces accurate, credible death investigation products" (p. 5). As of January 2011, there existed 2,342 medical examiner/coroner offices that served 3,343 counties nationwide. Of these offices, only 45 were accredited by the National Association of Medical Examiners, the primary accrediting body for the field (National Science and Technology Council, 2014). An additional nine offices were accredited by the International Association of Coroners and Medical Examiners (National Science and Technology Council, 2014).

Within the general field of forensic science professional accrediting organizations have been in existence for over 30 years. While accreditation is not a panacea for perfection, it nonetheless provides a measure of confidence in the reliability, validity, and accuracy of conclusions reached and services rendered. Laboratory accreditation was noted by the 2014 report as providing a "thorough evaluation of laboratory policies and procedures as measured against international standards" (National Science and Technology Council, 2014, p. 3). Laboratories make use of the accreditation process to seek guidance into their level of performance through external audits while simultaneously strengthening their methods of operations. This provides valuable information to agencies seeking forensic services and serves to identify those laboratories, which have demonstrated compliance with established standards.

The 2014 report outlined three potential avenues to the implementation of laboratory accreditation. The first suggestion entailed passing formal Congressional legislation requiring "a broad but well defined set of forensic service providers" to acquire accreditation by a specified period of time. The means of enforcing such accreditation requirements was identified but not made public by the interagency working group charged with researching the myriad issues involved with forensic accreditation procedures.

A second approach to accreditation postulated the drafting of law mandating accreditation, predicated on the goal of having state governments adopt such tenets nationwide. State-level stakeholders such as the Uniform Law Commission, the National Governors Association, the National Committee of State Legislators, and those within the forensic science community would work in conjunction with one another to craft a meaningful and enforceable legal standard (National Science and Technology Council, 2014).

The third, and perhaps simplest approach, was to have the Department of Justice mandate accreditation for all laboratories under its direct control. For those laboratories working outside of direct control of the Department of Justice, accreditation would become mandatory for certain financial benefits or training opportunities (National Science and Technology Council, 2014).

The final suggestion offered by the council was the utilization of "portions of some or all of the first three approaches in a well-integrated fashion in collaboration with Federal, state and local stakeholders" (National Science and Technology Council, 2014, p. 8). The council ended their suggestions with the caveat that mandatory funding would require additional study and consideration and questions regarding funding availability would have to be addressed. While accreditation might appear benign on the surface, there remain a plethora of questions in need of sufficient answers.

Analyst certification

While the process of accreditation for laboratories might seem a daunting prospect, obtaining universal certification for forensic science analysts poses even more formidable challenges. One of the biggest obstacles facing analyst certification lies in the fact that those working within forensic science come from a wide array of various academic disciplines. For analysts to become certified would necessitate the task of drafting suitable certification standards capable of encompassing many, varies disciplines. However, this is not taking into consideration standards that might be necessary for differing levels of certification such as those for managers, analysts, or technical support.

A related challenge standing in opposition to analyst certification is the fact that such standards might not be cost effective or even practical. There are those specialties existing within the forensic sciences for which few practitioners exist. The costs associated with requiring certification for those few individuals might pose financial management issues. As noted by the council's 2014 report, "As a general rule in the certification community, it is difficult to create and maintain a certification program for fields with fewer than 50 practitioners" (National Science and Technology Council, 2014, p. 11).

Related to the many issues brought up by the consideration of financial matters is the possibility that curricula and educational materials would need to be developed for those fields that do not have such resources available. For those practitioners working in fields for which certification does not exist they would be in need of training tools to provide them with the necessary educational background to allow for the pursuit of full certification.

Certification for analysts would also necessitate organizational change that could prove troublesome or perhaps even insurmountable. Agencies providing forensic science services would need time and resources to accommodate certification requirement mandates. If certification of analysts became a formal requirement, personnel would need time to obtain full credentialing before they could return to their laboratory work stations. In addition, an agency would also likely need to modify pay scales to recognize those analysts holding full certification.

Apart from purely organizational or agency related matters is the fact that the current store of certification bodies would need to amend their existing requirements in order to be in full compliance with a universal standard of certification. Laboratory accrediting bodies would also need to maintain records of those analysts holding certification as well as being able to monitor agencies' personnel to ensure analysts maintain their certification credentials.

Conclusion: role of forensic science

One of the more prominent challenges facing the field of forensic science pertains to the ever changing role of science within criminal justice and criminalistics. The current state of the field is in transition as stated by Tjin-A-Tsoi (2013),

"New insights, technologies, and customers, combined with falling costs and increasing capabilities cause the sector to grow rapidly" (p. 2).

In previous years forensic science acted in a supporting role according to the directives of police detectives, often finding itself at the mercy of available physical evidence. As noted by Tjin-A-Tsoi (2013), the field of forensic science is no longer required to remain behind the scenes. The field and its agents are increasingly taking a prominent role in criminal investigations as fast and reliable results are now possible.

This rapid growth and the nature of the role change that have occurred in the field have created a situation in which forensic science has grown into a full-fledged profession. No longer is it sufficient to have laboratory settings in which "individual experience and expertise of practitioners play a dominant role" (Tjin-A-Tsoi, 2013, p. 2). Contemporary laboratories require a highly educated, well trained staff who often work in collaboration in a bureaucratic infrastructure utilizing cutting edge scientific technology and technique.

While growth in any profession is to be lauded, advancement and the accompanying bureaucratic infrastructure bring a range of salient issues and challenges that must be addressed. One of the biggest hurdles facing the field is the fact that the basic structure of laboratories existing within the United States has remained virtually unchanged, yet the number of laboratories and personnel performing scientific analyses has increased. Funding requirements are often inadequate to accommodate the amount of work demanded of analysts and laboratory facilities.

In addition, the need for accreditation and certification standards can be understood by examining the place of forensic science within the larger field of criminal justice. The administration of justice is, by necessity, an adversarial process. Powerful competing forces impact both laboratories and forensic practitioners. As a result of such interactions, the need for established ethical and scientific guidelines has never been more essential. Through the processes of accreditation and certification, forensic practitioners can be guided by a sense of normalcy as they find balance in their place within the larger field of criminal justice.

Key terms and definitions

Innocence Project: A national litigation and public policy organization dedicated to exonerating wrongfully convicted individuals through DNA testing and reforming the criminal justice system to prevent future injustice.

Exoneration: The proof that a formerly convicted individual was not guilty of the offense for which they were convicted.

Backlog: An accumulation of uncompleted work.

Conviction: The judgment of a court, based on the verdict of a jury or judicial officer or on the guilty or nolo contendre plea of a defendant, that the latter is guilty of the offense with which he or she has been charged.

Ethics: Standards of moral and official conduct.

Protocol: A system of rules that explain the correct conduct and procedures to be followed in formal situations.

Chain of custody: A formal, written process recording the persons having custody of evidence from initial point of receipt to final disposition. American Society of Crime Laboratory Directors Laboratory Accreditation Board:

Forensic Quality Services: A nongovernmental organization providing accreditation services to public and private sector forensic science facilities.

American Association for Laboratory Accreditation: A nonprofit, nongovernmental, public service, membership society providing accreditation, and training services for forensic scientists.

References

Bolt, R.H., Cooper, F.S., David, E., Dens, P.B., Pickett, J.M., & Stevens, K.N. (1970). Speaker identification by speech spectrograms: A scientist's view of its reliability for legal purposes. *Journal of the Acoustical Society of America, 47*, 597–612.

Breithaupt v. Abram, 352 U.S. 432 (1957).

Daubert v. Merrell Dow Pharmaceuticals, Inc., 509 U.S. 579 (1993).

DiFonzo, J.H. (2005). The crimes of crime labs. *Hofstra Law Review, 34*(1), 1–11.

Donnelly, S. (2012). Forensic science in a human rights framework. *Australian Journal of Forensic Sciences, 44*(1), 93–103.

Doyle, A.C. (1905). *The original illustrated Sherlock Holmes*. Edison, NJ: Castle.

Escobedo v Illinois, 378 U.S. 478 (1964).

Feige, D. (2004). Printing problems: The inexact science of fingerprint analysis. Slate, May 27.

Frye v. United States, 293 F. 1013 (1923).

Giannelli, P.C. (1997). The abuse of scientific evidence in criminal cases: The need for independent crime laboratories. *Virginia Journal of Social Policy and Law, 4*, 439–478.

Giannelli, P.C., & Imwinkelried, E.J. (2002). *Encyclopedia of crime and justice*. New York: Macmillan Reference, USA.

Goldstein, R. (2011). Improving forensic science through state oversight. *Texas Law Review, 90*(1), 225–258.

Goodwin, R.J., & Gurulé, J. (2009). *Criminal and forensic evidence: Cases, materials, problems*. New York: Matthew Bender & Company, Inc.

Hansen, M. (2013). Crime labs under the microscope after a string of shoddy, suspect and fraudulent results. *American Bar Association Journal*, September, www.abajournal.com/magazine/article/crime_labs_under_the_microscope_after_a_string_of_shoddy_suspect_and_fraudu.

Hsu, S.S. (2012). Convicted defendants left of uninformed of forensic laws found by Justice Department. The Washington Post, April 16.

Ingram, J.L. (2015). *Criminal evidence* (12th ed.). Waltham, MA: Anderson Publishing.

Innocence Project (2015). *The cases: DNA exoneree profiles*. www.innocenceproject.org.

Johnson, H.A. (1988). *History of criminal justice*. Cincinnati, OH: Anderson Publishing.

Kaye, D.H. (1993). DNA evidence: Probability, population genetics, and the courts. *Harvard Journal of Law and Technology, 7*, 101–172.

Koppl, R. (2005). How to improve forensic science. *European Journal of Law and Economics, 20*(3), 255–286.

Lander, E.S., & Budowle, B. (1994). DNA fingerprinting dispute laid to rest. *Nature, 371*, 735–812.

McCartney, C. (2006). *Forensic identification and criminal justice: Forensic science, justice, and risk*. Cullompton: Willan Publishing.

McCormick, C.T. (1954). *Law of evidence*. St. Paul, MN: West Publishing Company.

McDonald, R. (1998). Juries and crime labs: Correcting the weak links in the DNA chain. *American Journal of Law and Medicine, 24*(2), 345–363.

National Academy of Sciences (1979). *Strengthening the forensic sciences: A path forward*. Washington, DC: Department of Justice.

National Research Council (2009). *Strengthening forensic science in the United States: A path forward*. Washington, DC: National Academy Press.

National Research Council (1996). *The evaluation of forensic DNA evidence*. Washington, DC: National Academy Press.

National Research Council (1992). *DNA technology in forensic science*. Washington, DC: National Academy Press.

National Science and Technology Council (2014). *Strengthening the forensic sciences*. Washington, DC: Office of the President.

Newton, D. (2008). *DNA evidence and forensic science*. New York: Infobase Publishing.

O'Brien, J. (2013). *The scientific Sherlock Holmes: Cracking the case with science and forensics*. Oxford: Oxford University Press.

Office of the Inspector General (2014). *An assessment of the 1996 Department of Justice Task Force review of the FBI laboratory*. Washington, DC: U.S. Department of Justice.

Office of Technology Assessment (1990). *Genetic witness: Forensic uses of DNA tests, OTA BA-438*. Washington, DC: U.S. Government Printing Office.

People v. Castro, 545 N.Y.S.2d 985 (1989).

People v. Wesley, 533 N.Y.S.2d 643 (Co. Ct. 1988).

Possley, M., Mills, S., & McRoberts, F. (2004). Scandal touches even elite labs: Flawed work, resistance to scrutiny seen across U.S. Chicago Tribune, October 21.

Pyrek, K. (2007). *Forensic science under siege: The challenges of forensic laboratories and the medico-legal investigation system*. Burlington, MA: Elsevier Academic Press.

Saferstein, R. (2011). *Forensic science: An introduction*. New York: Pearson.

Schmerber v. California, 384 U.S. 757 (1966).

Shreffler, P.A. (1989). *Sherlock Holmes by gas lamp*. Zionsville, IN: The Baker Street Irregulars.

Solomon, J. (2003). *FBI's DNA lab subject of probe*. Toronto Star, April 29, p. A2.

State v. Coon, 974 P.2d 386 (Alaska 1999).

Thompson, W.C., Ford, S., Doom, T., Raymer, M., & Krane, D. (2003). Evaluating forensic DNA evidence: Essential elements of a competent defense review. *Champion, 27*(3), 1–24.

Tjin-A-Tsoi, T.B.P.M. (2013). *Trends, challenges and strategy in the forensic science sector*. Amsterdam: The Netherlands Forensic Institute.

United States Code (2000). Pub. L. 93–595, §1, Jan. 2, 1975, 88 Stat. 1937; Apr. 17, 2000, eff. Dec. 1, 2000.

United States Department of Justice (2012). *Census of publicly funded forensic crime laboratories, 2009*. Washington, DC: Office of Justice Programs.

United States Department of Justice (2008). *Census of publicly funded forensic crime laboratories, 2005*. Washington, DC: Office of Justice Programs.
United States Department of Justice (2005). *Census of publicly funded forensic crime laboratories, 2002*. Washington, DC: Office of Justice Programs.
United States v. Jakobetz, 747 F. Supp. 250 (1990).
Warden v. Hayden, 387 U.S. 294 (1967).
Wigmore, J.H., & McNaughton, J.T. (1961). *Evidence*. Boston, MA: Little Brown.

Appendix 14.1

Forensic science associations and professional organizations

American Academy of Forensic Sciences
American Board of Criminalistics
American Board of Forensic Anthropology
American Board of Forensic Odontology
American Board of Forensic Toxicology
American Society for Quality
American Society for Testing and Materials
American Society of Crime Laboratory Directors
American Society of Questioned Document Examiners AOAC International
Association of Firearm & Tool Marks Examiners
Association of Forensic Quality Assurance Managers California Association of Criminalistics
Canadian Society of Forensic Sciences
Council of Federal Forensic Crime Laboratory Directors Forensic Science Society
International Association for Identification
International Association of Arson Investigators International Association of Bloodstain Pattern Analysts
International Association of Coroners and Medical Examiners
International Association of Forensic Nurses
International Association of Forensic Toxicologists
Mid-Atlantic Association of Forensic Scientists
Midwestern Association of Forensic Scientists
National Association of Medical Examiners
National Center of Forensic Science
National Forensic Science Technology Center
New Jersey Association of Forensic Scientists
Northeastern Association of Forensic Scientists
Northwest Association of Forensic Scientists
Society of Forensic Toxicologists
Southern Association of Forensic Science
Southwestern Association of Forensic Scientists
Wisconsin Association for Identification

Source: National Research Council (2009).

15 The ethics of forensic science

Proceed with caution

Randall Grometstein

Introduction

In this chapter we will survey the reasons why forensic science practitioners of the early 21st century should proceed with caution. First, we will describe how the scientific basis for most forensic science techniques has been called into question; second, we will discuss how the nature of forensic investigation involves both subjective and objective elements, which means that investigators' own unconscious biases may influence the results; third, we will discuss the common ethical framework people often employ for dealing with these issues that leads to noble cause corruption. We will conclude with observations on the future of forensic science practice, including the paradigm change provided by the Justice Department's Sentinel Event Review.

Twentieth century forensic science: from the age of discovery to the age of doubt

Modern forensic investigative techniques were developed in the 20th century. Medical and laboratory science was primitive in 1892, when Andrew and Abby Borden were found hacked to death in their home in Fall River, Massachusetts, one hot August morning. Thirty-two-year-old Lizzie Borden was tried for the murders of her father and stepmother.[1] The prosecution claimed to have found the murder weapon in the Bordens' basement and presented testimony that, on the day before her death, Mrs. Borden had complained to her neighbor, Dr. Bowen, that she feared she was being poisoned (Brown, 1992, p. 62). The defense called an expert witness to rebut this testimony. A chemist from Harvard University testified that none of the axes and hatchets found in the Bordens' basement had human blood on them, and that the contents of Mrs. Borden's stomach showed no evidence of poison (Brown, 1992, p. 212). The case against Lizzie Borden depended largely on circumstantial evidence and the testimony of witnesses. She was acquitted in 1893, and no one else was ever charged with the murders. A century later, scholarly debate rages over whether the jury's verdict was correct, while a modern reader wonders how the chemist made his determinations. The Borden case took place at a moment in time when modern forensic techniques were just being developed. (For example, fingerprint examination was adopted in 1901 by the London Metropolitan Police and in 1902 by the New York City Police Department. The first blood groupings were identified in 1902.) The Harvard chemist undoubtedly applied the laboratory techniques of the day and interpreted the results to the best of his ability. This example, however, serves to remind us of the limitations of the forensic sciences.

The 20th century will be remembered as the Age of Discovery in forensic science. From fingerprints to DNA identification, forensic examination techniques proliferated, along with the application of insights from fields such as anthropology, psychology, entomology, pathology, photography, and psychiatry. The prospect of a scientific, objective basis for crime detection beckoned.

Doubts emerge in the courts and the scientific community

By the 1990s, however, doubts were emerging about the accuracy of most forensic science analyses and the ability of the criminal justice system to administer them accurately. Ironically, it was the newest forensic technique, DNA identification, which cast doubt on the validity of the others. Genetic profiling was developed in the 1980s, and a decade later, the Innocence Project helped a number of defendants appeal their convictions on the basis that DNA identification, not available at the time of their trial, could show that their convictions were wrongful. These defendants had been convicted with the aid of forensic evidence that had appeared to indicate their guilt, as interpreted by experts. Now, DNA evidence suggested two disturbing conclusions: these older techniques may not be reliable, and experts may have misinterpreted or exaggerated the results of forensic tests. The number of people exonerated by DNA is currently 358, while the number of all cases recorded by the National Registry of Exonerations is 2,240.[2] In 2009, the National Academies of Science released a report calling for a major overhaul in how evidence is defined, gathered, handled, and interpreted (National Academies of Science, 2009). In a parallel development, the federal courts were refining the standards for expert testimony about scientific evidence from the so-called "Frye standard" (Frye v. United States, D.C. Circuit, 1923)[3] to the Daubert standard (Daubert v. Merrell Dow Pharmaceuticals, Supreme Court, 1993).[4] The Daubert case requires a scientific expert to demonstrate the scientific basis for his or her testimony. A few years later, the Supreme Court ruled in Kumho Tire Company v. Carmichael (1999)[5] that a federal court must determine the reliability of expert testimony by a nonscientist before admitting it. The effects of the standards in *Daubert* and *Kumho Tire* were felt in the federal courts immediately. In 2001, a federal judge in Philadelphia refused to allow an FBI fingerprint expert to state on the witness stand that the latent prints found were those of the defendant (Slobodzian, 2002). Although the judge later allowed the testimony, the case was a harbinger of further challenges to the scientific basis for fingerprint identification.[6] Not all states followed the example of the federal courts in adopting the Daubert standard for expert testimony, but those that did have seen a wave of successful challenges to the introduction of many types of 20th-century forensic evidence, including bite marks, hair analysis, handwriting analysis, ballistics analysis, and arson investigation.[7] Thus, research into the causes of wrongful convictions, as well as courtroom challenges to traditional types of forensic evidence, have changed the landscape for forensic science. The report of the National Academies of Science (2009) suggests that the prospect of a scientific, objective basis for crime detection is as elusive as ever.

To the scientist, it will not be surprising to learn that forensic science is imprecise. Science proceeds by trial and error, conclusions are provisional, and doubt is an ever-present companion on the journey.[8] But the criminal justice system is less comfortable with doubt and error due to the application of both legal and moral standards. Legal judgments are final, and it is difficult to make conditional moral judgments ("If there is sufficient evidence to prove it, the defendant may have committed a heinous act"). Criminal justice practitioners straddle these two epistemological worlds – the real world of imperfect tests with margins of error, in addition to the absolute world of legal and moral standards of the community – and face the constant temptation to fit the available evidence to the Procrustean bed of legal and moral judgment. This temptation, of course, is a basic ethical dilemma for the justice system. We aspire to the greatest certainty possible in making legal and moral judgments, so we strive to minimize factual uncertainty. Even as we do this, however, we must recognize the limitations of our current knowledge and of the evidence available. The ethical forensic investigator must faithfully serve two different and exacting masters – the imprecise real world and the absolute moral world – without sacrificing one for the other.

If the only issue facing the criminal justice system with regard to forensic investigation were the science itself, that would be a formidable enough hurdle to overcome. But the issues exposed in recent years also concern the practice of forensic science. Issues involving crime laboratories include: the struggle to keep up with steadily mounting caseloads, which can lead to "dry-labbing" by some crime lab personnel who pretend to test drug samples but issue reports without testing, poorly trained lab personnel, and labs with no professional accreditation. There are issues involving evidence collection, storage and retention, including the untested rape kits piling up in many police departments.[9] Police officers, prosecutors, and investigators need ongoing training and continuing education to understand the evolving field of forensic investigation. Additionally, jurors may report to the courthouse with expectations about the role of forensic investigation as portrayed on television crime dramas, only to discover the difference between reality and television (the so-called "CSI effect"). These are unsettled times for the criminal justice system. The purpose of this chapter is to review the ethical challenges posed by forensic evidence as the 21st century gets underway.

Evidence: objective or subjective?

The important role played by evidence in our criminal justice system can be seen in the data on case clearance.[10] In 2013, for example, fewer than 50% of all violent offenses, as well as 20% of property offenses, were cleared by arrest or by exceptional means.[11] The clearance rate is also an indicator of how many crimes cannot be solved due to lack of evidence, that is, slightly more than 50% of violent offenses and 80% of property offenses. During much of the 20th century, improvements in forensic techniques suggested that it might be possible to increase the amount of evidence available to investigators. More evidence would lead to better legal judgments of guilt and innocence on the part of courts. However, along with doubts about the scientific basis for a number of forensic techniques came the disquieting realization that "evidence" has both an objective and a subjective component. Improved forensic techniques can help only with the objective component.

The subjective component of evidence is hinted at by the law school professor's admonition to her class that "Evidence does not speak for itself." We apply the term "evidence" once we have determined that an object in the real world is connected with the event we are investigating. If a bloody bandana was found 100 yards away from the house where the victim was killed, does it constitute evidence of the crime? The bandana does not become evidence until an investigator determines it to be such, and in the Texas case of Michael Morton, the bandana was determined not to be evidence, a decision which resulted in the wrongful conviction of an innocent man and a 25-year delay in identifying the real culprit.[12] Similarly, an investigator sends a piece of stained clothing to a laboratory for determination as to the type of stain. On the basis of a laboratory report, the investigator determines whether the stain is evidence in the investigation. Two levels of human judgment are involved in this second case, the test and the interpretation of the result. Thus, the label "evidence" indicates a conclusion, not a fact. Conclusions, of course, are subjective, not objective, since they are a matter of human judgment. A recent Supreme Court decision recognized this subjective element when the court ruled that a prosecutor cannot simply introduce a lab report into a trial without making available the person who performed the test to explain the results.[13] Thus, the judgment of the expert who interprets the evidence is as necessary as the evidence itself.

Cognitive biases

In recent years, research by cognitive psychologists and neuroscientists on how the human mind operates offers a disquieting portrait at variance with our belief that we are rational thinkers (Benforado, 2015). We are poor at assessing risk, harbor unconscious biases, and employ a variety of cognitive shortcuts that undoubtedly promoted our survival in the distant past but are ill-suited to the task of rational thinking. Additionally, beginning with Solomon Asch (Asch, 1955), Stanley Milgram (Milgram, 1963), and Philip Zimbardo (Haney et al., 1973), social psychologists have been exploring the cognitive demands placed on individuals by their position within social networks. Far from relying on their own judgment, people can be influenced by peers, by persons in authority, and by the role they occupy. These experiments suggest that criminal justice personnel, such as forensic investigators, may also be influenced by peers, people in authority, and role demands. If our goal is to make the criminal justice process more objective, we will have to confront the cognitive biases of criminal justice personnel as well as the ways in which criminal justice operations may influence their findings. Moral panic theory suggests that shocking and unusual cases often result in innovations that appear necessary at the time but are found to be regrettable in hindsight (Grometstein, 2008). Routine forensic science procedures should also be examined to see whether there are ways to minimize unconscious bias and other sources of contamination, similar to the danger of investigator bias in simultaneous lineups, for which sequential lineups appear to be a better alternative (Wells, 2014). As with lineups, it will be possible to identify ways in which routine forensic procedures might be better designed to minimize investigator bias. Forensic practitioners may need to adopt blind examination techniques for the same reasons of investigator bias that medical researchers acknowledge in conducting trials of new medications (Economist, 2012). Knowledge of the context in which the evidence was found may influence forensic science practitioners to interpret ambiguous evidence in favor of the prosecution (ibid.). Similarly, when laboratories have unmanageable caseloads, it is not uncommon for a prosecutor to contact a lab chemist personally to plead that an urgent sample be processed quickly, as happened in the Massachusetts drug lab scandal of 2012.[14] Objective results are difficult to achieve in the context of personal relationships between laboratory personnel and law enforcement. Thus, we must examine forensic practices to identify and mitigate sources of unconscious investigator bias. This examination will have to be done by researchers, rather than by forensic science personnel themselves, in part because forensic science practitioners have a conflict of interest in this regard.

The National Academies (2009) report notes that forensic science disciplines fall into two distinct groups. Some disciplines – for example, biological/serological screening, including DNA analysis, and toxicology – developed within a scientific framework and are based on a body of scientific knowledge. Other forensic science disciplines – for example, fingerprints and bite mark analysis – consist of empirical applications of science and are performed exclusively by law enforcement personnel. The latter group is the one that is challenged most significantly by the Daubert standard, and the NAS report expresses the hope that a scientific basis for these disciplines may be developed (p. 37). But what if a discipline cannot be shown to have a scientific foundation? The NAS report described the testimony of several forensic science representatives who did not acknowledge that their methods were less than 100% accurate (pp. 13–14). People are most likely to resist change when they perceive a threat to their professional standing and their livelihoods. Thus, a bias in favor of the status quo may be the most difficult one for forensic science personnel to identify and deal with.[15]

Finally, there are structural biases built into the system. Many crime labs receive state funding for each conviction obtained (Koppl and Sacks, 2013). In the twelve states with provisions of this type, forensic science personnel might be pardoned for feeling as though they are on the prosecution's team, since the funding for their employer depends upon the success of their analyses. This structural bias, of course, is likely to affect the practitioner's findings.

Resource challenges in forensic science practice

Resource challenges for forensic science practice have mounted over the past decades. These challenges include mastering new laboratory techniques such as DNA analysis as well as the increasing use of forensic evidence in court. Resources such as personnel, training costs and budgets for forensic labs have not kept pace with the demand.[16] Many state forensics labs are not accredited by any professional association. Resource challenges lead to predictable results, including so-called "dry-labbing," as discussed earlier, case backlogs, demoralized forensics personnel, the hiring of untrained or undereducated personnel, and may contribute to the rape kit backlog mentioned earlier. Additionally, many lab personnel receive insufficient supervision, and lab results are rarely audited. In the Massachusetts case described earlier, for example, an expert stated that the average chemist can process 2,500 samples per year, while the Massachusetts chemist who falsified thousands of test results had reported processing as many as 11,000 samples per year (Harmon, 2012).

Resource challenges thus are both financial and administrative. Resource issues must be addressed at the level of the state, not the agency. The types of errors discussed here should be considered organizational accidents or "sentinel events" (Justice Department, Mending Justice: Sentinel Event Reviews, 2014) alerting people to the existence of possible flaws that threaten the integrity of the system as a whole.[17] Rather than trying to blame the responsible parties, participants in a Sentinel Event Review try to identify weaknesses in the system that affect system outcomes. How might a Sentinel Event Review have been conducted following the discovery that a Massachusetts lab chemist had been dry-labbing? In actuality, the Inspector General conducted a comprehensive review of the case that combined a trenchant analysis of the failures of named individuals with a host of recommendations for improvements in how the lab should function. The lab, previously under the purview of the Public Health Department, was assigned to the Massachusetts State Police. By contrast, a Sentinel Event Review might have expanded the review's focus from the lab where the chemist worked, to a broader view of the system as expressed by representatives from law enforcement, prosecutors' offices, and the courts. Ideally, suggestions would have been put forward that would increase the oversight of the process by key players in the entire system.

Ethics of the noble cause

From the vantage point of the forensic science system as a whole, we have looked at a variety of reasons for caution in forensic practice, ranging from the need to demonstrate the scientific basis for many forensic practices, to the subjective nature of evidence itself, to the cognitive biases that affect people's interpretation of forensic processes and results, to structural biases in the system, and resource challenges. Now we turn to the perspective of the individual forensic practitioner by examining an issue that any training program for forensic personnel needs to address: the ethical standard applied by a forensic practitioner in the course of his or her work. Thus, we will discuss some basic forms of ethical reasoning, beginning with an example. A chemist gets a request for an expedited drug test but realizes that such an action would violate lab protocols. The chemist must weigh the alternatives. On the one hand, following lab protocols represents good practice and organizational policy, but, on the other hand, the chemist may view the proposed violation as achieving an important goal (assisting the prosecution).

Broadly speaking, there are several ways to make the decision. One is to focus on the rightness or wrongness of the contemplated action (following or violating lab protocols) and make the decision on that basis. This type of reasoning is categorized as means-based or deontological. If the chemist applied this line of reasoning, in all likelihood she would decide to follow protocol and refuse the request for expedited processing. Another way to make the decision is to say that it is all right to violate protocol for a good purpose, such as assisting the prosecutor who made the request. This is called ends-based or teleological reasoning. If the chemist uses ends-based reasoning, she will decide to violate lab protocol. In the Massachusetts case, the chemist appears to have employed ends-based reasoning.

Means-based and ends-based reasoning are tools we all use to resolve dilemmas. Like all tools, they can be used appropriately or inappropriately. Organizations encourage the use of means-based reasoning by developing codes of ethics, manuals of procedure, and policy statements, but organizations rarely provide on-going ethical training to assist employees in using these tools appropriately.[18] A difference in ethical perspective between an employee tempted to use ends-based reasoning and a supervisor interested in upholding organizational standards is a common organizational problem. However, in the case of criminal justice agencies, there is an additional impetus toward ends-based or teleological reasoning on the part of employees. Criminal justice agencies are engaged in the noble cause of getting the bad guys off the street.

The noble cause is explained by Michael Caldero and John Crank as "the moral commitment to make the world a safer place to live... by getting the bad guys off the street" (Caldero and Crank, 2004). Many criminal justice personnel are idealists dedicated to the cause of combating crime. This attitude can lead to what Caldero and Crank call noble-cause corruption, in which a criminal justice practitioner bends or breaks the rules in order to arrest or convict a "bad guy."[19] Examples from law enforcement include "testilying" (lying on the witness stand), writing reports that reflect what should have happened rather than what did happen, and planting evidence on a suspect because the officer believes him to be guilty. While economic corruption is motivated by self-interest and is thus easier to recognize and condemn, noble-cause corruption is motivated by idealism rather than self-interest and thus inspires an ambivalent response in most observers (Grometstein, 2007). But noble-cause corruption poses a grave threat to the integrity of a criminal justice system committed to constitutional notions of due process. Earlier we considered the example of the chemist deciding whether to perform an expedited examination of a suspected drug sample. The opportunity to assist in the prosecution of a drug offender may tempt the chemist to break the rules of the lab, even though a bribe would not provide the same temptation. The nobility of the cause being served can blind the chemist to the fact that her decision leads to the same result as if she had accepted payment for her findings.

What can an organization do to discourage noble-cause corruption? Organizational leaders need to set a good example and to raise the subject with employees in the context of performance evaluations. Take the example of two district attorney's offices. One office sends a message to its employees by rewarding them for cases won. Supervisors do not appear to be monitoring noble-cause corruption. A different office emphasizes the integrity of the criminal justice system and the importance of each person's role in maintaining that integrity.[20] Employees are rewarded for doing a good job, regardless of the outcome of the case. The first office encourages, and the second office discourages, noble-cause corruption.

Noble-cause corruption and evidence

The process of collecting, testing, and interpreting of evidence is rife with opportunities for noble-cause corruption. Because the collecting and testing of evidence involves many different people, including those involved in the chain of custody, as well as those in the lab, there are numerous opportunities for mistakes or omissions that could render the results useless. Investigators may be tempted to believe that the ends (getting the bad guys off the street) justify the means (failing to report errors or omissions) or they may be biased by their insider's acquaintance with criminal justice personnel and their theory of who did it. As a result, they tend not to report errors or omissions in the process. The final stage is the interpretation of the evidence and its role in the criminal justice process. An examiner writes the report for the prosecutor's office. Is it likely at this point in the process that the examiner will express doubts about the validity of available bite mark evidence, for example, when the analysis has passed through so many steps already? Or doubts about the collection and analysis of bite mark evidence in general? Then, when the examiner takes the stand to explain the results, the examiner will naturally downplay any uncertainty in the results. To do any less would imply that his or her colleagues and coworkers are incompetent or engaged in a fool's errand. This example contrasts strongly with the scientific tradition, which holds that it is the duty of every participant to be critical of the results of any scientific experiment or procedure. The scholarly author points out the weaknesses of any conclusions tentatively reached as well as questions that the work cannot answer. In forensic investigation, participants are doing their job by supporting the results of a process, while in the scientific endeavor, participants see posing questions as part of their job.

Conclusion

Forensic science practice is undergoing significant change at this time. For much of the 20th century, pioneers explored new territory and pushed back the frontiers of forensic science. Starting in the 1990s, however, the frontiers began to close, as consolidation and evaluation raised questions about the claims staked out by the pioneers. In the 1990s and early 2000s, courts grappled with new standards for the admission of forensic testimony as well as the steady flow of cases of wrongful conviction based on forensic results. In the second decade of the 21st century, we can see that the forensic science system is suffering from a host of challenges, from the epistemological (doubts that forensic techniques yield valid results, disputes over professional credentials), to resource challenges (insufficient laboratory facilities to process all the drug samples generated in the war on drugs), to the findings of psychologists researching cognitive bias (the role of investigator bias in routine forensic science practices), to the need for ongoing training for forensic science practitioners. Additionally, we see that ethical issues require addressing, because of the pervasiveness of noble-cause corruption.

The Justice Department's Sentinel Event Review suggests a way forward. This type of review is intended to focus on identification of the system error or errors that must logically have occurred, and does not focus on efforts to blame individuals. The SER represents a new paradigm. Representatives of all parts of the system that might be responsible for the event in question must be present, and all have input into designing acceptable solutions. As a consequence, the voice with which they speak is a louder voice, one more likely to be heard.

We will close with an interesting prediction by Judge Donald Shelton (Shelton, 2011). Shelton notes the conservatism and resistance to change on the part of judges and prosecutors. He looks to the jury as the source of pressure to change current practice:

> Jurors are well aware that DNA can identify perpetrators with certainty....In the end, what may cause reform in this area is acquittals. Jurors, not judges, may be the source of the strongest message to the government about the vulnerability of some evidence prosecutors try to use to gain convictions.
>
> (2011, p. 140)

The National Academies report (2009, pp. 48–49) cites research to support the view that the CSI effect is real, although it is not clear what effect it will have on courtroom practice.

Thus, the forensic science practitioner of the early 21st century is advised to proceed cautiously, keeping in mind the uncertain terrain he is crossing. The practitioner should avoid sweeping claims of certainty and keep in the forefront of her mind the possibility of error and uncertainty.

Key terms and definitions

Ends-based reasoning: Deciding whether or not to take a particular action based on the anticipated outcome of the action.

Means-based reasoning: Deciding whether or not to take a particular action based on the rightness or wrongness of the action itself.

Noble cause corruption: Justifying a decision to bend or break the rules in order to get the "bad guy" off the street.

Cognitive bias: An unconscious bias on the part of the investigator in favor of the suspect's guilt.

Daubert standard: The requirement that a scientific expert demonstrate the scientific basis for the testimony he or she proposes to give in court.

Sentinel event review: An organization-wide review that seeks to identify sources of error and weakness within the organization that may lead to errors. This approach is an alternative to the traditional after-the-fact inquiry that tries to assign blame for the error.

Notes

1 There is a large literature on the Lizzie Borden case, both popular and scholarly. See, for example, Williams et al. (1980), Brown (1992) and Kent (1992).

2 See www.innocenceproject.org/cases-false-imprisonment, as well as www.law.umich.edu/special/exoneration/Pages/about.aspx. Accessed June 27, 2018.

3 Frye v. United States, 293 F. 1013 (D.C. Circuit, 1923).

4 Daubert v. Merrell Dow Pharmaceuticals, 509 U.S. 579 (U.S. Supreme Court, 1994).
5 Kumho Tire Company v. Carmichael, 526 U.S. 137 (1999) (Daubert standard applied to expert testimony from non-scientists).
6 In another example, in 2006 a Baltimore County Circuit judge refused to allow a fingerprint expert to testify that a latent print was made by the defendant in a death penalty case. See State of Maryland v. Bryan Rose, Baltimore County case number K06-545 (cited by National Academies of Science 2009, p. 43).
7 Judge Donald E. Shelton, for one, doubts that judges as a group have been able to grapple with the challenge of being the scientific gatekeeper of evidence. See Shelton (2011). He attributes this difficulty to the pro-prosecution bias of most judges, many of whom are former prosecutors (Shelton 2011, pp. 136–137).
8 DNA identification techniques have been changing rapidly, perhaps more rapidly than practitioners can keep up with.
9 See USA Today, Tens of thousands of rape kits go untested rape across USA, July 7, 2015, available at www.usatoday.com/story/news/2015/07/16/untested-rape-kits-evidence-across-usa/29902199/. Accessed June 27, 2018.
10 See Uniform Crime Reports available at www.fbi.gov. On the page entitled "Crime in the U.S." for any particular year, click on the link for clearances.
11 www.fbi.gov/about-us/cjis/ucr/crime-in-the-u.s/2013/crime-in-the-u.s.-2013/offenses-known-to-law-enforcement/clearances/clearancetopic_final. Accessed October 31, 2015.
12 The bandana was not tested, nor was its existence disclosed to the defense. After 24 years in prison, Morton won the right to have the evidence tested, and the testing led his release. www.chron.com/exonerees/stories/michael-morton/. Accessed November 1, 2015.
13 In Melendez-Diaz v. Commonwealth of Massachusetts, 557 U.S. 305 (2009), the U.S. Supreme Court ruled that the Sixth Amendment's confrontation clause requires a prosecutor to make available the person who performed a chemical drug test when she introduces a chemical drug test report into evidence. Alternatively, under a "notice-and-demand" statute, a prosecutor can notify the defendant of an intention to submit a chemical drug test report without the testimony of the person who performed the test, provided that the defendant has an opportunity to object and demand that testimony.
14 Andrea Estes and Scott Allen, Chemist often called, wrote to prosecutor. Boston Globe, October 27, 2012, page A1.
15 An example of this kind of investigator bias – the human desire to keep one's job when it is threatened by the advent of a new and better technology – appeared in an interview with the supervisor of the New York Police Department crime lab's hair-and-fibers unit (Toobin, 2007).

The supervisor described her attempts to carve out a role for her unit in an era in which DNA analysis is the gold standard for forensic evidence: "I know everyone wants to go straight to the DNA lab these days, and I don't blame them," she said.

> But they have to come to us first, and for good reason. Ninety percent of the hair recovered at a crime scene doesn't have roots in the right stage of growth. You need roots for nuclear DNA. You might ask, Why is microscopic trace examination necessary? Why not submit all hairs for DNA analysis immediately? But you cannot go to the medical examiner with two hundred hairs. We have so many crimes and hairs. Hairs are everywhere. We couldn't possibly give every hair to the m.e. We have to screen it first, see if there is a possibility of any kind of DNA test, and then we can send it over.

16 Scandals at drug labs have led to the overturning of many convictions in recent years. Perhaps the most influential was the investigation into the FBI's crime lab in the 1990s which cast doubt on the investigation of many significant cases. See *The FBI Laboratory: An Investigation into Laboratory Practices and Alleged Misconduct in Explosives-Related and Other Cases (April 1997),* by the Office of the Inspector General, U.S. Department of Justice, available at https://oig.justice.gov/special/9704a/. Accessed November 6, 2015. In 2015, the FBI announced that nearly all of its hair examiners gave flawed testimony in court about the results of their tests for years (Hsu, 2015).

State and city labs experiencing similar scandals include Fred Zain of the West Virginia State Police lab (Clines, 2001) and the Houston crime lab (Liptak, 2003). More recently, problems have been revealed at labs in Colorado, Philadelphia, Detroit, and North Carolina (Friedersdorf, 2015).

17 See www.nij.gov/topics/justice-system/pages/mending-justice.aspx. Accessed November 7, 2015.

18 Codes of ethics have been developed for many of the forensic science fields, for example, the Association of Crime Laboratory Directors, www.ascld-lab.org/guiding-principles/. Accessed November 11, 2015. Codes of ethics generally attempt to set standards for practice; these standards are among the "means" discussed in the chapter.

19 If criminals are the "bad guys," then those who pursue them are obviously the "good guys." Viewing one's own motivations as good tends to provide one with the assurance that one is doing the right thing. Sissela Bok, discussing the ethics of lying, reminds us that, when we lie, we assure ourselves of our good intentions, however questionable this assumption would be to an observer (Bok, 1989).

A philosophical tradition that began with Plato held the "noble lie" to be among the categories of permissible lies. Originating with a story showing that God ordained a social structure in which the upper class was superior to the lower classes (Bok, 1989, p. 167), the "noble lie" is one told by a member of the ruling class authorized to deceive the public for its own good: "In other words, lying is excusable when undertaken for 'noble' ends by those trained to discern these purposes." While today society is no longer ruled by a titled aristocracy, many public servants continue to believe that they know best, and the public does not need to know how they perform their jobs. Bok describes lies for the public good as the most dangerous kind of lies (1989, p. 166), because the liar believes in his own altruism and therefore underestimates the considerable damage to the social fabric caused by the lie.

20 As expressed by an inscription on the walls of the Justice Department in Washington, D.C.: "The United States wins its case whenever justice is done one of its citizens in the courts."

References

Asch, S. E. (1955). Opinions and social pressure. *Scientific American* **193**: 31–35.

Benforado, A. (2015). *Unfair: The New Science of Criminal Injustice*. New York, Crown Publishers.

Bok, S. (1989). *Lying: Moral Choice in Public and Private Life*. New York, Vintage.

Brown, A. R. (1992). *Lizzie Borden: The Legend, the Truth, the Final Chapter*. Nashville, TN, Rutledge Hill Press.

Caldero, M. A. and J. P. Crank (2004). *Police Ethics: The Corruption of Noble Cause*. Cincinnati, OH, Anderson Publishing.

Clines, F. X. (2001). Work by expert witness is now on trial. *New York Times,* Sept. 5, 2001.

Economist (2012). Ignorance is bliss: Forensic scientists know too much about the cases they investigate. *The Economist*, Jan. 21, 2012.

Friedersdorf, C. (2015). CSI is a lie: America's forensic-investigation system is overdue for sweeping reform. *The Atlantic Monthly,* April 2015.

Grometstein, R. (2007). Prosecutorial misconduct and noble cause corruption. *Criminal Law Bulletin* **43**(1): 63–75.

Grometstein, R. (2008). Wrongful conviction and the moral panic about organized child abuse: National and international perspectives. *Wrongful Conviction: International Perspectives on Miscarriages of Justice.* C. R. Huff and M. Kilias. Philadelphia, PA, Temple University Press, pp. 11–32.

Haney, C., et al. (1973). Interpersonal dynamics in a simulated prison. *International Journal of Criminology and Penology* **1**: 69–97.

Harmon, L. (2012). Black mark on the state drug lab. *Boston Globe,* Sept. 14, 2012.

Hsu, S. S. (2015). FBI admits flaws in hair analysis over decades. *Washington Post,* April 18, 2015.

Kent, D. (1992). *The Lizzie Borden Sourcebook.* Boston, MA, Brandon Publishing.

Koppl, R. and M. Sacks (2013). The criminal justice system creates incentives for false convictions. *Criminal Justice Ethics* **32**: 126–162.

Liptak, A. (2003). Worst crime lab in the country: Or is Houston typical? *New York Times,* March 11, 2003.

Milgram, S. (1963). Behavioral study of obedience. *Journal of Abnormal and Social Psychology* **67**: 371–378.

National Academies of Science (2009). *Strengthening Forensic Science in the United States: A Path Forward.* Washington, D.C., The National Academies Press. https://doi.org/10.17226/12589.

Shelton, D. E. (2011). *Forensic Science in Court: Challenges in the Twenty-First Century.* Lanham, MD, Rowman & Littlefield.

Slobodzian, J. A. (2002). Fingerprint experts are pointed toward the courthouse door; prosecutors worry about a federal judge's ruling in Philadelphia. *Philadelphia Inquirer,* Jan. 22, 2002, C03.

Toobin, J. (2007). The CSI effect: The truth about forensic science. *The New Yorker* **83**: 30.

Wells, G. L. (2014). Eyewitness identification: Probative value, criterion shifts, and policy regarding the sequential lineup. *Current Directions in Psychological Science* **23**(1): 11–16.

Williams, J., et al. (1980). *Lizzie Borden, a Case Book of Family and Crime in the 1890s.* Bloomington, IN, T.I.S. Publications.

Index

Note: Boldface page numbers refer to tables; italic page numbers refer to figures.